Space, Time and Quanta

An Introduction to
Contemporary Physics

DEDICATION

To

Angie, Alan, Seth, Shawn, John, Dave, Douglas, Nils, Mike,
Kirsten, Jon, Jon, Martin, Andy, Mary, Gordon, Jeff, Barry,
Kevin, Kim, Ben, Brant, Karen, Dan, Jenny, Jay, Mark, David,
Sarah, Aaron, Frank, Ann, Rob

Williams College, Physics 141, Autumn 1985

Matt, Scott, Levi, Bob, Arthur, Will, Ed, Dave, John, Alan, Dan,
Mike, Matt, Brad, Spencer, Dale, Dara, Caryl, Renee, Jim,
Laura, Nick, Mike, Leanna, Craig, Dave, Judy, Brian, Oscar,
Andres, Tom, Dan, Maggie, Tricia, Ed, Russ, Teresa, Bill,
Warner

OSU, Physics H131–2–3, 1989–1990

Yogi, Mary, Kevin, David, James, Jeff, George, Matt, Bill, Jason,
Bill, Quique, Pete, Brian, Cameron, Eric, Eric, Archy, Andrew,
Kevin, Tobin, Phil, Bruce, Walter, Will, Eric, Shanc, Attiya,
Scott, Chet, Brian, Preston, Russ, Raj, Gary, Chris, Andy, Jack,
Bill, Cong, Chris, Sergey

OSU, Physics H131–2–3, 1990–1991

Space, Time and Quanta

*An Introduction to
Contemporary Physics*

Robert Mills
The Ohio State University

W. H. Freeman and Company
New York

Library of Congress Cataloging-in-Publication Data

Mills, Robert, 1927–
    Space, time and quanta : an introduction to
    contemporary physics / Robert Mills
        p.cm.
    Includes index.
    ISBN 0-7167-2436-7
    1. Physics. I. Title.
QC21.2.M57   1994
530—dc20   93-33825
                                                       CIP

Printed in the United States of America

1 2 3 4 5 6 7 8 9 0   VB   9 9 8 7 6 5 4

CONTENTS

Contents

Contents

Contents

PREFACE

In the fall of 1985, while I was a visitor at Williams College, I was asked to teach a course entitled "Twentieth-Century Physics." The course was intended to introduce first-year students to the delights and wonders of physics (so physicists view them, of course) and to serve as a feeder course into the major for those who were so inclined. The bread-and-butter courses in seventeenth- to nineteenth-century physics were to come later, after enthusiasm and commitment had been suitably aroused.

The students were bright, with a good high school math background, and many—but not all—had had a high school physics course. There appeared to be no text at the right level, assuming mathematical competence but no prior college physics, so I wrote out notes as I went along and had them duplicated for distribution to the class. When the term was over, David Park, noting the absence of suitable texts for such a course and being as enthusiastic as I had become about this way of introducing prospective majors to the subject, encouraged me to expand the notes into a book. David, who had taught the course for a number of years and to whom I am *most* grateful, took the time to go over the notes carefully and make astute comments in the margins; he subsequently served as the most diligent and helpful of reviewers for the manuscript during the extensive revision process.

The book is intended for first-year college students who are considering a science or engineering major. It assumes competence in high school math—specifically algebra and geometry, and perhaps some analytic geometry; it introduces calculus as needed. It is intended to be self-contained in this regard, although a concurrent calculus course would be ideal. Many of the concepts of nineteenth-century physics are introduced at appropriate times to provide the necessary background and context, but the motivation is always centered in the physics of today.

Problems are an important part of the book. I want to stress, however, that the emphasis is not on problem solving but on conceptual understanding. Many problems are numerical—aimed at giving a feel for the orders of magnitude involved in the different areas of physics and getting the reader used to unfamiliar concepts; many are analytical; many are qualitative and conceptual, with the aim throughout of conveying the logic that ties the concepts together.

The book is intended for use as the primary text in a one- (or two-) semester introductory course of the sort that was offered at Williams, but it could be used in a variety of other contexts:

- As a modern physics supplement in a more conventional physics course where the primary text gives insufficient coverage of one or more of the main areas of contemporary physics, such as special relativity, cosmology, or quantum physics (see comments that follow on my own experience with this mode of use)

- As a text (in whole or in part) for the physics side of an interdisciplinary course designed for well-prepared students in other disciplines

- As a source for supplementary readings in a variety of contexts, both in physics courses for non-physics majors and in more general introductory courses dealing with the scientific/cultural setting of our times

- As a text (primary or supplementary) for an advanced placement course in physics for well-prepared high school students

It is also my hope that the book will provide a solid introduction to the world of contemporary physics for educated readers outside the academic subculture. I hope that such readers will dig into the math as much as they can but will skim boldly. A lot of the substance of the book lies in concepts rather than in math, though in the end the math is inescapable. Physics is the real world, not mathematical structures, but the models we construct to understand the physical world are unavoidably mathematical.

During the academic years 1989–1990 and 1990–1991, I used a draft of this book as a supplementary text in one of the honors physics sections of the three-quarter introductory physics course at Ohio State University. We spent around 40 percent of the course on this book, covering in the remainder of the time the essential topics of a conventional engineering physics course. The students' enthusiasm and their retention level were substantially greater than in previous years when I followed a more conventional pattern, and for reasonably bright students who go on in physics, I don't believe the necessary omissions in conventional physics will be hard to fill in as necessary.

Part I of the book, "Space and Time," deals with relativity and gravitation (general relativity) and with the special problems associated with time. Part II, "Quantum Physics," delves into the quantum world, including some discussion of what it means to be an observer. In the original version of the book there was also a Part III, "Particles and Fields," covering both the quantum characteristics of ordinary particles and fields—electrons, protons, neutrons, and the electromagnetic field—and the modern picture of quarks, leptons, and gauge fields as the fundamental constituents of nature. Part III has been omitted to help limit the size and cost of the book, but a bound duplicated copy of the manuscript of Part III can be obtained from the author for $15.00, including mailing within the contiguous 48 states.

Please write to Robert L. Mills, Department of Physics, Ohio State University, Columbus, OH 43210.

Parts I and II of the book can be used quite independently of each other. I ask the reader's patience with a certain lack of uniformity in the mathematical level; some subjects lend themselves readily to mathematical development, whereas others can be treated in only the most qualitative manner. Indented paragraphs consist of parenthetical material, sometimes a bit more technical, and can usually be skipped or skimmed.

The sections into which the chapters are divided are the key to getting around in the book. Within each section, the equations and figures are numbered beginning with 1 and are referred to by that number alone. References to equations and figures in other sections include the chapter and section numbers in decimal notation. Each page shows the section number and each right-hand page shows the section title for easy hunting.

The glossary is an important and integral part of the book. It contains over 400 entries and is extensively cross-referenced. I believe that a free-associating trip through the glossary would constitute a minicourse in physics in its own right.

Let me call attention, too, to the appendices: one on mathematical facts, one a brief summary of Newtonian mechanics, and one a collection of miscellaneous information on units, physical constants, and astronomical facts.

My gratitude and debt to Williams College are great, specifically to Ballard Pierce for the initial invitation and for unstinting encouragement and assistance during my stay, to David Park for the reasons I have mentioned and for much support and help with the course, and to Karen Drickamer, who provided invaluable logistical support. I want to express warm appreciation to Wayne Armbrust for his careful scrutiny of early draft manuscript and for his very helpful comments, to Stuart Raby and Terry Walker for valuable briefings on particle physics and cosmology, and to John Josephson for referring me to the Yoga Sutras of Patanjali, which are referred to in §15.4.

I find myself remarkably grateful to the publisher's reviewers, Royal Albridge, James Donaghy, J. P. Draayer, Dick Henry, Glenn Julian, David Park, and Lyle Roelofs, whose thoughtful and detailed comments gave rise to a large number of improvements, both small and great. I helped myself shamelessly to any number of their suggestions and ideas.

I am especially indebted to the students in Physics 141 at Williams, who contributed much more than they know to the course itself and to the book that the course has spawned; and to the students of Honors Physics H131-2-3 at Ohio State, whose enthusiastic criticisms and suggestions were of great value.

I must express, too, my warm gratitude to the editors and staff at W. H. Freeman, especially to Jerry Lyons, who had faith in the idea and who

Preface

guided this book from its amorphous beginnings to its final, much improved, form, and to Penny Hull, for her enthusiastic and creative oversight of the production process. I am indebted also to Connie Day, the copy editor, who corrected all sorts of gaffes and awkwardnesses and contributed (through our marginal debates) to my own understanding of a number of nuances of the English language, and to Ellen Murray, the indexer, who brought an astonishing degree of order to the chaos of ideas running through the text.

My special thanks are due also to Helen Mills, whose wisdom, experience, and encouragement were of incalculable value in every part of the process.

Finally, the whole project would not have been possible without the unstinting support and counsel of my wife, Lee, for which—and for whom—I am more grateful than I can possibly say.

Robert Mills
March 1994

Introduction

 THE PICTURE GETS FUZZY

The world of nineteenth-century physics was a world of certainties. It consisted entirely of matter—solid, liquid, or gaseous. For philosophers, solid matter was the very epitome of the "real": A thing was real if you could grasp it or hurt your toe kicking it, and abstractions like love and beauty—and energy for that matter—had a most doubtful claim to reality. Matter was everything. The idea of a force field such as the electric or magnetic field, something existing at each point in space and causing forces on electric charges or currents, had been developed by Michael Faraday and James Clerk* Maxwell, but such a field was still understood very much in mechanical terms, as the displacement or movement of a

---

*Pronounced like "Clark."

material medium, the *ether*. This ether was supposed to fill all of space, even where it was completely empty of any ordinary kind of matter—what you'd call a perfect vacuum. To a present-day physicist, Maxwell's theory of the electromagnetic field seems to give a powerful description of something completely *non*material whose reality has nothing at all to do with matter as we know it, and yet Maxwell himself was apparently unable to understand his fields except in terms of a material ether.

A few other abstractions were beginning to enter physics—and were causing distress in the process. Two important examples are *energy* and *entropy*, which were key ingredients of the developing theory of heat, known as thermodynamics. It was becoming clear in the last century that the amount of energy in the universe is conserved—energy can't be created or destroyed—but it's very hard to hold it in your hands and say, "This is something real." Entropy, too, which measures the amount of disorder in the universe and plays an important role in the irreversible flow of time, was even more difficult than energy to understand as something real.

The behavior of matter was then believed to be governed with complete precision and complete predictability by various "laws of motion." The motion of physical objects was described by Newton's laws, which dated from the seventeenth century, whereas the electric and magnetic force fields, conceived as I said entirely in terms of a material ether, were elegantly described by Maxwell's equations, developed in the 1860s. Chemical and thermal processes were not so well understood, but here also scientists felt they were well on the way to an equally complete—and equally mechanical—description.

Space and time were believed to be absolute, and the flow of time was thought to move the universe and the human race forward from one moment to the next at a universal and inexorable pace. Isaac Newton's view still held in the 1800s: "Absolute, true and mathematical time, of itself, flows equably without relation to anything external ...." The laws of physics, the very general statements that describe the behavior of matter and the forces between different bits of matter, were being refined and verified with ever-increasing precision, until it seemed that the only remaining task was to measure the constants of nature to more and more decimal places—and to tidy up a few loose ends, a few little nagging inconsistencies.

In this century we have tugged at the loose ends—and have seen all of nineteenth-century physics unravel.

The giants in this enterprise are clearly Albert Einstein and Niels Bohr, though many others played key roles in tearing apart the old picture and building a new one. Matter has been supplanted by fields as the more fundamental concept, time has lost its unique and absolute status, and our faith in physics itself as an exact science has been badly shaken by the fuzziness and indeterminacy of quantum theory. The fact that physicists are still doing physics, however, is not simply a tribute to their pigheadedness but is rather due to the extremely high level of exactness and predictability that still remains, despite the fuzziness, when we probe the behavior of things in the proper fashion—when we ask the right kinds of question.

If you don't examine it too closely, the new picture looks just like the old one. When you do look closely, it's like a newspaper photograph, with the thousands of dots that make it up corresponding only roughly to the picture you thought you saw. In the case of physics, our close look reveals a structure that's radically different from the nineteenth-century picture, and that seriously violates nearly all our common-sense impressions of how things behave. These common-sense notions, of course, like the nineteenth-century picture of the physical world, correspond to the newspaper photograph viewed from a distance. One of things I want to do in this book is explain, in terms of the new physics, why the old-fashioned picture works so well—and I hope you will wonder, as I do, why in fact the little dots arrange themselves into a picture at all.

I.2 CENTRAL THEMES

What are the most distinctive characteristics of twentieth-century physics? I would say the loss of objective certainty, the fundamental role of symmetry, and the movement toward unification.

Conceptually, we've seen a loss of absolutes and an increasing appreciation of the important role of the observer in defining reality. We've seen the loss of complete predictability—a hallmark of classical physics—and we've seen a peculiar fuzziness in how well we can describe the state of a physical system. We have also seen a greatly increased level of mathematical abstraction in the theoretical description of nature: The mathematical structures we use often have no direct physical meaning, and sophisticated rules of interpretation are required before they can be related to experimental observations. We've had to deal with changes like these both in Einstein's theories of relativity and in quantum theory.

In terms of practical knowledge, we've probed both the infinitesimal and the infinite—the microscopic and the cosmic—and in both directions have found structures unlike anything that scientists of the nineteenth century could have imagined: exotic objects such as quarks and gluons in the small, and curved spacetime, black holes, and the Big Bang in the large. The theories we've been developing to describe these structures embody all the characteristics mentioned: loss of objectivity, laws of physics governed by symmetry requirements, and always the pursuit of unity.

I want to elaborate now on these central themes, which run through all the parts of this book.

Loss of Objective Certainty

The relativity of truth in Einstein's view of space and time and the uncertainties associated with quantum theory have led to some serious soul-searching about the nature of the *real*. We are no longer sure what it means for a thing to be real, and we don't know what it means to *know*. Indeed this epistemological need, this need to understand what it is to "know" a thing, has become more urgent as the observer has taken center stage in physics. Facts seem not to have an independent objective existence but rather are mediated in every case through an observer and are stated and known only relative to that observer. Absolutes remain, however. At the immediate level it's relationships that we deal with, but the description of these relationships (as far as we can tell) has absolute validity! Only relative velocities have meaning, for example, but we can make firm statements about the different relative velocities in some experimental situation—statements that are true from *every* point of view.

Einstein insisted, and we've come to believe him, that space itself—the space through which the earth and the stars move—cannot be thought of either as being at rest or as having any definite state of motion. Not only has it become impossible to answer the question "How fast is the earth moving through space?" but the very question no longer has any absolute meaning. Even the gravitational force, although it seems such a real part of our everyday life, appears now to depend on your point of view. For example, the pull of gravity seems to vanish when you're in a state of free fall—in an elevator whose cable has snapped, for example, or in an orbiting spacecraft. (Einstein claimed that the point of view of someone in the falling elevator, in which it appears that the vehicle is at rest and that there's no gravitational force at all, is just as valid as the point of view in which there *is* a force and the vehicle, along with its contents, is accelerating downward because of that force.) The sense that we have lost our solid foundations and that everything is *relative* was taken by many people as applying to other things besides physics, sending ripples of both apprehension and liberation through our Western culture and elevating Einstein himself to a kind of mystical eminence. Note that this kind of cultural use of the idea of relativity is very vague and has no foundation in physics. And note too, as I said just now, that certain kinds of statements in physics *do* have absolute validity.

Quantum theory too has left a trail of insecurity and uncertainty as it has developed in the twentieth century. Even more than relativity, it has shaken our belief in the possibility of objective knowledge about reality—and our belief in that reality itself. It has robbed physics, at a very fundamental level, of its power to make precise predictions about future events. I think the shaking of our beliefs has taken place only at the philosophical level (though I think that's important enough). For one thing, these uncertainties and unpredictabilities become negligible in the world of big objects like people and machines. In addition, we find we can make precise and reliable statements *about* the uncertainties, which enable us to do high-precision physics as long as we ask only meaningful questions. For example, we've discovered that it's not meaningful to speak of the position of a particle in the absence of a direct observation but that it *is* meaningful to inquire into the *probability* of finding it at one location or another. The laws of quantum physics, in fact, cannot predict events; they are reduced to predicting the *odds* that different events might occur.

This replacement of certainties by probabilities is basic to our quantum world. It's not like a coin toss or a roll of dice, which is unpredictable simply because you don't have precise enough information about what is in fact a completely determined motion. In quantum physics there doesn't really seem to be *any* real objective state of affairs at a time when a system is not under observation. Different possible sequences of events, even though they're mutually exclusive according to classical physics, all contribute to the final outcome; that is, *all* the possible histories leading up to the moment of observation play a role in determining the probabilities for that observation. This interference among the different possible histories is in fact closely related to the interference of classical waves such as sound waves: There are "dead spots" and "live spots" in an auditorium because the sound waves from the performer follow different paths as they bounce off the walls of the room and can either cancel or reinforce each other as they arrive at the location of the listener.

Much of the early shaping of this new way of looking at the world and much of the battling for this new view were done by Niels Bohr and the group that he gathered around him at his institute in Copenhagen in the early years of this century.

Symmetry

Another thread that runs through the physics of this century is the notion of symmetry. From being an accidental property of certain things, nice but not of fundamental importance, symmetry has come to play a central role in physics, apparently controlling the very structure of the laws of physics and the number and character of the elementary particles of nature. I need to explain what I mean here by symmetry. You can speak of the symmetry of an object, or you can speak—as we do in physics and as I'll explain in a moment—of the symmetry of physical laws. It is the symmetry revealed in the *laws* of physics that has become so important in recent years.

We say that an object is symmetrical if it has a certain *invariance* property—that is, if the object is left unchanged by some operation on it. A cylinder, for example, is invariant under rotations about its axis, and a sphere, showing a greater degree of symmetry, is invariant under rotations about any axis through its center. (Each of these examples has also a *reflection symmetry:* It looks the same in a mirror.) [What other examples can you think of? Give an example of an object that has no symmetry at all.]

The symmetries that I need to talk about are the symmetries of the laws of physics themselves. For example, you may repeat some experiment in different locations or in different orientations, and you'll find that everything behaves in exactly the same way—according to the same rules. You can even look at things from a moving frame of reference. For example, if you get on a high-speed train and try to walk or play ping-pong, you find that the same rules apply as if the train were at rest (assuming no vibration, of course). The reflexes and muscular skills that you learned on the ground work just as well as ever. An even more striking example is the fact that we have no sense at all that (in old-fashioned terms) the earth is moving through space at a prodigious speed (about 20 miles per second relative to the sun and around three times faster than that relative to the center of our galaxy). The laws that govern the motion of our bodies and the objects around us work exactly the same as if we were at rest. In physicists' language, the laws of physics are invariant with respect to a variety of possible choices of reference frame.

The first indication of the power of symmetry to dictate physical laws was presented in 1905 by Einstein, who overthrew the nineteenth-century understanding of space and time and rewrote the laws of physics on the strength of a symmetry principle, the *principle of relativity*. His vision was so clear, and so beautiful, that after the initial shock scarcely anyone doubted that he had it right. This principle of relativity is the requirement referred to earlier:

> *The laws of physics—including the behavior of light—must be exactly the same for any two observers moving with constant velocity relative to each other.*

It's called the principle of relativity because it makes the idea of any absolute stationary reference frame meaningless in practice and says, rather, that we can describe events only relative to some observer and that all uniformly moving observers are equivalent. We continue to say that things are "stationary" or "moving," and it's usually clear what we mean, but to be precise we have to specify the reference frame *relative to which* a thing is stationary or moving. This principle leads to drastic changes in the description of nature devised by Newton. The new description goes by the name of the *special theory of relativity,* or just *special relativity.*

It turns out, surprisingly, that when we relate the views of two observers moving relative to each other, we find that the dimensions of space and time get mixed—that what looks like pure space to one observer looks like an admixture of time and space when viewed by the other. This mixing is closely analogous to what happens when the coordinate axes are rotated on a plane or in ordinary space. As a consequence, we have to think of space and time together as a single four-dimensional space (three dimensions of space, plus time), for which we use the word *spacetime.*

Einstein's vision of the deep equivalence of different points of view led him to pursue even further the idea that the laws of physics should appear exactly the same to all possible observers: He now insisted on including observers who are accelerating or rotating. This further step was suggested by the fact that the inertial effects of acceleration or rotation—like the forces experienced by astronauts during blast-off and like the centrifugal force that throws you to one side during a sharp turn in a car—are indistinguishable from gravitational effects. That is, you can attribute such effects to your own motion or, equally well, to the presence of different gravitational forces.

Einstein's inspiration was to extend this simple kind of equivalence to all kinds of motion and to state it as a universal rule, referred to as the *principle of equivalence:*

> *The laws of gravity must be such that the apparent forces due to any possible kind of motion are indistinguishable from gravitational forces.*

This would not make sense if it weren't possible for gravitational forces to change with time and (as you'll see) to depend on the velocity of the object they're acting on. This vision led Einstein, by a difficult and tortuous path, to a general theory of gravity that not only far transcends Newton's in power and sophistication but also leads us to a new view of space and time as *curved*. We live apparently in a curved universe, and the forces that we attribute to gravity turn out not to be forces at all but simply a necessary consequence of that curvature. It's as if you were a two-dimensional creature sliding smoothly along some irregular surface with dimples distributed here and there and were being deflected in your path every time you passed near one of the dimples.

This curved universe can be said to have a shape, either open and infinitely extended or closed like the surface of a balloon—we don't know which. Every star or planet sits in the middle of one of the dimples, and the dimple can become infinitely deep in the case of a star that has collapsed and become a black hole.

The theory that describes this curved spacetime is called the *general theory of relativity,* or just *general relativity.* It gives to spacetime itself an active, dynamical role, and it foreshadows, in a curious way, the modern concept of gauge fields (see below). Note that general relativity is again based on a symmetry principle—a requirement of invariance: the insistence that the laws should be the same for all observers, including even those in accelerating and rotating frames of reference.

The next step, the beatification of symmetry, you might say, was Noether's theorem (1918), which gives a relation between symmetries and conservation laws in physics. A conservation law is any rule, derived from the basic laws of physics, that says the total amount of some quantity is constant and doesn't change with time. A notable example is energy, which can be neither created nor destroyed but only transformed from one form to another. What Emmy Noether* showed (with some exceptions that don't concern us here) was that for every symmetry of the laws of physics there is a corresponding conservation law. We now know also, though it wasn't part of her original theorem, that the converse is true: Every conservation law must be associated with a corresponding symme-

---

*German, Nöther: pronounced like "neater" but with the lips pursed as for "ooh."

try. The proof of the theorem depended on the fact that all the laws then known were of a certain type—what we call derivable from a Lagrangian. It is astonishing to me, given the tremendous changes that have been wrought in physics since 1918, that Noether's theorem still applies and has become, in fact, much more general than it was to begin with. It's so easy to invent universes where it doesn't apply that I take it to be a fundamental *principle* of our universe, rather than just a theorem, that symmetries and conservation laws go hand in hand. For this reason I would like to promote Noether's theorem to *Noether's principle:*

> *The laws of physics must be such that every symmetry of nature corresponds to a conservation law, and vice versa.*

The next stage—what could be called the canonization of symmetry—was the discovery during the last 35 years that many, and maybe all, of the laws of physics themselves can be generated from symmetry principles. It now seems that all the interactions of physics are caused by a special kind of field called a *gauge field,* whose structure and behavior are completely dictated by a new symmetry requirement: the requirement of local symmetry. What's a local symmetry? Recall that symmetry, as I've been talking about it, is the equivalence of the laws of physics from different related points of view—broadly speaking, different frames of reference. The laws show a *local symmetry* if the equivalence persists even when you choose a different point of view at every point in space and at every possible time.

It seems likely that every continuous symmetry of nature is associated in this way with a gauge field, and in each case the conserved quantity corresponding to that symmetry (by Noether's principle) is exactly the thing that interacts with the gauge field. The familiar fact that gravity interacts with mass is an important example; mass is a form of energy, and energy is one of the conserved quantities related to the symmetries of space and time. (In fact, gravity interacts with all forms of energy, though often in a less obvious way than we're used to.) The electromagnetic field interacting with electric charge is a second example, and finally we have several new kinds of gauge field, interacting with exotic new kinds of "charge," that extend the idea to include all the interactions we know of.

Unification

The third characteristic—not new to this century, to be sure, but representing more and more a major theme of physics—is *unification.* Unification means showing that what may look like very different areas of physics are in fact governed by the same fundamental laws, or showing that what look like a large number of different particles are in fact made up of a small number of fundamental ones. We have seen unification in previous centuries: when sound and heat, for example, were identified as mechanical motion at a microscopic scale, governed by the same mechanical laws that Newton had devised for larger objects; and again when light was found to be an electromagnetic wave, governed by exactly the same equations that describe static or slowly varying electric and magnetic fields.

In this century, we've seen the unification of space and time, which are understood now as being both of the same fundamental character. We have seen the further unification of spacetime with gravity, which is no longer thought to be an arbitrarily added force but is believed to be the result of spacetime itself being curved. We have seen all the matter and forces that make up the world as we know it unified through the ideas of *quantum field theory:* All of the forces are directly due to fields like the gravitational and electromagnetic fields, whereas each of the particles, both familiar and exotic, that populate the universe is seen as a kind of quantum "ripple" of one of the fundamental fields. In the case of the electromagnetic field, these ripples, or *quanta* (the singular is *quantum*), of the field are called *photons*. Everything is now fields. (Gravity is still the troublemaker—like a field but somehow different.) I have to admit that it's hard to describe just what a field is; I'll try to do so in §6.2 and §11.5. This idea of a field is just one example of the increased abstractness of the mathematics that we need to use to describe the physics.

Gravity, incidentally, is just one example of our constant struggle with semantics. The words we use to describe phenomena reflect the mode of thinking that we're in at the moment. Many times it's completely natural to speak of a "gravitational field," a force field pervading space and producing the familiar gravitational forces, but at other times we need to use the geometrical picture of a curved spacetime, in which the idea of a gravitational field has no place. The actual physical effects are the same, but the language is very different.

We have seen the different elementary particles and fields of nature fall into families, forming patterns whose symmetries reflect the fundamental symmetries I mentioned. The electric and magnetic fields are related in this way to each other, and the combined "electromagnetic field" is in turn related, by further symmetries, to the fields responsible for the weak interactions—the interactions that produce certain kinds of radioactivity. The proton, the neutron, and the pi meson, members of the family of strongly interacting particles called *hadrons,* are now known to be composed of particles more elementary still, known as *quarks,* fewer in number and related again by appropriate symmetries. We believe now that there are six types of quark and six types of *lepton* (the electron and its family of related particles), along with three or four types of gauge field that are responsible for all the forces of nature. Physicists dream of removing the apparent arbitrariness of these different families and of deriving them all from a single grand group of symmetries. Many schemes have been proposed for performing this ultimate unification, but we're not there yet. I believe myself that a radical revolution at the deepest conceptual level may still be necessary before the pieces all fit together.

Part I of this book belongs to Albert Einstein. A separate introduction to Part I would make too many introductions before we get down to work, so I hope you will glance again at the various references in this introduction to Einstein and to his vision of a universe with no fixed framework and no meaningful distinction between "moving" and "stationary" as applied to both objects and observers. I hope to make clear in what follows what fueled this vision and how far it has led us.

SPACE AND TIME

See the little light ray moving
Space and Time are worth improving

SAMPLER MOTTO FOR OUR TIME

CHAPTER 1

Is the Earth Standing Still?

1.1   MOVING OBSERVERS AND LIGHT

The key idea of special relativity, that the laws of physics appear exactly the same to two observers moving uniformly relative to each other, was around long before Einstein. This equivalence of different observers worked beautifully for the motion of mechanical objects and all sorts of other things, but there was one little loose end: It didn't seem possible for it to be true for the behavior of light. According to Maxwell's theory of light as an electromagnetic wave, and according to very careful observations, light waves in empty space have a single definite speed c—a constant of nature. To an observer moving very fast alongside a light signal as it travels through space, on the other hand, the speed of the signal should appear less, just as when a car passes you at 65 miles per hour while you are going 55, and the car moves ahead of you at a relative speed of 10 mi/h. This would be natural enough, but it spoils the equivalence of different observers. If this description is right, the laws that describe electromagnetic radiation should be different for moving observers than for stationary ones.

Until Michelson and Morley's famous experiment in the 1880s, no one had ever looked for a shift like this in the speed of light: The speed of light is so great, and terrestrial speeds so small in comparison, that the effect would be too small to see in any normal observation. In the nineteenth century questions of invariance or symmetry were not a big issue, and no one was disturbed by a possible inequivalence between moving and stationary observers.

The story unfolds in two directions: On the one side was the experiment of Michelson and Morley, designed not to test any fundamental principles but merely to exploit the expected shift in the apparent speed of light in order to detect the motion of the earth through space—and finding quite unexpectedly that there is no shift at all, as if the earth were standing still. On the other side was Einstein, who, quite independently of any experimental observations, had become deeply committed to the principle that all observers should see the same physics, including especially the behavior of light. The consequences of extending the principle to light turned out to be earth-shaking. They start off with an explanation of the odd result observed by Michelson and Morley, and they lead finally to the need to rewrite all of physics and reshape our basic conceptions about space and time.

The initial focus of our discussion must be light, which is one kind of electromagnetic (EM) wave. Visible light is the most familiar example of EM waves; other common varieties of EM radiation are radio waves, microwaves, x-rays, and infrared and ultraviolet radiation. The speed of light in empty space, universally represented by the letter c, plays a central role in Part I of this book. Its value, to a very good approximation, is 3×10^8 m/s.*

The equations for the electromagnetic field that Maxwell developed in the 1860s give an elegant and satisfactory description of everything then known about EM theory, including a prediction of the speed of EM waves in empty space that agreed nicely with the observed speed of light, as well as it was then known. From Maxwell's equations, that speed turned out to be a universal constant independent of the wavelength

---

*The speed of light is in fact *exactly* 299,792,458 m/s, by a recent international agreement to use this exact value for c to redefine the meter, the basic unit of length.

and independent of any possible motion of the source. This would have to be a wave's speed as seen by a *stationary* observer, of course, because an observer moving in the same direction as the traveling wave would, as I said, expect to see it traveling at a reduced speed. In an extreme case that Einstein had thought about a lot, and that had begun worrying him by the age of sixteen, the observer might be moving *at* the speed of light, and the light wave should then appear to be standing still. Now a stationary light wave is definitely not allowed by Maxwell's equations because they predict a definite speed for the waves, and that's c, not zero. Thus, according to the laws of physics as they were then understood, Maxwell's equations apply *only* for a truly stationary reference frame.

This idea disturbed Einstein greatly because of his commitment to the idea that all frames of reference should be equivalent and that the speed of light should therefore appear the same to all uniformly moving observers. The term frame of reference is used a lot in this kind of discussion. It denotes the basis a particular observer uses to describe events, and it typically refers to a system of space and time coordinates used by that observer. You can think of the *reference frame* as the observer's laboratory, supplied with clocks, rulers, and other measuring devices, moving with the observer and providing a framework for all possible measurements. (See §2.1 for a more detailed discussion of reference frames.)

There is every indication that when Einstein resolved this problem in 1905 with the special theory of relativity, he had taken no particular notice of the Michelson–Morley experiment, a crucial and delicate experiment that seemed to show this equivalence of different frames of reference very clearly. As Einstein said in 1931, the Michelson–Morley experiment "uncovered an insidious defect in the ether theory of light, as it then existed, and stimulated the ideas of H. A. Lorentz and FitzGerald out of which the special theory of relativity developed."*

---

*Quoted by A. Pais in *Subtle is the Lord, The Science and the Life of Albert Einstein,* 1983, Oxford University Press, New York, p. 116. The big issue at the time was the motion of the earth relative to the *ether,* the supposed material substance thought to fill all of space and to act as a medium for the propagation of light, analogous to the role of air with sound waves. Experiments such as Michelson and Morley's were known as ether-drift experiments and were interpreted, though with severe difficulties, as indicating that the ether is dragged along with the earth. The concept of an ether was buried by Einstein (it was already pretty sick by then), and I shall not discuss its history any further here. As we now understand, EM waves require no medium, and they exist and propagate even in completely empty space.

Einstein's way of thinking, which has deeply influenced later generations of physicists, is that nothing can be said to be truly "at rest"—that it's physically meaningless to speak of a thing being at rest or moving *except in relation to something else,* an idea that people have very naturally found it difficult to get used to.

The logical development that I'd like to take you through in the next few chapters unfolds in three stages:

(1)   How the Michelson–Morley experiment works, and how it forces us to believe that in the real experimental world, the speed of light always comes out the same no matter who measures it, as expressed also by Einstein in his principle of relativity.

(2)   Why this equivalence of differently moving observers forces us to conclude that moving clocks and rulers must necessarily function in an unexpected way, and what sort of changes are in fact required.

(3)   How this unexpected behavior of measuring implements leads to a new and non-Euclidean view of the geometrical structure of space and time. This new view is embodied in the idea of *Lorentz invariance,* which is the mathematical statement of the relativity principle—the principle that all uniformly moving observers see the laws of physics working in exactly the same way.

PROBLEMS

1.   The speed of light is now defined to be exactly 299,792,458 m/s.
  (a)   What is the approximate relative error in taking $c = 3 \times 10^8$ m/s, or 300,000 km/s, as we do in most of our discussion? (Relative error is the error expressed as a fraction or percentage of the actual value.)
  (b)   Calculate how many miles light travels in a millisecond— 1/1000 of a second. How many times around the earth can light travel in a second? (See Appendix C, §5, for astronomical information, and imagine an occasional mirror to keep the light beam on track.) The nearest star outside the solar system is 4 light-years away (that is, it takes 4 years for light to get to us from there). How many miles is it to that star?*

---

*This is the star called α Centauri, or Alpha Centauri, which means (α being the first letter in the Greek alphabet) that it's the brightest star in the constellation Centaurus, the Centaur.

(c) What difference would it make in the time for light to get to us from α Centauri if the figure for c differed by 1 m/s—that is, if it were 299,792,457 m/s instead of 299,792,458 m/s?

(d) A light signal traveling at the correct speed, 299792458 m/s, and some other signal traveling at 299,792,457 "m/s" are sent from New York to San Francisco, a distance of about 4800 km. How far is the light signal ahead of the other signal when they arrive in San Francisco? (This experiment has to be done in an evacuated tube, because the slowing down of the light signal due to air is about 90,000 times bigger than the effect you're looking at.)

2. The radius of the earth's orbit about the sun is 1.49×10^{11} m. Find its period in seconds, and evaluate v/c, its speed relative to the speed of light. (One year is 365.24 days.)

3. The basic rate equation for velocities is distance = velocity × time, or

$$\Delta x = v \Delta t$$

where $\Delta x \equiv$ "*change in x*" = *distance traveled in time interval* Δt. An observer S′ traveling at speed v_0 tries to measure the speed of an object that is traveling at speed v, greater than v_0, in the same direction.

(a) Find how far the object travels relative to S′ in a time Δt—that is, the difference between the distance the object travels, Δx, and the distance S′ travels, Δx_0—and thereby show that the apparent velocity of the object as seen by S′ is

$$v' = v - v_0$$

Explain whether your reasoning, and hence this formula, continue to be valid if $v_0 > v$ or if $v_0 < 0$.

(b) The speed of the earth relative to the sun is roughly 1/10,000 of the speed of light. Find the difference Δt that the motion of the earth should make in the time it takes a light signal to travel from New York to San Francisco [prob. 1(d)]. How far does a light signal travel in the time Δt?

1.2 MICHELSON, MORLEY, AND EINSTEIN

The first clue that there was something wrong with our understanding of space and time appeared, as I said, in the 1880s, twenty years before

Einstein's theory of relativity. The Michelson–Morley experiment was intended to measure the speed at which the earth moves through space by exploiting the small variations in the apparent speed of light that I referred to in §1.1. The experiment was accurate enough that it could detect speeds much less than the known speed of the earth's motion around the sun, and Michelson*—working first alone and then with E. W. Morley in 1887—expected no difficulty in seeing results. The idea was to look for differences, due to the motion of the apparatus, in the time it takes for light to travel in different directions. To the great astonishment and distress of the experimenters, a *null* result was obtained. That is, no such differences were observed, implying that the earth is standing still! The experiment was repeated in subsequent years by many other scientists looking for possible explanations of this apparently tremendous coincidence, and the null result persisted. Something was apparently wrong with even the very simple idea that the motion of an object should appear slower (or faster) to an observer moving in the same (or in the opposite) direction.

To get a feeling for how the problem is resolved, we need to think carefully about how we make measurements in space and time. We have finally come to realize that the only way to explain the Michelson–Morley result is to suppose that our measuring devices—our clocks and rulers—must somehow be functioning in a peculiar way. Physicists in the late 1800s tried very hard to explain the effect in various other ways. (For example, they proposed a change in the behavior of light near the moving earth: If light travels through the ether, and if the ether is dragged along by the earth, then that might explain the effect.) Eventually such explanations had to be abandoned. Others before Einstein—FitzGerald, Lorentz, and Poincaré in particular—realized that a distortion of moving objects, including rulers, would explain results like those of Michelson and Morley, but they could see no reason why there should be such a distortion nor why it should so precisely conspire to produce the null result that Michelson and Morley observed.

It was Einstein who got the perspective right. It was he who recognized the equivalence of all frames of reference as a fundamental principle of nature, and the apparent misbehavior of clocks and rulers as the necessary consequence of that principle. One very important reason Einstein's point of view has been so widely accepted is that physicists have learned not to believe in coincidences. It is generally much more

---

*A. A. Michelson—pronounced like "Michaelson."

likely that an apparent coincidence is due to some underlying cause than just happenstance, and being suspicious of coincidence almost invariably leads to richer physics and deeper understanding. The alternative point of view, remember, is that the apparent conspiracy of clocks and rulers *is* just a big coincidence. We would then have to believe that there is no fundamental principle involved, that there really is just one unique reference frame that is truly at rest, that clocks and rulers function in this strange way for some entirely extraneous reason, and that this strange behavior just happens to conceal the motion of the observer.

The idea that there is no preferred frame of reference—that it's meaningless to speak of any observer as being truly at rest—is difficult to get used to. To be accurate, you have to keep adding "relative to so-and-so" whenever you speak of a stationary or moving object or observer and when you talk about the peculiar behavior of moving rulers and clocks. In this book I'll generally speak as if the earth is at rest and try to remind you from time to time that any other perspective would have done equally well.

PROBLEM

1. A fossilized bug is embedded in a rock, undisturbed for as many years as necessary for this discussion. Describe the motion of the bug relative to (a) the rock, (b) the center of the earth, (c) the sun, (d) the center of our galaxy.

1.3 INTERFERENCE OF LIGHT WAVES

The device used by Michelson and Morley is called an *interferometer* and was designed by Michelson specifically for the purpose of performing this experiment. The interferometer has found a great many other valuable applications over time and is still in use today in modern forms. There are simple interferometers in most physics departments—they're not hard to make.

By making use of the phenomenon of interference of light waves, the interferometer lets us compare very precisely the times it takes for light to travel in two different directions in the apparatus. The simple logic involved in interference is this: If light can arrive at a particular point by two different paths, then the two waves may arrive in step or out of step or somewhere in between. If they arrive in step the two waves reinforce each other and if they arrive out of step they cancel each other; both of these effects go by the name interference. The fact that light shows interference was, historically, the important piece of evidence that light is indeed some kind of wave.

Waves

Because of the key role light plays in helping us understand relativity, and also because we need to understand the role that wave interference plays in the Michelson–Morley experiment, I believe we should now pause and look briefly at the nature of waves.

One of the most basic facts about waves is that they typically fill up all the space available. Consider some common examples: The waves on a pond or ocean go on and on in all directions, as far as the water extends. They can be smooth and regular or rough and irregular; they can vary in character from one location to another. Even if the entire pond appears still, the surface is never perfectly motionless, so there is always at least a little bit of wave motion across the whole pond.

Another important example is that of sound waves in a room. Again the wave motion—the pattern of pressure variations that constitutes sound—fills the entire room, and again it can vary in character from one place to another. Note, however, that the water waves extend across a two-dimensional surface, while the sound waves fill a three-dimensional volume.

In both of these cases there is a "medium," the water surface or the air in the room, whose vibrations at different points make up the wave motion of the whole. The medium can also be one-dimensional, as in the case of vibrational waves on a stretched guitar string.

Let me now be a little more specific about some of the facts I've been describing, and let me start with a definition:

▌ *A wave is a spatial pattern that varies with time.*

Think, for example, about the waves on a pond. If you take a snapshot or better yet push the pause button on a video of the scene, you see the spatial pattern at a single moment of time, and the elevation of the water surface is different at each point on the surface. If you now release the pause button and allow the picture to go forward, the whole pattern is set in motion. This variation of the picture with time is characteristic of wave motion.

There are several important points that you should think about carefully:

(1) A wave is not the same as an oscillation. The word *oscillation* refers to a motion that has no spatial pattern but is simply a variation in time, like the motion of the bob on a pendulum or the day-by-day variation of the temperature reading on a thermometer. A snapshot of an oscillation reveals no pattern at all. A typical oscillation consists of some simple motion repeated over and over many times. A single repetition of this motion is called a *cycle*—one single back-and-forth swing of the pendulum, for example.

We have several words to describe different aspects of this cyclical motion: The *period* of the oscillation is the length of a single cycle, measured in seconds or days or years. The period is often written as T. The *frequency* is the rate of oscillation—the number of cycles per second—and is written as f. The frequency is simply the inverse of the period,

$$f = \frac{1}{T} \tag{1}$$

and the unit used for frequency is the hertz, abbreviated Hz. A hertz is just one cycle per second. One reason the unit was introduced was to allow the use of scientific prefixes; a kilohertz (kHz) is a thousand cycles per second, and a megahertz (MHz) is a million cycles per second. The inverse relation (eq. 1) is straightforward when you think about it: If each cycle takes $1/20$ of a second—that is, $T = 0.05$ *s*—then there are 20 cycles each second and $f = 20$ Hz. If each cycle lasts 100 seconds ($T = 100$ s), on the other hand, then there's only $1/100$ of a cycle each second and $f = 0.01$ Hz.

A wave, by contrast, consists of an infinite number of oscillatory motions, one at every point in space and all generally different. When you take a snapshot, you see the pattern made by all these different oscillations at that moment of time.

(2) A wave is very often, and typically, oscillatory in character, but it need not be. Examples of non-oscillatory waves: the wave thrown off by the bow of a speedboat, the sonic boom from a supersonic plane, the sound wave emitted from a single gunshot or handclap. In each of these examples, there's a single big disturbance that passes any one point for just a short time, and only tiny motions are seen anywhere else. Such a localized disturbance, passing just once like the bow wave, is called a *wave pulse.*

(3)   The physical quantity that exhibits the wave pattern, which is called the *field variable,* can be any locally defined quantity measured in almost any sort of unit. In the case of water waves it's the elevation, measured in meters or inches; in the case of sound waves it can be the pressure, measured in newtons per square meter (N/m^2) or pounds per square inch (lb/in^2). (The newton is the metric unit of force.) For light waves and radio waves it turns out to be the electric and magnetic field strengths. The field variable oscillates about some central value that's normally taken as zero. The actual value of this variable at some point in space and time is referred to as the *displacement* of the wave at that point, and the maximum value of this displacement is called the *amplitude* of the wave.

The graph of a typical wave at a single moment of time is shown in fig. 1. The figure shows the particularly simple and smooth type of wave known as a sine wave (explained more fully in Chapter 11); a wave can in fact have an arbitrary shape. The *wavelength* of a wave pattern is illustrated in fig. 1; it's the length of a single complete unit of the repeating pattern, written as λ and measured in meters or feet.

(4)   Waves can be one-, two-, or three-dimensional, depending on how many spatial variables are involved. Thus a wave in a stretched rope is one-dimensional, and its displacement—y, say—is a function only of x

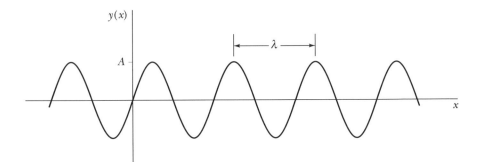

Fig. 1   A sine wave. A is the amplitude and λ is the wavelength.

and t: $y(x, t)$, where x is the distance along the rope. A sound wave, on the other hand, is a three-dimensional wave, because it's a pattern of pressure variations over all three dimensions of space: $p(x,y,z,t)$ or $p(\underline{r},t)$. The symbol $\underline{r}$, which is called a *vector*, represents the location of a point in our three-dimensional space. You can think of $\underline{r}$ as just a short-hand way of indicating the three coordinates x, y, and z. Note that the wave in a horizontal rope is still one-dimensional even if it's allowed to oscillate in both the vertical and horizontal planes, because the rope itself extends only in one direction. If y and z represent the displacement of the rope in the two directions perpendicular to its length, then the wave is described by the two functions $y(x,t)$ and $z(x,t)$, still with just the one spatial variable x.

The math needed for a complete description of wave motion is very complicated because of the diversity of types of wave and the complexity of wave motion, which consists of an infinite number of oscillations at all the different points in space. Chapter 10, especially §10.1, deals with this in more detail.

Interference

A familiar example of the interference of light waves is the appearance of colors on a soap bubble or an oil slick. You probably know that ordinary white light is really a mixture of many different colors, each corresponding to a different wavelength. What is happening at some particular spot on a soap film or oil slick is that the colors are being reflected selectively—some colors are being reflected and others are not. If there were only one wavelength of light shining on the surface, just one pure spectral color, then you would see dark and light patches all in that same color rather than areas of different color. In each of these examples, the soap film and the oil slick, the light encounters a very thin film of transparent material. As shown in fig. 2, the light may be reflected off the first surface it encounters, or it may go on through that first surface, be reflected off the second surface, and then pass back through the first surface again. (Most of the light just goes on through the film without being reflected at all, but we're interested here only in the part that's reflected.) There are thus two different ways for the light to be reflected off the film and return to the eye of the observer, and consequently there is the possibility of interference. If the thickness of the film is just right, the two reflected waves can arrive at your eye *in step* with

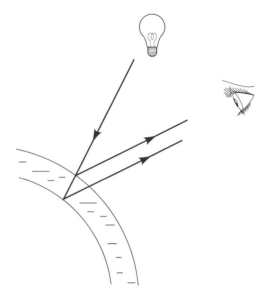

Fig. 2 Interference from a soap film. (The thickness of the film is exaggerated.)

each other, as shown in fig. 3, and you see the colored light reflected from that portion of the film. This is *constructive interference*. At another spot, where the film is a quarter-wavelength thicker, say, the light that enters the film has to travel a half-wavelength more on its two-way trip

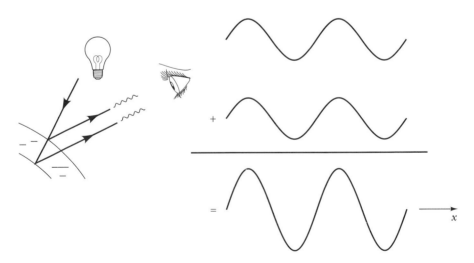

Fig. 3 Constructive interference. The waves reinforce each other.

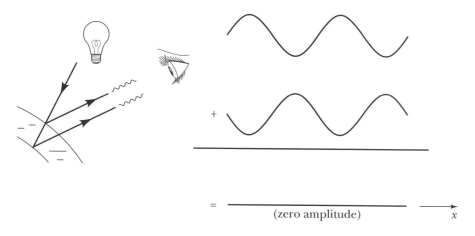

Fig. 4   Destructive interference. The waves cancel each other.

through the film, and thus it gets exactly *out of step* with the light that bounced off the first surface. The two waves then cancel each other, as shown in fig. 4, and that portion of the film looks dark to you. This is called *destructive interference.* Thus you get light and dark areas that depend on the varying thickness of the film in relation to the wavelength of the light. If you use light of a different wavelength, you get a different pattern.

Now, because white light consists of many different wavelengths, when the film is illuminated with white light each of the different wavelengths produces a different interference pattern. At a place where the film has a particular thickness, there is constructive interference for some wavelengths and destructive interference for others, and the wavelengths that are reinforced give the colors you see. Of course, if all colors were reflected equally, they'd add up to white light again, but where only some colors are reflected and not others, you see the light as colored. What color you see depends on the thickness of the film at that point.

The Michelson interferometer

Michelson's interferometer is a device that uses this same principle to allow a precise comparison of two different light paths, with the added feature that you can control the lengths of those paths. It is illustrated very schematically in fig. 5. I've shown a laser as the light source, because

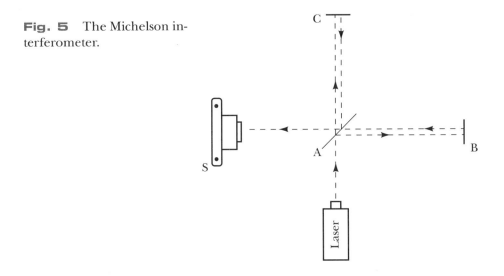

Fig. 5   The Michelson interferometer.

it's important that a monochromatic light source—one that produces light of a single wavelength—be used. Michelson had to use other less effective devices. Mirror A is half-silvered, which means that it has only a very thin reflecting coating, so about half the light is reflected and half goes on through. After being reflected from mirrors B and C, the light can get to observer S by again passing through or being reflected by the half-silvered mirror A. (Observer S is now a camera, in order to remove any risk of eye damage from the laser.) The two possible light paths, shown separately in fig. 6, can produce constructive or destructive inter-ference, depending on the difference in the lengths of the two paths. Fig. 6(a) shows light that's reflected first at mirror A and again at mirror B and then passes through mirror A to arrive at the observer S. Fig. 6(b) shows light that passes through mirror A, is reflected off mirror C, and finally is reflected off mirror A to reach S. If the two legs AB and AC are exactly the same length, then the light waves following the two different paths arrive in step at S and reinforce each other. Moving either mirror B or mirror C by a quarter-wavelength would change that light path by a half-wavelength, causing the waves to arrive out of step and cancel each other.

For use in measuring lengths with precision, the interferometer would have one mirror movable and controlled by a screw drive. As the mirror is moved, the field of view at the telescope would go dark and bright successively, with one cycle of dark and bright for each half-wavelength

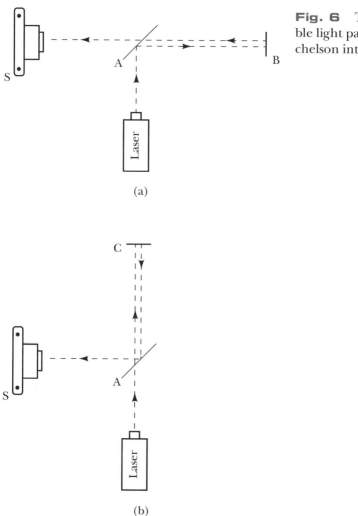

Fig. 6 The two possible light paths in the Michelson interferometer.

(a)

(b)

that the mirror moves. A standard procedure for allowing a more precise estimate of fractions of a cycle is to tip one of the mirrors slightly: If the telescope is focused on that mirror, then some parts of the field of view are dark and others bright, in parallel bands known as *fringes*. As the movable mirror is advanced, the fringes appear to move across the field of view, and a displacement of the fringes corresponding to a small fraction of a single cycle can be estimated with accuracy.

The wavelengths of visible light range from 400 to 700 nm (1 nm = 1 nanometer = 10^{-9} m), or around 1/20,000 of a centimeter, so this device allows us to make extremely precise length measurements. Michelson and Morley, in fact, were able to detect with considerable confidence a change in one of the two legs of as little as 1/100 of a wavelength.

PROBLEMS

1. The colors on a film of oil on water are produced by interference between light reflected from the upper surface of the film and light reflected from the lower surface. The distances the two rays of light travel differ by $\Delta\ell$, which is twice the thickness of the film.
    (a)   If the thickness of the film is 1.0 μm (1 μm = one micron = 10^{-6} m), find the wavelengths in the visible spectrum (0.4–0.7 μm) for which the two rays exactly reinforce each other ($\Delta\ell$ is an integer number of wavelengths) and the wavelengths for which they exactly cancel each other ($\Delta\ell$ is an integer number of wavelengths plus half a wavelength). Ignore the fact that the wavelength of light inside the film is modified by the reduction of the speed of light in oil.
    (b)   Find the minimum film thickness for which 0.5-μm light will be exactly reinforced and 0.6-μm light will be exactly canceled.
    (c)   Make an order-of-magnitude estimate (to within a factor of 5, say) of the film thickness at which the colors will merge into a uniform white.
2. If you place a hair between two very flat pieces of glass, the air gap varies in thickness from zero where the two pieces of glass are in contact to the thickness of the hair at the location of the hair. When this arrangement is viewed under monochromatic light of wavelength λ, the air gap shows a large number of light and dark fringes. Find a formula for the thickness of the hair in terms of the number n of dark fringes and the wavelength of the light used. How many fringes will be seen if the hair thickness is 0.08 mm and the wavelength of the light is 0.6 μm?
3. The interferometer shown in fig. 5 is used with light of wavelength λ = 0.540 μm to measure the displacement of the movable mirror. The number of fringes counted as the mirror is moved is 92.7.
    (a)   Find the displacement of the mirror.
    (b)   How many fringes would be counted if light of wavelength 0.650 μm were used?.

   (c) How many fringes would be observed for a displacement of 1 cm, using $\lambda = 0.540$ μm?

4. **(a)** Show that the time delay corresponding to an increase of 1/100 of a wavelength in the length of one of the interferometer legs, a change such as Michelson and Morley could have detected, is of the order of 10^{-16} s.

   (b) To get an idea of how small a time this is compared to one second, find the time interval that is the same fraction, 10^{-16}, of the age of the universe, which is around ten billion years.

5. When you use laser light in the interferometer, the fringes all look just the same regardless of the difference in path lengths. Suggest a way to find the "central fringe" corresponding to a path difference $\Delta\ell = 0$.

1.4 MEASURING THE SPEED OF THE EARTH: THE MICHELSON–MORLEY EXPERIMENT

How would you use the interferometer described in §1.3 to detect the motion of the earth? Well, if the apparatus is moving through space, then the distance the light has to travel is different for one leg than for the other, even though the two legs are exactly the same length. This difference has just the same effect as a difference between the lengths of the legs and should therefore be observable as a shift in the interference pattern* in just the same way, as I'll explain in detail in this section. The effect can also be described in terms of an expected difference in the apparent speed of light that results from the motion of the apparatus. This way of looking at it will be of interest to us later because the apparent speed of light is one of our main concerns in the theory of relativity, but as a practical matter it's easier to describe the effect in terms of the difference in the light paths, and hence in the time it takes for light to travel along the two paths.

---

*There's a slight complication in that you can really measure only *changes* in the interference: The pattern you see doesn't tell you how many wavelengths difference there is between the two path lengths, but changes in the pattern can be observed with accuracy. You get around this difficulty by watching for a shift in the pattern as you rotate the apparatus through 90° (see prob. 5).

The null result that Michelson and Morley obtained, then, was that they saw no shift at all! The precision with which they could detect shifts in the pattern should have enabled them to get an observable effect for a speed of about 10 km/s. The speed of the earth in its orbit around the sun is about 30 km/s, and (as we now know) the speed of the solar system in its motion around the center of our galaxy is over 90 km/s, so Michelson and Morley should have had no difficulty at all in seeing the effect. Is it possible that Copernicus was wrong and Ptolemy was right: that the earth is standing still and the heavens are revolving around it?*

To Michelson and Morley, and to many others, the result was mystifying because the earth was obviously *not* at rest, and they could see nothing at all wrong with the logic that led to the predicted result. Someone like Einstein, on the other hand, who believed deeply in the equivalence of different frames of reference, would have been convinced that the moving observer *must* see exactly the same physics as a supposedly stationary one. Thus not only would Michelson and Morley get a null result, just as a stationary observer would, but any experimenters who tried to make a direct measurement of the speed of light would inevitably get exactly the value c regardless of how they might be moving. If the apparatus has to function in an unexpected way in order to get such results, then so be it. Much of the theory of relativity amounts to a description of this odd behavior, which we have to get used to because it is indeed the way things are.

Let's look more carefully at the experiment and see first how it was supposed to work without regard to relativity theory. The object here is to figure out the difference between the two light paths in the apparatus in fig. 1.3.5, in terms of the speed v at which the apparatus is moving through space, so that we can then turn the result around and *deduce v* from the interference effect caused by that path difference. Let's assume for this discussion that the direction of motion is in the direction of the leg AB—that is, to the right in the diagram of fig. 1.3.5.

What we want, then, is the difference between ℓ_{ABA}, the light path from A to B and back to A [fig. 1.3.6(a)], and ℓ_{ACA}, the light path from A to C and back to A [fig. 1.3.6(b)]. (The paths from the light source to A

---

*Nicolaus Copernicus, a Polish astronomer (1473–1543), got into trouble for proposing that the earth moves in orbit around the sun. Claudius Ptolemaeus was a Greek mathematician and astronomer (127–151) whose complicated scheme for describing the motion of the stars and planets around the earth was accepted for more than a thousand years.

and from A to the observer are the same in both cases, so we can ignore them in our discussion.) When the light is traveling from A to B, it's going in the same direction as the apparatus, so its speed v_{rel} relative to the apparatus (see §1.1, especially prob. 1.1.3) is the difference $c - v$, and the time it takes to get there, which I'll call t_{AB}, is longer than if v were zero:

$$t_{AB} = \frac{L}{v_{rel}} = \frac{L}{c - v} \tag{1}$$

Here L is the distance from A to B, (the length of one leg of the interferometer), and the time is equal to the distance divided by the relative speed. The distance the light signal travels in that time, ℓ_{AB}, is the light speed times the time:

$$\ell_{AB} = ct_{AB} = \frac{Lc}{c - v} \tag{2}$$

For the return trip, from B to A, the relative speed is $c + v$, because the light is now traveling in the opposite direction to the apparatus. The time is

$$t_{BA} = \frac{L}{v_{rel}} = \frac{L}{c + v} \tag{3}$$

and the light path is

$$\ell_{BA} = ct_{BA} = \frac{Lc}{c + v} \tag{4}$$

We now combine the two distances ℓ_{AB} and ℓ_{BA} to get the total light path A → B → A:

$$\ell_{ABA} = \ell_{AB} + \ell_{BA} = \frac{Lc}{c - v} + \frac{Lc}{c + v} \tag{5}$$

$$= \frac{2Lc^2}{c^2 - v^2} \tag{6}$$

This is the path length that we'll want to compare with the light path for the route ACA. It's useful to introduce here a quantity γ, defined as

33

$$\gamma = \frac{1}{\sqrt{1 - v^2/c^2}} \qquad (7)$$

and to write the light path ℓ_{ABA} as

$$\ell_{ABA} = \frac{2L}{1 - v^2/c^2} = 2L\gamma^2 \qquad (8)$$

where you'll note that the factor $2L$ would be the length of the light path if the apparatus were at rest. The factor γ^2, then, is just a way of describing the increase in path length that is due to the motion of the apparatus.

The quantity γ (lower-case Greek letter gamma), introduced here as a shorthand symbol for the expression on the right side of eq. 7, turns out to play a very important role in the theory of relativity, and it crops up everywhere. You'll see as we proceed that it's the most natural measure of the degree of relativistic effects of all sorts, which is why I'd like you to get used to it. A graph of the relativistic factor γ versus v is shown in fig. 1.

Note the following properties of γ:

$$v = 0: \quad \gamma = 1 \qquad (9)$$

$$|v| \ll c: \quad \gamma \approx 1 + v^2/2c^2 \qquad (10)$$

$$v \to \pm c: \quad \gamma \to \infty \qquad (11)$$

Fig. 1   The relativistic factor γ.

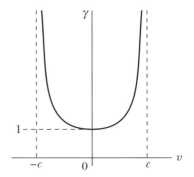

Mathematical note: Eq. 10 is an example of the very useful approximation

$$(1 + \varepsilon)^n \approx 1 + n\varepsilon, \quad \text{for } |\varepsilon| \ll 1 \text{ and } |n\varepsilon| \ll 1 \qquad (12)$$

Eq. 12 is true for any n and is easy to check for $n = 0, 1, 2, \ldots$ by using the binomial theorem and neglecting ε^2 (see prob. 2 at the end of this section). Eq. 12 is easy to derive exactly using calculus. To get eq. 10, substitute the values $\varepsilon = -v^2/c^2$ and $n = -1/2$.

You should now examine what you've found out about the light path ℓ_{ABA}. First, note that the path is longer when $v \neq 0$ than when $v = 0$ and that as v approaches the speed of light c, the factor γ becomes very large and the path becomes very large also—we say that $\ell_{ABA} \to \infty$. This is due to the first leg ℓ_{AB}, given by eq. 2, where the denominator goes to zero as $v \to c$, reflecting the fact that the light has to travel a very long distance to catch up with mirror B if the mirror is going at nearly the speed of light. Second, note that for the actual Michelson–Morley experiment, the speed v is very much less than c. For the earth's motion around the sun, $v/c \approx 10^{-4}$, which means that we're very close to the center of the graph in fig. 1 and the approximate form of eq. 10 holds. The deviation of ℓ_{ABA} from its zero-velocity limit $2L$ is proportional to the square of v/c and is therefore of the order of a part in 10^8, or a millionth of 1%.

Now let's calculate ℓ_{ACA}, the length of the light path A $\to$ C $\to$ A along the other leg of the interferometer, in order to compare it with ℓ_{ABA}. Again the light has to travel a distance greater than $2L$, but the picture is now different. The path is shown in fig. 2. The logic is the same, but the formulas are different. Because the apparatus is moving to the right, the light signal has to travel along a diagonal path to get to mirror C, and it reaches that mirror when it has moved to the location C$'$. The light has to move at an oblique angle θ, as shown,* and the velocity vector for the light signal is therefore oriented at the same angle

---

*You don't need to worry about why the light signal goes at an oblique angle, as shown. For one thing, the angle is extremely tiny; for another, the motion of the mirror at A has just the effect of reflecting the signal at the correct angle even if the speed of the apparatus is large; and finally, you always make the little adjustments to the mirrors needed to get the light to go where you want it.

Fig. 2   Light path for fig. 1.3.6(b).

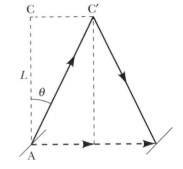

θ relative to leg AC of the interferometer, as shown in fig. 3. The speed of the light signal relative to the apparatus, v_{rel}, is therefore

$$v_{rel} = c\cos\theta \tag{13}$$

the component along the direction AC. (See Appendix A, §2, for a review of vectors.)

   The angle θ is determined by trigonometry (see Appendix A, §1): The light is traveling at speed c along the diagonal in fig. 2, and mirror C is traveling at speed v along the top, so that

$$\sin\theta = \frac{v}{c} \tag{14}$$

and eq. 13 then gives

Fig. 3   The velocity vector of the light signal of fig. 2. Its components are v, the speed of the apparatus, to the right, and v_{rel}, the speed of the light signal relative to the apparatus.

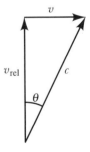

$$v_{rel} = c\cos\theta = c\sqrt{1 - \sin^2\theta} \qquad (15)$$

$$= c\sqrt{1 - \frac{v^2}{c^2}} = \frac{c}{\gamma} \qquad (16)$$

We get the light path by the same routine as before (eqs. 1–4):

$$\ell_{AC} = ct_{AC} = \frac{Lc}{v_{rel}} = L\gamma \qquad (17)$$

The return trip is just the same because the slope of the diagonal path is the same, so the light path for the complete trip $A \to C \to A$ is

$$\ell_{ACA} = 2\ell_{AC} = 2L\gamma \qquad (18)$$

Recall that $2L$ is the length of the light path when the apparatus is at rest, and note that the extra factor is now γ, compared with γ^2 for the path $A \to B \to A$.

Now it's the difference between the two paths ℓ_{ABA} and ℓ_{ACA} that will determine the interference, so we must compare these two path lengths, eqs. 8 and 18. The same factor γ appears in both of these, but it is squared in one and not in the other. The effect (the increase in path length due to the motion of the apparatus) is extremely tiny in both cases, as I mentioned, but is about twice as great for ℓ_{ABA} as for ℓ_{ACA}. To estimate the effect, you can use the formula of eq. 12 to get approximations for both γ and γ^2. We used it to get the approximate value $1 + v^2/2c^2$ for γ (eq. 10), and we can use it again, with $\varepsilon = -v^2/c^2$ and $n = -1$, to get

$$\gamma^2 \approx 1 + \frac{v^2}{c^2} \qquad (19)$$

So you see that ℓ_{ABA} is greater than ℓ_{ACA} and that the difference, finally, is

$$\ell_{ABA} - \ell_{ACA} = 2L(\gamma^2 - \gamma) \approx (2L)\left(\frac{v^2}{2c^2}\right) = \frac{Lv^2}{c^2} \qquad (20)$$

Let me remind you of how you would expect to observe this path difference. What you're able to see with an interferometer is not just a state of affairs, like this path difference, but a *change* in the state of affairs. If you rotate the apparatus through 90°, then leg AB will be perpendicular to the direction of motion and leg AC will be parallel to it, and their roles will be reversed. The path ℓ_{ACA} will now be the longer one, and you'll have

$$\ell_{ABA} - \ell_{ACA} \approx \frac{-Lv^2}{c^2} \tag{21}$$

In the process of rotating the apparatus, the path difference changes by twice the difference shown here, or $2Lv^2/c^2$, which would then be observable as a shift in the interference pattern, just exactly as if one of the mirrors were simply moved through half that distance.

This is the effect that Michelson and Morley were hoping to observe, and thereby to deduce the value of v, the speed of the earth through space—through the "ether," as they understood it at that time. The important result, as I've said, is that *they failed to see any effect*. We now have to consider what can be wrong with the argument that I've described to you. I've presented this argument in a somewhat casual way, but the logic is actually quite clean. In such a case we have to reexamine our assumptions, which we'll do with some care in the next section.

PROBLEMS

1. Write out in detail the algebraic steps needed to go from eq. 5 to eq. 8, $\ell_{ABA} = 2L\gamma^2$.
2. (a) Check the approximation formula (eq.12)

$$(1 + \varepsilon)^n \approx 1 + n\varepsilon, \text{ where } \varepsilon \ll 1$$

for the cases $n = 0, 1, 2, 3$, using the binomial theorem where appropriate and neglecting ε^2 and higher powers of ε.
   (b) Check the approximation formula with your calculator for at least five other values of n, including negative and noninteger values, and construct a table to show how in each case the approximation gets better as ε gets smaller.
3. (a) Explain physically why $\ell_{ABA} \to \infty$ in eq. 8 when $v \to -c$.
   (b) Explain physically why $\ell_{ACA} \to \infty$ in eq. 18 when $v \to \pm c$.
4. Complete the derivation of eq. 17 from eq. 13, and explain carefully why the same formula applies for the return trip from C to A.
5. You've seen in this section that $\Delta\ell = \ell_{ABA} - \ell_{ACA}$, the difference between the light paths for reflection from mirror B and from mirror

C (fig. 1.3.5), is given by $\Delta\ell \approx L v^2 / c^2$ when the apparatus is moving with speed v in the direction of mirror B, as in the text.

(a) Find the change in $\Delta\ell$ as the apparatus is rotated by 90°, if the leg length L is 10 m and the speed v is $(8 \times 10^{-4}) c$, the speed of our solar system relative to the center of our galaxy.

(b) What fringe shift does this correspond to if the wavelength of the light used is 0.5 μm? (In this problem, we must assume that the odd effects due to relativity—not yet discussed, of course—do not occur.)

6. Suppose the Michelson interferometer has legs of slightly different lengths, L_1 and L_2, and that the experiment is done first with leg 1 parallel to the motion of the apparatus and then with leg 2 parallel to the motion.

(a) Show that the difference in the light paths in the two cases is given by

$$\text{Case a: } \Delta\ell_a = 2L_1 \gamma_0{}^2 - 2L_2 \gamma_0$$

and

$$\text{Case b: } \Delta\ell_b = 2l_1 \gamma_0 - 2L_2 \gamma_0{}^2$$

(b) Show that the difference, then, between $\Delta\ell_a$ and $\Delta\ell_b$—this difference being what is measureable—is approximately given by

$$\Delta(\Delta\ell) \equiv \Delta\ell_a - \Delta\ell_b \approx \frac{(L_1 + L_2) v_0{}^2}{c^2}$$

$$\approx \frac{2L v_0{}^2}{c^2}$$

where L is the average of the two lengths L_1 and L_2.*

7. In the Michelson–Morley experiment, using the interferometer of fig. 1.3.5, suppose the length of leg AB is L_1, the length of leg AC is L_2, and the device is moving rigidly in the direction of leg AB with speed v. Assuming that length L_2 is not affected by the motion of

---

*Note that the quantity $\Delta(\Delta\ell)$ is here expressed not as the difference of two large quantities—which would require that you know them to great precision in order to get the difference right—but simply as a small factor, $(v_0/c)^2$, times the average length L. This means that L need not be known to great precision. You typically need to calculate $\Delta(\Delta\ell)$ to two- or three-figure accuracy at most, so you need L only to that same accuracy, even though the path difference you're interested in may be of order $10^{-8} L$. Thus you can normally just use L_1 or L_2 instead of the average L, provided they don't differ by too much.

the apparatus, determine how length L_1 needs to vary with v in order to make the path lengths ℓ_{ABA} and ℓ_{ACA} equal (which is what Michelson and Morley found).

1.5 MOVING OBJECTS SHRINK— THE PRINCIPLE OF RELATIVITY

In the last section we worked out what Michelson and Morley should have seen, according to the straightforward logic of classical physics. Now we must confront the fact that they didn't see the effect they were looking for at all and must think more carefully about what that implies.

There are four assumptions that enter into the argument of the previous section and that need to be questioned. It turns out that three of them seem thoroughly reliable, leaving just one as the necessary culprit. First, it's assumed that there's no moving medium that would affect the speed of the light signal relative to the assumed stationary reference frame. After much argument and many experiments to test for a moving medium (the ether, which we've discussed), this possibility has been effectively ruled out. Second, it's assumed that light has a speed that doesn't depend on the motion of the source (otherwise, because the source of the light is moving with the earth, the argument I've shown you would be invalid). The very fact that light is known to be a wave motion essentially rules out any dependence on the motion of the source; it's the nature of wave motion to behave in a way that depends only on the local environment, not on any distant source. Nevertheless, the experiment was repeated using starlight, just to be sure, and the null result persisted even though the source of light was detached from the earth. Third, it's assumed that a small dependence of the speed of light on its wavelength would not cancel the effect of a difference in path length. The wavelength as measured on earth is different from the wavelength as seen from the assumed stationary reference frame, and the waves traveling in different directions in the apparatus would all have different wavelengths, and therefore different velocities, as seen in the stationary frame. It turns out that the speed of light would have to vary quite a lot for different wavelengths in order for that variation to cancel the effect of the difference in path length, so this possibility can be ruled out simply on experimental grounds: The speed of light in a vacuum is known experimentally to be constant to a high degree of accuracy over a wide range of wavelengths.

What is left to question is the assumption that the legs of the apparatus don't change in length during the 90° rotation that's supposed to reveal a shift in the interference pattern. Such a change in length would be far too tiny to have been seen in any other way, so the only real objection is that it would violate our common-sense idea that rigid objects retain the same dimensions when they're moving as when they're at rest so long as there are no stresses on them. Let's see what this distortion would have to be to account for the null result of the experiment.

What we have to suppose is that the length of a moving rod is different when it is aligned along the direction of motion than when it is at right angles to that direction, and that its length can actually change as it's rotated relative to the direction of motion. So let $L_\parallel$ be the length of the leg that is parallel to the motion—leg AB in the original setup described in §1.4—and let $L_\perp$ be the length of the leg that's perpendicular to the motion—leg AC in our derivation. There's no change in the reasoning, and the light paths that we calculated, eqs. 1.4.8 and 1.4.18, can simply be reexpressed in terms of these two different lengths, as in prob. 1.4.6:

$$\ell_{ABA} = 2L_\parallel \gamma^2 \tag{1}$$

$$\ell_{ACA} = 2L_\perp \gamma \tag{2}$$

It's now easy to see that the path lengths will be the same if only

$$L_\perp = L\gamma \tag{3}$$

This, then, is the condition we're looking for—the condition required by the null result of Michelson and Morley (see prob. 1.4.7). We can't tell yet whether $L_\parallel$ is reduced or $L_\perp$ is enlarged or both, but some distortion has to take place so that eq. 3 is satisfied. When the apparatus is rotated the distortion will change; leg AB will be perpendicular and leg AC will be parallel, but the path difference will still be zero provided the leg that is now parallel to the motion is again shorter than the one perpendicular to it.

Now, where does Einstein come into all this? As I mentioned earlier, Einstein figured out about this very rule for distortion without being particularly aware of or concerned about the Michelson–Morley experiment. His argument rested on his principle of relativity, stated (in my own language) in the introduction and repeated here for convenience:

> The principle of relativity: *The laws of physics—including the
> behavior of light—must be exactly the same for any two observers
> moving with constant velocity relative to each other.*

This is a kind of principle of democracy in physics, a fundamental
rule that all observers are equal—and it's a consequence of this principle that the laws of nature must conspire to make it impossible to tell
whether a system is moving or at rest.

At this point Einstein was applying his idea of the equivalence of
different observers only to uniformly moving reference frames: frames
moving at constant velocity. The odd effects of acceleration are dealt
with only when we get to general relativity—Einstein's theory of gravity.
This is a tricky point, though, because you can't say whether a reference
frame S′ is moving at constant velocity except relative to some other
frame S, and how do you know S isn't accelerating? To deal with this
problem, you have to back off and rephrase the principle a little more
carefully, something like this: There exists a set of reference frames,
uniformly moving relative to each other with all possible relative velocities, such that the laws of physics appear exactly the same for all of them.
These reference frames are then referred to as inertial reference frames,
because in some sense they're not accelerating.

Einstein wasn't thinking about the Michelson–Morley setup, of
course, but that example will still do very well to show how the principle
of relativity works—how it can be used, in fact, to determine how the
Michelson–Morley apparatus is altered by its motion through space and
to give a perspective from which that behavior, odd as it is, can be understood.

In the first place, the principle of relativity answers our question
about whether leg AC gets longer or leg AB gets shorter as the apparatus
moves along the direction of AB. The principle tells us that there must
be complete equivalence between the behavior of things as seen by an
observer moving with the apparatus and by an observer fixed with respect to the distant stars, let's say, and with respect to whom the earth
and the apparatus are moving in the way Michelson and Morley supposed. With this in mind, consider the motion of leg AC, the leg assumed to be perpendicular to the direction of motion. Let's take the leg
to be just one meter long as judged by an observer, whom I'll call S′ (you
say "ess prime"), who is moving with the apparatus. Suppose also that
there's a "stationary" observer S (stationary, that is, relative to the standard reference frame in which the distant stars are at rest on the average) and that S is holding a meter stick parallel to leg AC as it goes by, in
order to see whether there's a change in its length. Both observers will

agree that the two ends A and C pass the meter stick simultaneously, and they can't help but agree on whether the points A and C pass across points on the meter stick or miss the meter stick altogether—that is, whether AC is shorter or longer than the meter stick. S and S' could, for example, attach pieces of chalk to the apparatus at A and C and see whether they mark the meter stick as they go by; then S and S' will certainly agree on whether chalk marks were made or not. If there are chalk marks, they must agree that $L_\perp$ is less than 1 m, and if there are no chalk marks, they must agree that $L_\perp$ is greater than 1 m. The point now is that if leg AC looks long to S, then the meter stick looks short to S. Each of them sees a rod moving in a direction perpendicular to its orientation, and for one of them the moving rod appears to get longer while for the other it appears to get shorter, in violation of the relativity principle, which requires that both observers see physics working in the same way.

To satisfy the principle, then, we have to suppose that the length of leg AC is unaltered:

$$L_\perp = L \tag{4}$$

and therefore that the null effect is entirely due, according to eq. 3, to a shortening of the other leg, AB, the one that's moving parallel to its own orientation:

$$L_\| = \frac{L}{\gamma} \equiv L \sqrt{1 - \frac{v^2}{c^2}} \tag{5}$$

This is the famous Lorentz contraction, about which I'll have a great deal more to say as we go on.

You may well wonder why the same argument I gave you about leg AC doesn't apply here. If the moving leg looks short to a "stationary" observer like S, won't a stationary meter stick look long to the moving observer S'? It turns out that the situation is different. In the first place, using pieces of chalk doesn't work, because in this case leg AB is sliding along the stationary meter stick and leaves chalk lines all along it. In the second place, and more fundamentally, there is a real uncertainty about *when* the two ends A and B pass the two ends of the meter stick. As you'll see in §2.4, the two observers will actually disagree on which of these two events (A passing one end of the meter stick and B passing the other end) takes place first, and this in turn causes them to disagree on which rod is shorter. For both of these reasons, the previous argument doesn't apply, and there's no logical difficulty in supposing that leg AB is contracted according to eq. 5, as observed by the "stationary" observer S'.

You may have noticed that this contraction of moving objects along their direction of motion is called the Lorentz contraction rather than the Einstein contraction. Several physicists, particularly G. F. FitzGerald in 1889 and H. A. Lorentz,* in 1892, had noticed that the only way to explain the Michelson–Morley result without making strange assumptions about the behavior of the ether is to suppose that leg AB of the moving apparatus is contracted in the way I've described. Actually, neither FitzGerald nor Lorentz was thinking in terms of the exact factor γ: FitzGerald gave no explicit factor at all, and Lorentz gave the factor as $1 - v^2/2c^2$, corresponding exactly to the approximation of eq. 1.4.12, which is valid because v^2/c^2 is of the order 10^{-8} for the motion of the earth around the sun.

The significance of Einstein's contribution is that he put this effect into the context of a fundamental principle: If democracy among observers is to be preserved, the laws of physics *must* be such that the atoms of a moving object arrange themselves more compactly along the direction of motion—a strange, but by no means impossible, prediction that is borne out by the Michelson–Morley result and, since then, in a host of other experimental situations in physics.

Is this contraction real? I think we have to say that it is. In any given reference frame, the laws of motion do predict that a moving atom will shrink to a flattened shape. Maxwell's equations for the electromagnetic interaction and Newton's laws of motion for the electrons, suitably modified by relativity and quantum theory, really do say that the average distribution of the electrons around the nucleus will be spheroidal instead of spherical, flattened exactly by a factor γ along the direction of motion of the atom. It is this flattening of atoms, in fact, that is responsible for the contraction of rulers and other material objects.

How can I be sure about this flattening of atoms without having solved the really complicated equations that describe the problem? I know it by the following logic: I know that those equations all satisfy exactly the principle of relativity, which means that the solutions in two different reference frames must be appropriately related by the Lorentz contraction. You can therefore get precisely the correct answer by first solving the equations in the reference frame in which the atom is at rest (the "rest frame" of the atom) and then transforming the answer to what

---

*Pronounced like "Lawrence."

it looks like in the "lab frame" (which is what physicists call the original reference frame in which the atom is seen as moving). I know from symmetry requirements that in the atom's rest frame, the electron distribution in the ground state of the atom is exactly spherical (for a closed-shell atom like neon or argon, say), and I know that the transformation to the lab frame produces a Lorentz contraction of exactly γ. The important point here is that I used the principle of relativity only as a device for solving the equations and that I would have got exactly the same answer if I had solved the equations directly.

It's been noted recently that to a distant stationary observer a moving object appears to be rotated rather than flattened. This doesn't contradict my statement that the object is really flattened with respect to the stationary observer's reference frame. The apparent rotation is a subtle combination of two effects: the Lorentz contraction of the object and the difference in the time it takes for light rays to reach the observer's eye from different parts of the object. If the observer makes allowance for these time delays, or else uses instruments to make simultaneous on-the-spot measurements of the locations of the different parts of the object, he'll find indeed that the object is flattened, but not rotated, according to the exact standards of his own reference frame.*

PROBLEM

1. To estimate the relativistic factor γ for speeds very close to c, you can write $v/c = 1 - \varepsilon$, where ε again represents some very small number.
   (a) Show that in this case $\gamma \approx 1/\sqrt{2\varepsilon}$. Note that no matter how small ε is you can't neglect it, because v cannot equal c and it's the difference that counts.
   (b) Use the approximation of part (a) to find the value of ε, and to find the difference $c - v$ in meters per second, for the following values of γ: 10, 100, 1000. Find the error in the approximation in each case by reevaluating γ exactly, using the same value of ε.

---

*Whenever it makes sense to do so, I'll adopt the convention that stationary observers (stationary from some reasonable point of view) are male and that moving observers are female.

MOVING CLOCKS RUN SLOW

You've seen now how the Lorentz contraction, by reducing the distance that the light signal has to travel in one of the two alternative paths in the Michelson–Morley experiment, explains their null result and thereby satisfies the principle of relativity, which requires nature to conspire against any effort to determine absolute velocities. There is in fact another way in which you might determine the absolute velocity of the earth—a way that isn't frustrated by the Lorentz contraction. This is to ignore path ABA altogether and make a direct determination of the time it takes the light signal to follow path ACA in fig. 1.3.5. The light path, shown in fig. 1.4.2, is along two diagonals and is greater than $2L$ by the factor γ (eq. 1.4.18). The length of leg AC, as explained in §1.5, is not altered by the motion, so the time t_{ACA} is increased by that same factor γ over what it would be if the earth were at rest. A measurement of this time, combined with an accurate measurement of the length L, gives you γ and hence the speed v of the earth. Note that what I've described is essentially just a measurement of the speed of light, as it appears to the moving observer S′, and that the increase in the observed time for the light to make the round trip would be understood as a reduction in the observed value of c by the factor γ. Of course, observer S′ wouldn't know what the value of c was supposed to be unless she consulted with some other observer, such as S, who is presumed to be stationary. Note too that this measurement would have been impossible in practice at that time, because it would have required a determination of c to a part in 10^8, but is quite possible in principle (and today is possible in fact).

The point I'm trying to make is that such a difference in the observed value of c between two observers would again violate the principle of relativity, because it would amount to different laws of physics for a moving observer than for a stationary one. If you again examine the assumptions involved, you find that the only questionable feature of this experiment is the assumption that moving clocks read the same time as stationary ones. If your clock is wrong, the method doesn't work. All it takes to frustrate your effort to determine your absolute speed by this method is for moving clocks to run slow by a factor γ. This would precisely conceal the fact that the light signal takes longer to traverse the path. Using t for the time as measured by the "stationary" observer S and t' for the time according to the moving observer S′, you can write this as a formula:

$$t' = \frac{t}{\gamma} \tag{1}$$

In fact, all physical and biological processes must be slowed down by exactly this same factor; otherwise S and S′, by comparing notes, would be able in principle to observe the discrepancy. This is the famous relativistic time dilation. There was no experimental evidence for it then or for some years afterward; it was predicted solely as a consequence of the relativity principle. In modern times it has been amply and precisely verified, primarily in the observed behavior of particles traveling very close to the speed of light.

It's often difficult to speak accurately about such a confusing state of affairs. You are constantly having to take into account the variety of different points of view and to treat them all as being equally valid. One way to proceed—and you run into no contradictions by doing so—is to take the simple view that there's just one reference frame that is the correct one—that is in fact at rest. You can then predict from the equations of motion how moving rulers and clocks will be changed, thus conspiring to make the physics look all right in other reference frames, even though (from your point of view) they're not. Spatial contraction and time dilation are real effects, even though the detailed description of such an effect depends so much on which point of view you're adopting. A very familiar example is the "twin paradox," discussed in §2.6 and §3.3, where the twin who goes on a voyage through space and then returns is actually and literally younger when she returns to earth than the twin who stays behind. The description from the earth's frame is easy: You just apply the factor γ to the traveling twin and you get the correct answer; from some other point of view, such as that of the outward bound spaceship, the description is much less obvious, but it yields the same result.

I find myself tempted to say that the moving clocks and rulers misbehave, but this is clearly unfair because they're behaving just the way Einstein says they should. The world would have less elegance—less symmetry—if rulers and clocks didn't behave in this odd way, because in that case the laws of physics would take different forms in different frames instead of taking the same form in *all* uniformly moving frames, as they actually do.

What is hard to get used to, perhaps, is that the conspiracy is so exact that even though you may choose to believe in a true rest frame, there is absolutely no way for you to tell which frame that is. In practice the concept of an absolute rest frame is meaningless because of the impossibility of devising any empirical test to determine it. There is true democracy among reference frames. This idea has its own appeal, but it does leave you with a metaphysical question: How do you characterize the truth—the reality—that underlies a situation in which each point of view is different but all are equally valid? I don't believe you can do it by saying the effects are not real.

PROBLEM

1. Twin A takes a plane trip of round-trip distance 3000 km (3×10^6 m), traveling while in the air at a constant speed of 300 m/s (about 670 mi/h), while twin B stays at home. What is the age difference of the twins after the trip? Which twin has aged more?

CHAPTER 2

Clocks, Rulers, and the Universal Speed Limit

There is much more to the theory of relativity than just the two effects, Lorentz contraction and time dilation, that I talked about in Chapter 1. I want to emphasize in Chapters 2 and 3 those aspects that relate more directly to the revolution that relativity has led to in our view of the structure of space and time. For this reason I want to go back and think through with you the consequences of the relativity principle at a more fundamental level, taking a global view of space and time and emphasizing the relation between the reference frames used by different observers. Using spacetime diagrams as a way of visualizing events in space and time, we'll see why Newton's idea of absolute time and the related notion of absolute simultaneity of events have to be abandoned, and we'll revisit time dilation and the Lorentz contraction using our new tools. My main objective is to show you that spacetime must now be thought of as a *single unified space* with a geometry of its own that is very different from the Euclidean geometry you're used to.

 REFERENCE FRAMES

I want to start by introducing you more carefully to the use of reference frames, which allow you, as I've said, to describe events in space and time from the point of view of different observers. To specify the *location* of an event—that is, where it takes place in space—you have to start with some reference point, called an *origin*, and make measurements relative to that point. Say you drive a stake into the floor of your laboratory and make measurements from there. Then you have to specify three directions, normally at right angles to each other, in order to identify the different kinds of position measurement you'll make. It's conventional to call these the *x* direction, the *y* direction, and the *z* direction. Say you pick east as the *x* direction, north as the *y* direction, and straight up as the *z* direction. Then three distances, *x*, *y*, and *z*, will be enough to specify any point in your laboratory—or in the universe, for that matter. The value of *x* tells you how far east from the origin (or how far west, if *x* is negative), *y* tells you how far north or south, and *z* tells you how far up or down to go to get to the point you've chosen. It is conventional also to imagine three straight lines drawn through the origin, called the *coordinate axes*, to indicate these three directions. The one that runs east and west is the *x axis*, and so on. Any point in space will do for the origin, and any three mutually perpendicular lines will do for the coordinate axes. It might be better, though, to orient your axes with respect to the fixed stars rather than with respect to the rotating earth. The origin and the axes constitute a *spatial reference frame*.

Now you have to tell at what *time* the event takes place. All the same logic applies. You need a reference time, and you need a variable *t* to indicate how long before or after the reference time the particular event takes place. Events simultaneous with the reference time have the value $t = 0$. We conventionally take the positive direction as the future (it's confusing to do otherwise), so events later than $t = 0$ are assigned positive values of t, and earlier events have negative values. It may be a little hard to imagine a *line* drawn in the *t* direction to serve as a *t axis*, though when we draw diagrams of spacetime that's just what we'll do.

Now every possible event is precisely located in space and time by the four quantities *x*, *y*, *z*, and *t*, which we refer to as the *coordinates of the event*. The spatial origin *at the time* $t = 0$ constitutes a single point in spacetime and is called the *spacetime origin*. The three spatial axes, together with the time axis (which specifies how the spatial origin moves in time), constitute a complete reference frame. Since it takes four quantities to specify a given event, we say that spacetime is four-dimensional. There's no more mystery than that about the fourth dimension.

Note that the coordinates (the quantities x, y, z, and t,) cannot simply be numbers but must have *units* attached. A distance of 1 meter is the same as a distance of 100 centimeters, or 39.37 inches, so a number alone—1 or 100 or 39.37—can't be used to specify the distance. In the same way, t must always include a time unit. This is different from math, where a symbol like x or v or t often represents a number without any unit. We often call such a physical variable a *dimensioned variable* and say that "x has the dimension of a length" or "t has the dimension of a time." You can multiply and divide quantities with different dimensions—x/t, for example, has the dimension of length divided by time, or velocity*— but you can't add or subtract quantities with different dimensions.

PROBLEMS

1. Show that a *light-year*, the distance a light signal travels in one year, is 9.4605×10^{15} m, and determine this distance in miles. (A year is approximately 365.24 days.)
2. Find the coordinates of the sun in the local spatial coordinate frame described in the text (x is east, y is north, z is vertically up) for a person at $40°$ north latitude when the local time is (a) 6:00 a.m., (b) noon. Take the date to be the vernal equinox, when the tilt of the earth's axis can be ignored. Take the distance from the earth to the sun as 1.5×10^{11} m.
3. A boy sits in a chair.
    (a) In the local coordinate frame do his spatial coordinates change with time?
    (b) Do his spacetime coordinates change with time? Explain.
    (c) Describe how his spacetime coordinates change with time in a coordinate frame that's not rotating relative to the distant stars, with the origin at the center of the earth. (Have him sit in the chair for two days.)
4. Identify which of the following expressions make sense, and evaluate them in standard metric units:
    (a) $15 \text{ m} + 3 \text{ s}$

---

*Note that meters per second means exactly the same as m/s, meters divided by seconds, because (number of meters per second) × (number of seconds) = (total number of meters): (m/s) × s = m.

(b)   10 cm − 2 in.
(c)   1 lt-yr + 1 yr
(d)   0.001 lt-yr − 2×10^{13} m
(Treat the data as exact, and give answers to three significant figures.)

2.2   SPACETIME DIAGRAMS AND WORLD LINES

I now want to introduce *spacetime diagrams* (often referred to as *Minkowski diagrams*), one of the important tools for picturing events in space and time and for visualizing the relationships between different frames of reference. In its simple form, a spacetime diagram is simply an *x* axis and a *t* axis, as shown in fig. 1, on which you can plot points to represent events in spacetime.

There's a little problem connected with the fact that it takes four numbers *(x, y, z,* and *t)* to specify the location of an event in space and time, while the paper on which we draw our diagram is only two-dimensional. A lot of the time we'll be interested only in events that take place along a single line in space, for which the variable *x* alone will illustrate amply most of the physics we need to discuss. You have to remember that *y* and *z* are both equal to zero in that case and use the diagram just to represent the values of *x* and *t*.

If the spatial events you're looking at take place in a plane, with *z* always equal to zero, then you can display them in a three-dimensional spacetime diagram that can be visualized without too much difficulty, as in fig. 2. Another common way of using a spacetime diagram is to *imagine*

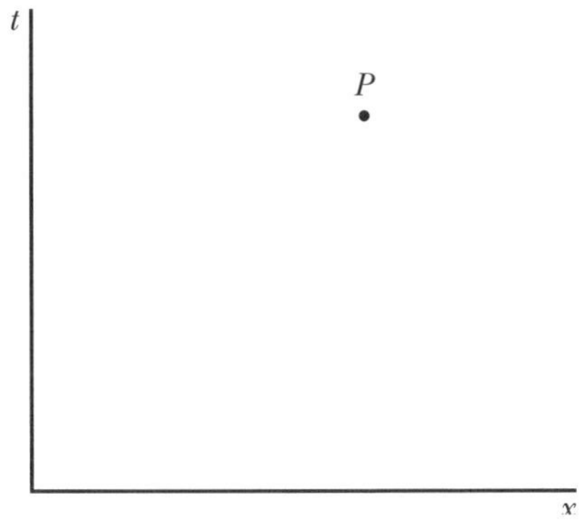

Fig. 1   Spacetime diagram.

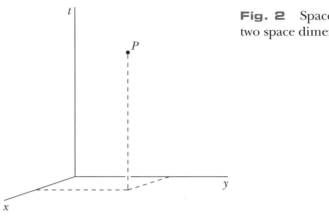

Fig. 2   Spacetime diagram for two space dimensions.

the additional dimensions, letting the single x axis in fig. 1 or the two spatial axes in fig. 2 symbolize the whole spatial world of three dimensions. (You'll see an example of this in fig. 7.)

Let's return to fig. 1. Any point P in this diagram has definite coordinates x and t and thus represents an event that takes place at location x at time t. In the diagram, x is the distance of the point from the t axis (the vertical line), and t is the distance from the x axis (the horizontal line). Appropriate units for x—meters, kilometers, or light-years, for example (for light-years, see prob. 2.1.1)—can be marked on the x axis to set the scale for measurements of x, and units for t can be marked on the t axis, as illustrated in fig. 3. Negative values of x, of course, correspond to points to the left of the t axis, and negative values of t to points below the x axis.

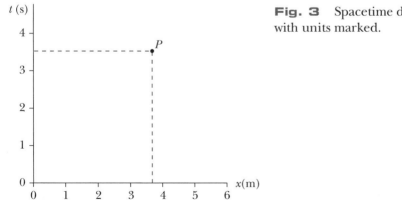

Fig. 3   Spacetime diagram with units marked.

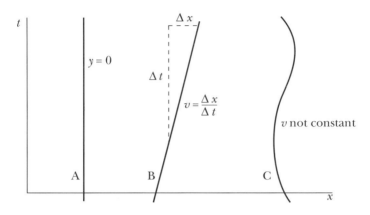

Fig. 4   World lines. Object A is stationary; object B moves at constant velocity; object C moves with variable velocity.

   The history of a particle can be thought of as consisting of a large number of events, the events being the particle's visits to different locations at different times. Because there's really a continuous infinity of different times, the history of the particle is represented by a continuous set of points that make up a line, either straight or curved, as shown in fig. 4. Each of the three lines is what's called the *world line* of a particle. Line A represents a particle at rest: The time varies, of course, but the position coordinate x is the same for all the points on the line [see prob. 2.1.3(a)]. Line B represents a particle moving in the $+x$ direction at constant speed, with velocity related to the slope of the line. The velocity is in fact $\Delta x/\Delta t$, which is the reciprocal of the slope as it's usually defined. I'll call this reciprocal slope the *inclination* of the line. The vertical line A, representing a particle with zero velocity, has zero inclination. The curve C represents a particle moving with variable speed, back and forth along a single straight line in space, first to the left, then to the right, and then to the left again.
   The world line of a light signal is a straight line with inclination $\pm c$, as shown in fig. 5. In drawing the diagram I've followed the common practice of choosing the scale so that light paths have inclination 1, thus making a 45° angle with the t axis. This is done in fig. 5 by letting the same distances on the diagram represent a year (in the t direction) and a light-year (in the x direction). In these units c has the numerical value 1; that is, light travels at a speed of just 1 light-year per year (see prob. 2.1.1). Note that this choice of the scale rather stretches things out of the

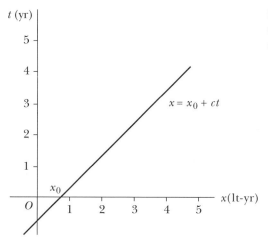

Fig. 5   World line for a light signal. The signal starts from position x_0 at time zero.

usual proportions: One second on the t axis is represented by the same distance on the graph as 186,000 miles on the x axis. However, for our purposes we'll often want to look at things that are traveling at speeds comparable to c, so this is a very suitable scale for us.

For another example, look at the spacetime diagram for the earth going in its approximately circular orbit around the sun, shown in fig. 6. Since the motion is in a plane—that is, in two space dimensions—we need a three-dimensional spacetime diagram. The vertical line labeled S

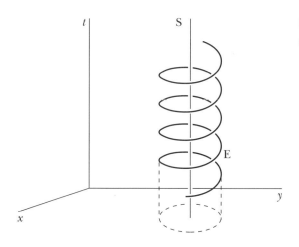

Fig. 6   World lines of the sun and the earth.

is the world line of the sun, which I've shown as standing still (same values of x and y for every value of t), and the helix around it is the world line of the earth. The values of x and y that the earth visits make up a circle with the sun at the center (shown as a projection onto the x–y plane). Each circular orbit takes one year, so when the path comes back to the same values of x and y, the time t has advanced by one year. I've drawn the axes to a different scale than in fig. 5.

For just one more example, try to imagine the spacetime diagram for a human being—yourself, say. Fig. 7 is a highly schematic version; a proper diagram would have to show all the events of your life—past, present, and future. You are not just a point particle but an assemblage of a large number of particles, so your spacetime diagram ought to show the world line of each particle. The result is something like a tube rather than a single line, like an oddly shaped piece of spaghetti extending over a time of 80 years, assuming that's how old you live to be. The thickness of the piece of spaghetti is just a foot or so, though it has to meander over many miles (even if you ignore the motion of the earth itself) in order to represent all the various travels you take in the course of your life. You can see what you look like at any given time by taking a cross-section, as at C or D. This spatial cross-section is shown as two-dimensional in the diagram, but it should really be three-dimensional, since it shows your complete appearance at that chosen time. Point A is the moment of conception, and a cross-section just after that time would show the fertilized ovum. At B you are born, and the cross-section at C shows you as you are right now (assuming you're of college age). All of your childhood is represented by the part of the tube between B and C, and your future life is the portion above C. The cross-section at D is supposed to represent you at age 50. At age 85, say, you die and the world lines of the different particles that go to make you up eventually disperse and get recycled in one way or another. I would argue that this is a much more complete picture of you than the usual one, which consists only of what you are at one single moment. In this way of looking at yourself, your past and your future are as much a part of you as your present.

As you can see, I really like spacetime diagrams. I think they give you a lot of insight into the meaning of time, once you get used to them.

PROBLEMS

1. Mark an appropriate scale, with appropriate units, on each of the three axes of fig. 6.

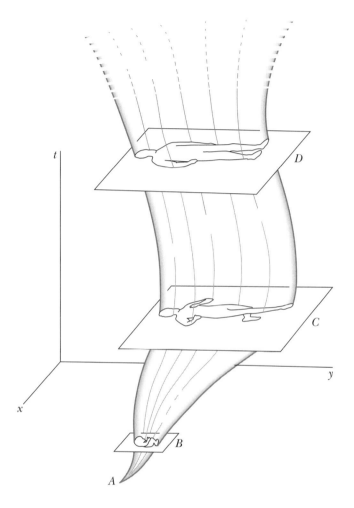

Fig. 7 World tube of a human being.

2. (a) Draw a schematic spacetime diagram for a few generations of your own family.
   (b) Write a verbal description of the spacetime diagram for the whole human race. Note the complete interconnectedness.
   (c) Estimate the ratio of length to thickness for the world tube of a human being, in units for which $c = 1$.

3. Using graph paper, make accurate spacetime diagrams for each of
    the following cases. Be sure to label the axes and indicate distance
    and time scales. Treat the woman and the train as pointlike objects
    with a precise position at any given time, and ignore relativistic ef-
    fects.
    (a) A woman walks along a straight road for four hours, starting at
         noon, at a speed of 3 mi/h. At 12:30, 2:00, and 3:15 she
         coughs. Indicate the coughs as events on the spacetime dia-
         gram.
    (b) A train leaves station A at time $t = 0$, travels for 1 h due east at
         60 mi/h to station B, stops at station B for 30 min, continues
         on for 1 h at 30 mi/h to station C, and then immediately re-
         turns directly to station A at 90 mi/h. Include—and label—the
         world lines of the train and of each of the three stations.
    (c) The train travels due east from station A to station B at 30
         mi/h and immediately returns to station A at the same speed.
         Show this in a stationary frame of reference and also in a
         frame of reference moving *west* at 30 mi/h. In each diagram,
         show the world lines of the train and of the two stations.
4. In world line C of fig. 4, mark the points where the speed is zero
    and the points where the speed has its greatest values (either to the
    right or to the left). If the particle had traveled at constant velocity
    from its initial location to its final location, would its velocity have
    been positive or negative? Draw the world line for that alternative
    motion.

THE SPEED LIMIT

In §2.1 and §2.2 you were introduced to some of the concepts and
methods that will be useful in the job before us, which is, you'll recall, to
work out carefully some of the consequences of the principle of relativity. We now return to that task.

The first thing to notice is that c, the speed of light, *has to be an*
absolute speed limit for the universe. No object can travel faster than c relative
to any observer. If an observer could travel as fast as c, then to her the
speed of a light ray traveling in the same direction would appear to be
zero. This would make her viewpoint, and hence the apparent laws of
physics from her perspective, very different from what we're used to. (As

I mentioned in Chapter 1, traveling observers are female and stationary observers are male.) Actually, this description of what she'd see is too simple because, as you saw in Chapter 1, rulers and clocks are distorted in their behavior. The distortion factor is γ in each case, and you'll recall that γ becomes infinitely large as $v \rightarrow c$, so the observer's rulers, and the observer herself, would be contracted to zero length at speed c while her clock would run infinitely slowly—a meaningless picture, really. It has to be impossible, then, for any actual massive object to reach the speed of light. You might think that you could keep pushing an object to higher and higher speeds until it reached speed c, but in fact, as you'll see in Chapters 4 and 5, this doesn't work. The harder you push the object the closer it comes to the speed of light, but it never quite gets there. It would take an infinite amount of energy to push an object to speed c. (The things that do go that fast, like light and gravity waves—and possibly neutrinos—are of a different character from ordinary matter and can never go any *slower* than c.) In fact, as I'll explain in due time, no information of any sort can be sent faster than c.

It is important to understand that the speed limit c doesn't have to do with just light. Light is nothing more than a familiar example of something that happens to hit the speed limit, and relativity would be true if our laws of physics didn't include electromagnetic waves at all. Note that the speed of light shouldn't be thought of as "the speed of a light signal through space" but rather as "the speed of a light signal as measured by an observer—any observer." That's the sense in which c is something absolute. It's one example of what physicists call an *invariant*, something that has the same value or form for all observers. We need to refer to observers in order to define it, but we don't need to specify which observer because the value is always the same.

PROBLEM

1. A traveler leaves the earth at $t = 0$, traveling at a uniform speed of 0.3 c.
    (a) After one hour a station on earth sends a brief light signal, which is reflected by a mirror on the spaceship and returns to the earth. Using graph paper, draw an accurate spacetime diagram for these events, including the world lines of the earth, the traveler, and the light signal, for each of the two reference frames, the earth's and the traveler's. Use hours for the time units and light-hours for the distance units.

(b) Instead of the light signal, a radio message lasting fifteen minutes is sent out from the earth, beginning one hour after the spaceship takes off, and is received by the ship in due time. Draw an appropriate spacetime diagram.

2.4 THE SIMULTANEITY SHIFT

The next thing we find—and this is very important and very deep—is that different observers won't be able to agree on whether or not two events happen at the same time: Events that appear simultaneous to one won't in general appear simultaneous to the other. This is again a consequence of the requirement that different frames of reference be completely equivalent. You get an indication of it in the Michelson–Morley experiment: For a stationary observer the light traveling parallel to the direction of motion of the apparatus takes longer to go one way than the other (eqs. 1.4.2 and 1.4.4), while for the observer moving with the apparatus the light takes the same time to go both ways. As you'll see, the two points of view can be reconciled only by allowing the different observers to have different definitions of simultaneity.

Look at this effect now using spacetime diagrams. Let S be the stationary observer whose reference frame consists of the x and t axes shown in fig. 1. Then draw the world line for an observer S′ moving with velocity v. The world line of S′ has inclination v, with $v < c$. Next consider (S′)'s observation of a light signal sent out and reflected back by a mirror at B, which might be thousands of miles away. The spacetime diagram is

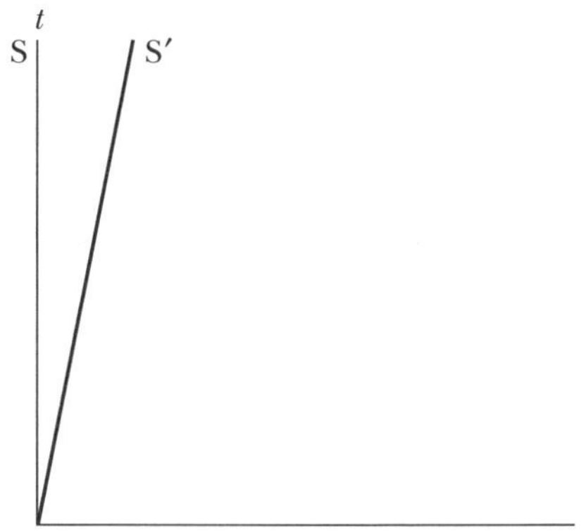

Fig. 1   World line for a moving observer S′.

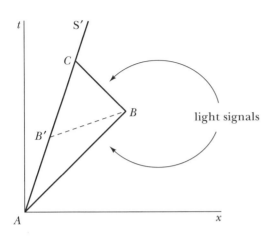

Fig. 2   Simultaneity for a moving observer.

shown in fig. 2. The light signal is sent out by S' at A, is reflected by the mirror at B, and returns to S' at C.

How do S and S' figure out when the light signal arrives at the mirror at B? They have to figure it out *indirectly*. They're not on the spot to see it arrive and have to deduce from far away the time when event B occurred. Each adopts the very reasonable point of view that he or she is at rest and that a light signal therefore travels at the same speed in both directions. These may be conflicting points of view, but neither can be regarded as wrong. From the point of view of S, the light takes longer to go from A to B than from B back to C, for the obvious reason that on the trip $A \rightarrow B$, the light has a greater distance to go than on the return trip $B \rightarrow C$, because S' is moving to the right. You can see from the figure that $t(B) - t(A)$ is greater than $t(C) - t(B)$. From the point of view of S' on the other hand, the light signal must take the same time to get from B to C as it did from A to B, because for her the two distances are the same. The time she assigns to event B must therefore be exactly halfway between the times she assigns to A and C (where she is on the spot, with watch in hand). The event point B', halfway between A and C, represents a tick of her watch at the moment that she regards as simultaneous with the event B. In fact, by a similar process using light signals, any of the event points on the dashed line $B'B$ would be regarded by S' as simultaneous with B'. Let's call the line $B'B$ a *simultaneity line* for S'.

The slope $\Delta t / \Delta x$ of this simultaneity line is given by the formula

$$\frac{\Delta t}{\Delta x} = \frac{v}{c^2} \qquad (1)$$

which tells us just how *non*simultaneous the events B and B' appear to S, given that they appear simultaneous to S'. (The slope of the line $B'B$, $\Delta t/\Delta x$, is more appropriate here than the inclination, $\Delta x/\Delta t$, because $B'B$ is not the world line of any object and because, in the simple case that $v \to 0$, Δt is zero and the slope is zero, while the inclination becomes infinite.)

Where does the eq. 1 come from? It's determined by the construction represented in fig. 2. Apart from the overall scale, the shape of the figure is entirely determined by the known inclinations of the three lines AB, AC, and BC, together with the fact that B' is the midpoint of line AC. The inclinations of these three lines are given simply by

$$AB: \quad \frac{x(B)}{t(B)} = c \ \text{(light signal)} \tag{2}$$

$$AC: \quad \frac{x(C)}{t(C)} = v \ \text{(velocity of S')} \tag{3}$$

$$BC: \quad \frac{x(C) - x(B)}{t(C) - t(B)} = -c \ \text{(light signal)} \tag{4}$$

The turnaround time $t(B)$ is arbitrary, but once $t(B)$ is fixed, the other locations and times are determined, as I'll now explain. I'll leave it to you (prob. 4) to complete the solution of these three equations for the three unknowns $x(B)$, $x(C)$, and $t(C)$ in terms of $t(B)$. The results are

$$x(B) = ct(B) \tag{5}$$

$$x(C) = 2x(B') = \frac{2cvt(B)}{c + v} \tag{6}$$

$$t(C) = 2t(B') = \frac{2ct(B)}{c + v} \tag{7}$$

I've also included here the resulting coordinates of the point B', because we need those for the final step. The slope of the simultaneity line $B'B$ is now obtained by simple substitution:

$$\frac{\Delta t}{\Delta x} = \frac{t(B) - t(B')}{x(B) - x(B')} \tag{8}$$

$$= \frac{v}{c^2} \tag{9}$$

Again, the missing steps are left for prob. 4. Note how $t(B)$ cancels out in the final result, because all we're calculating is a ratio.

Another, more practical way of stating this result is to say that if two events are perceived as simultaneous by the observer S′, then the observer S perceives them as separated by a time Δt given by

$$\Delta t = \frac{v\Delta x}{c^2} \tag{10}$$

where Δx is the distance between the events as measured by S and v is the velocity of S′ relative to S. (Note that this last statement is in "nonframist" language—that is, it avoids any bias toward a preferred, or "stationary," frame of reference.) I'll call the time separation Δt *the simultaneity shift*.

The faster S′ is traveling (the closer to c, that is), the more extreme the simultaneity shift becomes, as illustrated in fig. 3, where the inclination of the line AC is very close to c, and again B and B' are seen as simultaneous by S′.

One important consequence of our results about simultaneity is that the S′ coordinate axes get tipped in an unexpected way. First, (S′)'s time axis, called the t' axis, is the set of spacetime points for which $x' = 0$; it is just her world line, shown in fig. 1. This is consistent with the way the t axis is defined in the original coordinate system of S, as the set of points for which $x = 0$. What's more surprising, perhaps, is what happens to (S′)'s spatial axis, her x' axis. It must be the set of spacetime points for which $t' = 0$, which means that these points form a simultaneity line, with slope v/c^2 in the S coordinate system, as shown in fig. 4. This tipping of axes for a moving observer is the analog, in the funny geometry of spacetime, of a rigid rotation in Euclidean geometry.

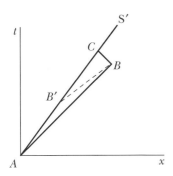

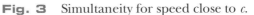

Fig. 3   Simultaneity for speed close to c.

Fig. 4   Reference frame for a moving observer.

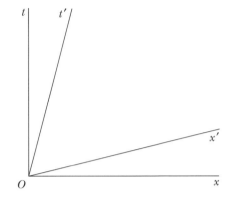

You can now see the difficulty with the idea of sending any kind of information faster than the speed of light. If, in fig. 3, observer S could send information about the result of a horse race from B' to B, say, which would be at speed $u = c^2/v$, then S' would think the information was being transmitted instantaneously, while another observer S'', going even faster than S', would believe that the information arrived earlier than it was sent. This isn't really a problem, because the notions of "later" and "earlier" are here just matters of interpretation. The real difficulty arises when we realize, from the complete equivalence of different observers, that observer S'' could send the information about the result of the race back to S in such a way that S would receive it *before* the race and could use it to bet on the winner. This scenario is illustrated in fig. 5, where the lines AA' and BB' are simultaneity lines for S and S', respectively; that is, the points A and A' are simultaneous in frame S, and B and B' are simultaneous in frame S'. (Note that the simultaneity line BB' is parallel to the x' axis, which is consistent with fig. 4.) The signal sent at A, just after the race, is received by S' at B' (which is later than it was sent according to S, but earlier according to S'). S' now relays the information back and it is received by S at C, before the race. This is later than it was sent (at B') according to S', but earlier than the original transmission at A by anybody's standard, in violation of common-sense ideas about cause and effect. Note that S, at point A, not only knows which horse won the race but also knows that he bet on it! What happens if he therefore decides not to bother with sending the message to S'? A particularly strange thing about this picture is that both S and S' are receiving messages that arrive *from the future* as they see it. In ordinary terms this means that S finds himself *sending* a message (from B' to A in

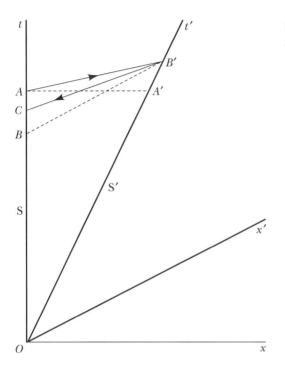

Fig. 5 S uses S′ to relay information backward in time.

his frame of reference) that he has no control over, and the content of that message is the information that he is receiving.

This violates all our beliefs about the way time works, because to know the future ahead of time gives rise to the possibility of altering it and thereby contradicting it. The possibility of a contradiction, and indeed the very idea of sending information in the first place, seem to me to rest on the assumption of *freedom of choice*, the ability to decide on your actions independently of preexisting conditions. I happen to believe in freedom of choice, but it is very difficult to incorporate into any mechanistic view of the world (see §15.4). Subtleties aside, the simple statement, almost universally accepted, is that if you could send signals faster than light then you could send them backward in time also, and that's not possible.

Leaving the human element out, physicists have noticed that physics alone doesn't forbid things to go faster than *c*, and they have taken the idea seriously enough to search carefully for what they call *tachyons* (from the Greek *tachy-*, which means "swift"), hypothesized particles that travel faster than c. The rules of relativity would prevent their ever being brought to rest or appearing to any observer to go slower than c. None have been found.

It's pretty well settled, then, that the fastest communication method is by signals at speed c, and an observer like S′ has no better way to establish the simultaneity of events than by means of light signals, in the way I've described.

You've had a glimpse in this section of the difficulty an observer has in making measurements and observations at different locations and times, far distant perhaps from the observer herself, and I've drawn you pictures of all sorts of different frames of reference. How does an observer in practice make measurements on events that are far away, and how do these measurements reflect the abstract relationships I've been talking about? A standard way of dealing with the problem—not very practical, to be sure, but concrete enough to remove any ambiguities—is for the observer to employ a large team of assistants, each with an assigned location in space and at rest relative to the observer. Each assistant is armed with a clock that's synchronized with the observer's clock by procedures like those described in this section, involving the exchange of light or radio signals. Then, after some event has been observed and recorded by the assistants on the spot, all the data can be sent back to headquarters, where the primary observer can evaluate them at leisure. Such a network of observers can be said to define a reference frame.

PROBLEMS

1. In the spacetime diagram shown in fig. 6, the line AC is the $t′$ axis— that is, the world line of observer S′, who is moving at velocity v relative to observer S. S′ checks the simultaneity of events 0 and B by sending a light signal from her spatial origin at time $-t_1$, at A, bouncing it back at point B, at position x and time t, and receiving it back at her origin at C, at time $+t_1$. She is assuming that the signal takes equal times to go and to return to her at C. The broken line ABC is the world line for the light signal. Prove that x and t, the coordinates of B, are related by $x = c^2 t/v$. (This is thus the equation of the $x′$ axis, the dashed line in the diagram, which is the set of event points for which $t′ = 0$, that is, points that S′ regards as simultaneous with the origin point O.) You may use the kind of argument given in eqs. 1–10, or you may use a geometrical argument based

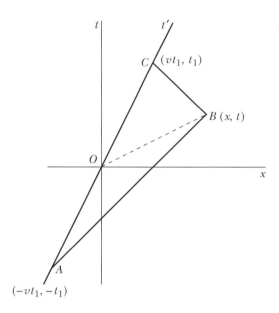

Fig. 6   (see prob. 1)

on the fact that light signals make an angle of 45° with the axes in the way I've scaled the diagram, so that triangles *AOB* and *BOC* are isosceles. (This argument doesn't work so well if different scales are used for the *x* and *t* axes, but the result is equally valid.)

2. Observer S′ is moving to the right relative to observer S with relative velocity $v_0 = 0.6c$. Two events *A* and *B* are seen by S as simultaneous, with *B* taking place to the right of *A*. S′ sees these same events as taking place 1 s apart in time.
    (a)   What is the spatial separation of the two events as seen by S?
    (b)   What is their spatial separation as seen by S′?
    (c)   Which event appears earlier to S′?

3. In fig. 7, a grid of light paths is used as the basis for a coordinate system. The grid is set up by sending out light rays in both directions from S's origin (which is the *t* axis) at equal time intervals, and by selecting those incoming light rays that arrive at the same equal intervals. The intersections of the light paths can be used to determine the coordinate grid, which will then consist of horizontal and vertical lines in the diagram. Draw in the coordinate grid.
    Now use the same procedure to set up a coordinate grid for S′, using light paths that intersect the *t′* axis at equal time intervals. In

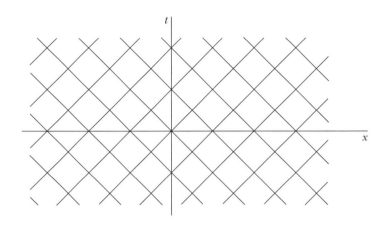

Fig. 7   (see prob. 3)

fig. 8 draw carefully a grid of light paths that have the same slopes as in the first figure but pass through the points marked on the t' axis. Then use the intersections of the light paths to construct the x' axis, and a few representative lines of constant x' and of constant t'. Keep $c = 1$ (light paths at $45°$) as in fig. 7.

4. Fill in all the missing steps in the derivation of eq. 10, the formula for the simultaneity shift.

Fig. 8   (see prob. 3)

2.5   WHOSE CLOCK IS SLOW? WHOSE RULER IS SHORT?

Clocks

Consider now how the two observers view each other's clocks. If a moving clock looks slow to a stationary observer, shouldn't the stationary clock look fast to the moving observer, in violation of the equivalence of all points of view? The key to the paradox lies in the mix-up about simultaneity, as I'll now try to explain. Suppose that each of our two observers S and S', shown in the spacetime diagram of fig. 1, has a clock that ticks once a second. The initial tick of each clock is at O, and the ticks 1 second later are at T and T', respectively. Letting $t(P)$ represent the time of any event P as measured by S and letting $t'(P)$ be the time of that event as measured by S', we can say that

$$t(T) = t'(T') = 1 \ s \tag{1}$$

According to what we learned in §1.6, (S')'s clock, which I'll call C', appears to S to tick more slowly by a factor γ than his own clock C, so in S's reference frame the event T' occurs later than the event T, as shown.

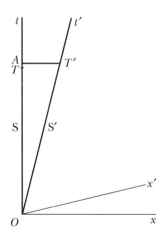

Fig. 1   Clock ticks, as seen by S and S'.

The point A represents clock C at the moment when clock C' ticks. In other words, A and T' are simultaneous according to S, so

$$t(A) = t(T') = \gamma \qquad \text{(in seconds)} \qquad (2)$$

Now we have to ask how these events look to S'. For symmetry between the two observers, T ought to occur later than T' from her perspective, which means that the S' simultaneity line through T must intersect (S')'s world line (the t' axis) at a point later than T', as shown in fig. 2. That is, A' represents the reading on clock C' at the moment when clock C is ticking at T, according to (S')'s standard of simultaneity. Fig. 3 represents the two clocks according to the two perspectives; fig. 3(a) is the picture according to S, at $t = \gamma$ (in seconds), and fig. 3(b) is the picture according to S', at $t' = \gamma$.

To make the argument exact, we have to find out whether S' finds the clock C to be running slow by the correct factor γ—that is, whether

$$t'(T) = \gamma \qquad (3)$$

To determine this, we have to locate point A', defined as the intersection of the simultaneity line through T and the world line of observer S', and

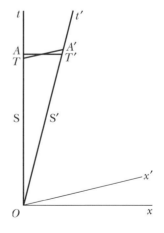

Fig. 2   Clock ticks at T and T'. Each clock appears slow to the other observer.

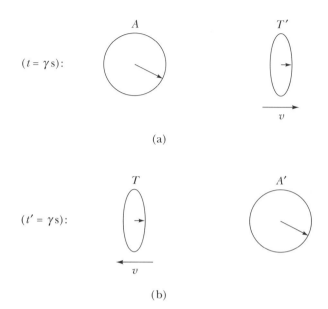

(a)

(b)

Fig. 3 Simultaneous views of the two clocks C and C′, (a) according to S, at $t = \gamma$, and (b) according to S′, at $t' = \gamma$. See fig. 2. Note: These clocks have second hands that make a full sweep every four seconds.

then see if the times at A' and T' are related by the correct factor γ. These two conditions give the equations

$$t(A') = t(T) + \frac{v}{c^2} x(A') \quad \text{(simultaneity shift—eq. 2.4.10)} \tag{4}$$

$$x(A') = vt(A') \quad \text{(velocity of S′)} \tag{5}$$

Solving these for $t(A')$ (see prob. 2), and setting $t(T) = 1$ (eq. 1), we get

$$t(A') = \gamma^2 \tag{6}$$

But $t(A') = \gamma t'(A')$, because we started with the fact that clock C' is slow by that factor, so we get

$$t'(A') = t'(T) = \gamma \tag{7}$$

which is the desired result (eq. 3). That is, the events A' and T are simultaneous according to S′, and T, the 1-second tick of clock C, is seen by S′ as occurring at time γ.

Note that physics is talking more and more about measurements by different observers, rather than about how things really *are*. There's no contradiction among the different measurements; it's just the inferences we draw that are likely to be contradictory. Is there then no meaning to "how things really are"? I think that ultimately there has to be, but it can't be seen at this level nor, perhaps, at any level we have reached thus far.

Rulers

We saw in connection with the Michelson–Morley experiment how the relativity principle—particularly the requirement of the invariance of the speed of light—leads to the necessity of the Lorentz contraction, which is the contraction of moving objects by a factor γ along the direction of motion. Just as before, we can ask how this effect manages to be symmetrical. If (S′)'s ruler looks short to S, how is it possible for S's ruler to look short to S′? And just as in the case of clocks, the answer lies in the relativity of simultaneity: the fact that S and S′ have different ideas about which events are to be regarded as simultaneous.

In §1.5 you saw that there can be no change in the length of a ruler if it's traveling sideways—that is, with the ruler at right angles to the direction of motion (like the wing of an airplane). We go on then to look at the case of a ruler traveling lengthwise, like an arrow, and try to compare it with a stationary ruler. What do the two rulers look like on a spacetime diagram? A ruler is an *extended* object, and every point of the ruler has its own world line (see the discussion of fig. 2.2.7). We can't conveniently draw them all, so let's just do the two ends. In fig. 4 I've drawn the world lines for the ends of a stationary ruler R of length L (taking S as the stationary observer again). The left end stays at x = 0, so its world line is simply the t axis, while the right end stays at x = L, a vertical line parallel to the t axis. The point B, which plays a role in the discussion to follow, marks the right end of the ruler at time t = 0. If R is a meter stick, then the centimeter subdivisions appear as 99 equally spaced vertical lines in between the two ends shown in the figure.

Next we look at the moving ruler R′. In fig. 5 we have a picture of R′, which is stationary in (S′)'s reference frame. The world lines of the two ends are straight lines of inclination v, the left one being the t′ axis. The point marked A is needed later in the discussion.

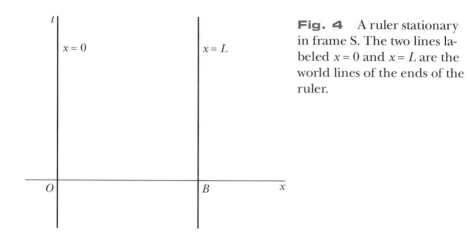

Fig. 4 A ruler stationary in frame S. The two lines labeled $x = 0$ and $x = L$ are the world lines of the ends of the ruler.

What happens now when S measures the moving ruler? He has to observe the two ends *simultaneously,* say at the time $t = 0$ (in his frame), corresponding to the points O and A in fig. 5. (Note that S′ *doesn't* see these events as simultaneous.) We have already found out that S will see the moving ruler as shortened by the factor γ. If we include S's ruler R in the picture it will look like fig. 6, with the end of ruler R′ (point A) to the left of the end of ruler R (point B) at $t = 0$, because of the Lorentz contraction.

When you look at it from (S′)'s point of view, on the other hand, you have to adopt her standard of simultaneity, indicated by the x' axis in fig. 6. Simultaneous observations by S′ of the two ends of ruler R are indi-

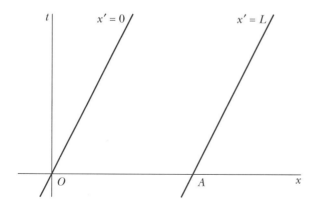

Fig. 5 World lines of the ends of a moving ruler.

Fig. 6   S and S′ view each other's rulers.

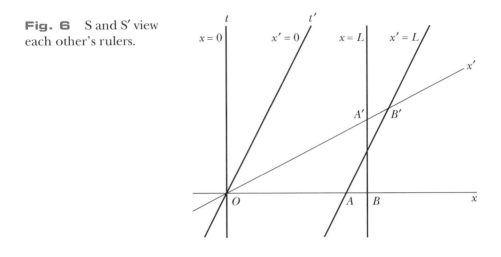

cated by points O and A', and at the same time the end of ruler R′ is at point B', so ruler R looks short to S′, again by the factor γ, as required by symmetry. (Note the close similarity to the diagrams and arguments on time dilation at the beginning of this section.) You can check this by writing the equations that give the inclinations of the two lines OB' and AB':

$$x(B') = \frac{c^2 t(B')}{v} \quad \text{(eq. 2.4.10)} \tag{8}$$

$$x(B') - x(A) = x(B') - \frac{L}{\gamma} = vt(B') \tag{9}$$

and solving for $x(B')$ by eliminating $t(B')$. After some steps (see prob. 3), this gives

$$x(B') = \gamma x(A') = \gamma L \tag{10}$$

as required.

This, then, confirms the equivalence of the two points of view and illustrates the consistency of the principle of relativity—that is, the fact that we don't run into contradictions as we look at the various odd consequences of assuming that different inertial frames of reference are equivalent.

PROBLEMS

1. Suppose there were no time dilation effect and (S')'s clock looked normal to S.
   (a) Draw a spacetime diagram similar to fig. 1, showing the ticks of the two clocks in this case, and then use a simultaneity line for S' to show that S's clock must necessarily look slow to S'.
   (b) Draw a second diagram to illustrate the case in which S's clock looks normal to S', and explain how the behavior of the clocks would look to S.
2. Perform the derivation of eq. 6, $t(A') = \gamma^2$, from eqs. 4 and 5.
3. Fill in the missing steps in the derivation of eq. 10: $x(B') = \gamma x(A') = \gamma L$.
4. Another way to look at the contraction of rulers is to suppose that the contraction factor γ isn't known to start with and then determine γ by requiring that each observer's ruler must look short to the other observer by the same factor. Write out the argument in this form, using the fact that in fig. 6 the slope of the simultaneity line OB' is v/c^2. That result, from §2.4, doesn't depend on knowing γ.
5. Dick and Jane are twins. Jane leaves on a space voyage, traveling at constant speed $c/3$. One year later, Dick takes off after her to show her his new puppy, Spot. He travels at speed $2c/3$ and finally catches up with her.
   (a) How long, in the earth's frame, does it take Dick to catch up with Jane?
   (b) How long does this seem to Dick?
   (c) What is the difference in their ages when he reaches her? Which twin is now older?
   (d) Find Dick's coordinates in Jane's reference frame at the moment he leaves the earth.
   (e) Draw careful spacetime diagrams, both in the earth's frame of reference and in Jane's, of these events.
6. A meter stick is flying by at speed v parallel to itself and therefore undergoes a relativistic contraction as seen by a stationary observer. Give the relativistic factor γ and the speed v for each of the following three cases (you'll need to use approximations in some cases, but not all).
   (a) The meter stick is shortened by one atomic diameter (say, 0.1 nm).
   (b) The meter stick is shortened to half its original length.
   (c) The meter stick is shortened to a length of 1 cm.

7. Observer S′ is traveling at her usual speed v relative to observer S,
    after parting at their common origin $x = t = 0$.
    (a) (S′)'s clock reads 1 s. Find the position and time of this event
         in S's frame (distances in light-seconds).
    (b) S′ "looks" at the clock that belongs to S at $t' = 1$ s (that is, one
         of (S′)'s team of observers, who is on the spot—see §2.4—
         looks at that clock at the moment when $t' = 1$ s). What does
         the clock read?
    (c) Event P takes place at $t = 0$, and $x' = 1$ lt-s. Find x and t'. Ex-
         plain how this illustrates the Lorentz contraction as seen by
         one of the observers.
    (d) What events (give their coordinates) can be used to illustrate
         the contraction as seen by the other observer?

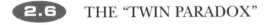 **2.6**   THE "TWIN PARADOX"

The curious effects we've been discussing, especially time dilation and
the Lorentz contraction, have a lot to do with the disagreements among
observers about how to synchronize clocks, and you might think therefore that they represent an illusion on the part of the observer. This is
not so, as I tried to explain in §1.6. I'd like to describe here a particular
case where you can see that the time dilation effect has to be real: the
famous twin paradox. One twin S stays on the earth (presumed to be at
rest), and the other twin, S′, travels away at a speed v and then returns at
the same speed. If S′ travels away for a time t' [in (S′)'s time] before
turning around, then the time t elapsed according to S at the moment of
turnaround is given (eq. 1.6.1) by

$$t = \gamma t' \qquad (1)$$

By the time S′ returns, S has aged by $2t$ and S′ has aged by $2t'$, so S′ is now
biologically younger than S by the factor $1/\gamma$. (Note that biological processes are governed by the laws of physics, so the rule for time dilation
must still apply. If it didn't, you could use biological processes to identify
an absolute rest frame, and the principle of relativity—the exact equivalence of different reference frames—would have to be seen as false.)
    This was thought to be a paradox (not by Einstein, though, who
knew better) because there appears to be a lack of proper equivalence
between the two twins. If S thinks that S′ is younger, shouldn't S′ think

that S is younger? However, there is no paradox (hence the quotation marks in the section title) because S' does not represent a single frame of reference, while S does. S' is not a uniformly moving observer, and the principle of equivalence doesn't apply. S' moves successively in two different frames, one with velocity v relative to S and one with velocity $-v$ relative to S. When S' changes direction, she goes into a frame with a *different simultaneity convention*. In fig. 1, the straight vertical line AC is S's world line, and the broken line APC is (S')'s world line. The two legs S'_1 and S'_2 represent the two successive reference frames of S'. According to frame S'_1, the turnaround point P is simultaneous with B_1, while according to frame S'_2, P is simultaneous with B_2. Thus, though it's true that $t(B_1) < t'(P)$ and that $t(C) - t(B_2) < t'(C) - t'(P)$, it's not correct to say that $t(C) < t'(C)$. This is because you've simply omitted from $t(C)$ the time interval $t(B_2) - t(B_1)$.

Does anything actually happen, when S' reverses direction at P, to cause the discontinuous shift from B_1 to B_2? *No, nothing physical.* Both of the points B_1 and B_2 are far away, and she can't observe them directly; it's simply that when she reverses her motion, she arbitrarily adopts a new criterion of simultaneity. In frame S'_2, events B_1 and P are simply not *treated* as simultaneous. It's very important to note, though, that when S and S' get together at C, they can make a *direct* comparison of their ages, and they must then agree on which has aged less or on which clock has lost time.

Let's look now at the whole scenario from the point of view of S'_1, which is shown by the spacetime diagram of fig. 2. The high-speed trip from P to C on the part of S' gives her such a big time dilation (slowing

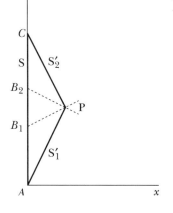

Fig. 1   The twin paradox. S' takes a trip and returns home.

Fig. 2   The same trip, viewed in frame S_1'.

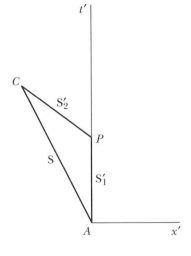

down of her biological clock) that it more than compensates for the slowing down of S's clock at his slower speed during the whole trip *AC.* That is, S′ is still younger than S at point *C,* regardless of which frame you choose. (Note that fig. 2 is just the scenario of prob. 2.5.5. You should go over that problem carefully now if you haven't already done so.)

The fact that the aging of S′—or the reading of her clock—is something that all observers have to agree on is very significant. I have called such a quantity, whose value is the same to all observers, an *invariant* (§2.3), and the invariant quantity in this case, the time elapsed according to the moving clock, is called the *proper time* and is represented by the symbol τ (the lower-case Greek letter tau). The proper time for S's path from *A* to *C* is

$$\tau_{AC} = 2t \tag{2}$$

and the proper time for (S′)'s path *APC* is

$$\tau_{APC} = \frac{2t}{\gamma} \tag{3}$$

Every traveler, no matter how complicated the journey, has a proper time τ determined by the world line of the traveler.

How much younger could a traveler manage to be? How far into the future could you take yourself, by means of time dilation, if you traveled far enough and fast enough and then returned to the earth? The limiting factors are time, acceleration, and fuel, as I'll explain in more detail in §3.3. The analysis is a bit difficult, but you can read about the results at the end of that section.

PROBLEMS

1. Draw the spacetime diagram for the twin paradox (figs. 1 and 2) in the frame S_2', the frame of the traveling twin as she returns to earth.
2. Show that the standard acceleration of gravity at the earth's surface, $g = 9.8$ m/s^2, can be written in astronomical units as $g \approx 1.03$ c/yr, or 1.03 lt-yr/yr^2. How long would it take to reach the speed of light at this rate of acceleration? (An answer to this question is not inconsistent with relativity. It's just that it's not physically possible to keep up this rate of acceleration as you approach speed c.)
3. How fast would a clock have to be traveling (in kilometers per second) to lose 1 hour per year? to lose 1 second per year?

CHAPTER 3

Moving Around in Spacetime

3.1 THE LORENTZ TRANSFORMATION

You learned in §2.1 that any given event is described in different ways by different observers. For every observer S there's a frame of reference (also referred to as S), with a set of four coordinates, x, y, z, and t, that give the position and time of that event relative to the reference frame S. A second observer S' uses a different set of coordinates, x', y', z', and t', to describe the same event. The values of the new coordinates are entirely determined by the results I've told you about in Chapter 2—Lorentz contraction, time dilation, and the simultaneity shift—and it's very useful to incorporate those results in a set of formulas that give the new coordinates explicitly in terms of the old. These formulas constitute what's called the *Lorentz transformation:* the mathematical rules for obtaining the coordinates of the given event P in reference frame S' from the coordinates in frame S. In particular, if S' is moving with speed v_0 in the x direction relative to S, then the coordinates of P in frame S' are given by

$$x' = \gamma_0 (x - v_0 t) \tag{1}$$

$$y' = y \tag{2}$$

$$z' = z \tag{3}$$

$$t' = \gamma_0 \left(t - \frac{v_0 x}{c^2} \right) \tag{4}$$

with

$$\gamma_0 = \frac{1}{\sqrt{1 - v_0^2/c^2}} \tag{5}$$

as usual. In the next few paragraphs I'll show you why these are the correct equations. (I'll stick to the case where S' is moving in the x direction relative to S; the equations get more complicated when the direction of relative motion is arbitrary, and it doesn't add anything useful. A reference frame may also be rotated in an arbitrary way—a complication of no importance to us here.)

As usual, for convenience, I've given S and S' the same spacetime origin. That is, an event at $x = y = z = 0$ and $t = 0$ in frame S is described by $x' = y' = z' = 0$ and $t' = 0$ in frame S'. Let's start by ignoring y and z, and consider only events occurring on the x axis in space (in the x–t plane in spacetime). The transformation on x and t can be understood from our previous results about the simultaneity shift (SS), time dilation (TD), and Lorentz contraction (LC), by referring to fig. 1. We're looking at the event-point P, and we want to express its coordinates (x',t') in S' in terms of its coordinates (x,t) in S. As usual, there are many different ways of deriving the formulas; I've found that it works out nicely to focus on the points A and B in the diagram and apply the different rules from Chapter 2, as I'll explain. From the simultaneity of A and P in S', the time $t(A)$ is given by the simultaneity shift rule (eq. 2.4.10):

$$t(A) = t - \frac{v_0 x}{c^2} \tag{6}$$

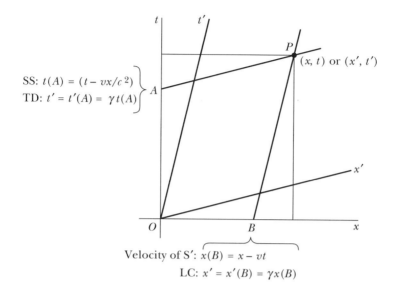

SS: $t(A) = (t - vx/c^2)$
TD: $t' = t'(A) = \gamma t(A)$

(x, t) or (x', t')

Velocity of S': $x(B) = x - vt$
LC: $x' = x'(B) = \gamma x(B)$

Fig. 1   Deriving the Lorentz transformation. The simultaneity shift (SS) and time dilation (TD) rules applied to point A yield t', and the Lorentz contraction (LC) and velocity formula applied to point B give x'.

But this is the time as read by S's clock, which is seen as slow in S' by the time dilation factor γ_0, so

$$t'(A) = \gamma_0 t(A) \tag{7}$$

from which eq. 4 follows because $t'(A)$ is just t'.

We now apply very similar logic to the point B: The value of x at B is

$$x(B) = x - v_0 t \tag{8}$$

because the line BP is the world line of a point moving with speed v_0. In fact, comparing this line with the corresponding line in fig. 2.5.5, we can think of it as the right end of a stick of length x' (the length as S' sees it) moving with speed v_0, which therefore looks contracted to S. But $x(B)$ is just the length of the stick as seen by S, so

$$x' = x'(B) = \gamma_0 x(B) \tag{9}$$

and this, together with eq. 8, gives us eq. 1.

The simple equalities for y' and z' follow from the argument of §1.5 that S and S′ always agree on the length of rulers transverse to the direction of motion.

Eqs. 1–4, the Lorentz transformation equations, are central to special relativity. They incorporate in compact form all the odd results we've been talking about in Chapters 1 and 2, and they can be used in all sorts of other situations as well. The same equations, with a few necessary modifications, also give the transformation rules for other physical quantities, such as momentum and energy, electric and magnetic fields, and so on. It's the essence of relativity to relate the views of different observers, and these transformation equations put the relationship between these different views in concise mathematical form.

The Lorentz transformation equations can be used in various ways. Let's look at a few examples.

(i)   If the equations are inverted—solved for x and t in terms of x' and t'—you get (prob. 1):

$$x = \gamma_0(x' + v_0 t') \tag{10}$$

$$t = \gamma_0\left(t' + \frac{v_0 x'}{c^2}\right) \tag{11}$$

These two equations are identical in form to eqs. 1 and 4, with the interchange of primed and unprimed coordinates and the replacement of v_0 by $-v_0$, corresponding to the fact that S is moving with velocity $-v_0$ with respect to S′. This symmetry between the views of S and S′ is a necessary consequence of the principle of relativity (§1.5) and includes as special cases the fact, discussed in §2.5, that each observer sees the same slowing of the other's clock and the same shortening of the other's ruler.

(ii)   S′ herself follows the path $x' = 0$: then $t' = \tau'$, where τ' is the proper time of S′ (§2.6), as read by her clock and corresponding to her subjective sense of the passage of time and her own biological processes. From eq. 1 we see that $x' = 0$ implies $x = v_0 t$, which simply says that S sees S′ moving with velocity v_0. From eq. 11 we see that

$$t = \gamma_0 t' = \gamma_0 \tau' \tag{12}$$

which is the standard time dilation.

(iii)   The Lorentz contraction too can be seen in these transformation equations. If S′ carries a rod of length L (as she sees it), then we can say that one end of the rod has $x' = 0$ and the other has $x' = L$ for any value of t'. For S to measure the length of the rod as it moves by, he has to observe both ends at $t = 0$—that is, at the same time in his frame. Then the end with $x' = 0$ has $x = 0$ at that moment (from eq. 1 with $x' = 0$ and $t = 0$), and the other end, also from eq. 1 but with $x' = L$, has

$$x = \frac{x'}{\gamma_0} = \frac{L}{\gamma_0} \tag{13}$$

so S sees the length as L/γ_0, which is shorter by the factor $1/\gamma_0$, the Lorentz contraction factor.

Both (ii) and (iii) can be illustrated by the example of an astronaut returning to earth from a long trip and traveling at speed $0.95c$. She is three light-years away and has only one year's supply of oxygen left. Will she make it? It will take her 3 lt-yr$/0.95c = 3.158$ yr to get here, and her relativistic γ factor is $[1 - (0.95)^2]^{-1/2} = 3.203$. In terms of her own proper time, then, it will take her $3.158/3.203 = 0.986$ years, and she'll just barely make it. (I haven't worried about how she's going to slow down and land without using up valuable time. I trust she has some stress-free braking system that will bring her to rest quickly without causing physical damage.) It's instructive to look at the process more closely from her point of view: The earth is coming toward her at $0.95c$, and its initial distance away, by her standard of simultaneity, is $3/\gamma = 0.937$ light-years. The earth will therefore reach her in less than one year of her time, which is consistent with our previous result.

A real experimental fact that shows the same logic as this example is the observation of muons that arrive at the earth's surface along with the cosmic rays from outer space. A muon is like a big brother of the electron, with a mass 207 times greater, that lives only 2.2×10^{-6} s on the average before falling apart. The muons we see are produced in the upper atmosphere in high-energy collisions between cosmic ray particles and air molecules. They're created at altitudes of the order of 10 kilometers, and even at speed c they would travel only about $(3 \times 10^8 \text{ m/s}) \times (2 \times 10^{-6}\text{s}) = 600$ m before decaying. Nevertheless, we see them here at ground level, thanks to the time dilation factor, just like the astronaut in our example. This observation of cosmic ray muons provided one of the first experimental confirmations of the reality of the relativistic time dilation effect.

(iv)   The next example is a light signal. As seen by S (fig. 2.2.5), each point on its path satisfies

$$x = ct \qquad (14)$$

and we can find out what S′ sees by substituting $x = ct$ into eqs. 1 and 4, which gives

$$x' = ct' \qquad (15)$$

(You should check this result; see prob. 2.) That is, both observers see it moving with the same speed c, as required by the principle of relativity.

(v)   As mentioned earlier, Newtonian physics corresponds to the limit $c \to \infty$. In this case $\gamma_0 = 1$, and the Lorentz transformation equations, eqs. 1–4, reduce to the following much simpler forms:

$$x' = x - v_0 t \qquad (16)$$

$$y' = y \qquad (17)$$

$$z' = z \qquad (18)$$

$$t' = t \qquad (19)$$

These equations describe what is called a Galilean transformation, the transformation to a moving reference frame in Newtonian physics. Time is absolute—the same in *every* reference frame—and the expression for x' simply reflects the fact that S′ measures x relative to her origin at location $v_0 t$. The Galilean transformation doesn't deal with light because we've set the speed limit, which is the speed of light, equal to ∞.

    The Lorentz transformation, as I've said, is the core of special relativity. You'll find that I refer to these transformation equations frequently, in the sections and chapters that follow, as the most compact way to work out the consequences of relativity in different physical situations.

PROBLEMS

1. Derive eqs. 10 and 11 for the inverse Lorentz transformation, expressing x and t in terms of x' and t'.

2. For a point P on the path of a light signal, with $x = ct$ in frame S, find the coordinates in frame S′ and show that $x' = ct'$. Note that S and S′ do not see the same coordinates for this point but do agree on the ratio x/t, which is the speed of the signal.

3. (a) Observer S′ is traveling in the x direction at velocity $-0.8c$ relative to observer S. Evaluate γ_0 and then determine the following, with all distances and times expressed in light-years and years, respectively.

   (i) The coordinates x and t of an event A for which $x' = 5$ and $t' = 4$

   (ii) The times t and t' for an event B for which $x = 4$ and $x' = 4$

   (iii) The times t and t' for an event C for which $x = -4$ and $x' = 4$

   (iv) The relative velocities v and v' for a particle P moving in such a way that t always equals t'

   (b) Carefully draw a spacetime diagram showing these events and the world line of particle P.

4. Mme. Curie (observer S′) is in a train traveling at speed $+0.5c$ relative to Dr. Einstein (observer S), who stays at home in Ulm. They both set their timers to zero at the moment Mme. Curie leaves Ulm. Mme. Curie observes an event that she interprets as taking place at $x' = 2$ lt-s (light-seconds) and $t' = 2$ s. Find the coordinates (x,t) of the event in Dr. Einstein's coordinate system.

 **3.2** COMBINING VELOCITIES

One notorious feature of relativity is the peculiar way velocities combine. Say you're in a railroad carriage traveling 88 ft/s (60 mi/h), and you throw a ball forward along the carriage with a speed of 50 ft/s. To the observer on the ground the speed of the ball should simply be the sum, 138 ft/s, because in 1 s the train travels 88 ft, and the ball travels that 88 ft plus an additional 50 ft, or 138 ft. But because of what I've told you about the speed of light, you know that has to be wrong. If you could throw the ball with speed c or, equivalently, shine a flash of light forward along the carriage, then the observer on the ground mustn't see the ball—or light—traveling faster than c. Thus you know you can't simply add the speed of the train to the speed of the ball. And you also know (taking the ground to be at rest) that your clocks and rulers behave oddly because of the motion of the train, so although the ball may seem to you to be going 50 ft/s faster than the train, that isn't the way it would be seen by an observer on the ground.

Let's use the rules for the behavior of clocks and rulers to get a formula for the speed of the ball from the point of view of the observer on the ground. The simplest way to do this is to use the inverse Lorentz transformation equations, eqs. 3.1.10 and 3.1.11. You are looking not at just one event but at a series of events—the set of points making up the world line of the moving object. In the S′ frame these points all satisfy

$$x' = v't' \tag{1}$$

which simply says that v' is the apparent velocity in the frame S′. If you plug this expression for x' into eqs. 3.1.10 and 3.1.11 and divide, you get

$$x = \gamma_0(v' + v_0)t' \tag{2}$$

$$t = \gamma_0\left(1 + \frac{v_0 v'}{c^2}\right)t' \tag{3}$$

and the quotient gives the velocity v according to S:

$$v = \frac{x}{t} = \frac{v_0 + v'}{1 + v_0 v'/c^2} \tag{4}$$

This is the famous formula for the composition of velocities in special relativity.

It's important to remember that there are three different participants in this problem, the two observers S and S′ and the object P that they are observing, as well as three different relative velocities to keep track of:

$$v_0 = \text{velocity of S}' \text{ relative to S}$$
$$v = \text{velocity of } P \text{ relative to S}$$
$$v' = \text{velocity of } P \text{ relative to S}'$$

If the railroad car mentioned previously is traveling at $0.5c$, for example, and S′ (that's you) throws the ball P at $0.6c$, then S (that's me) will see the ball traveling at $v = (0.5c + 0.6c) / (1 + 0.5 \times 0.6) = 0.846c$, which is still less than c.

The physical picture can be represented by a spacetime diagram, as shown in fig. 1. As usual, the t' axis is the world line for observer S′ moving with the train at speed v_0, and the x' axis sets her standard of simultaneity. She throws the ball at the time and place represented by the spacetime origin O, and the world line of the ball is the line OA. Between O and A, the time (according to S′) increases by t', which you can take to be 1 s, and the position according to S′ increases by x', which would be 50 ft in our example. The velocity v' in our example is thus $x'/t' = 50$ ft/s.

Now, in figuring out v, the velocity as seen by S, you have to take into account the three quirks of the measurement process: Lorentz contraction, time dilation, and the simultaneity shift. As it turns out, the Lorentz contraction and time dilation effects cancel, and the simultaneity shift alone is responsible for the crucial denominator in eq. 4. Note the spacetime point B in fig. 1, which is the point that according to S′ is at the same location as A but simultaneous with O. Thus B could represent the event of making a chalk mark on the floor of the carriage 50 ft away from S′ at the very moment that S′ throws the ball, according to (S′)'s standard of simultaneity. The line BA is then the world line of the chalk mark, and A is the event of the ball's passing that mark. Note that BA is parallel to the t' axis in the spacetime diagram. The simple point, now, is this: S′ is measuring only the time from B to A, while S is measuring it all the way

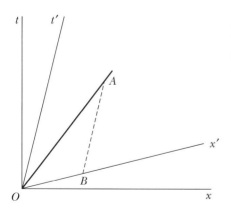

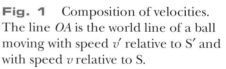

Fig. 1 Composition of velocities. The line OA is the world line of a ball moving with speed v' relative to S′ and with speed v relative to S.

from the x axis. The difference, corresponding to the simultaneity shift, is added to t and thus makes x/t smaller than you'd expect, by just the factor in the denominator of eq. 4. The argument in detail is more complicated than I want to go through here, and it's logically equivalent to the concise derivation that I gave using the Lorentz transformation equations. Note that the simultaneity shift is represented by the second term in the Lorentz transformation equation for t, eq. 3.1.11, which is the term responsible for the denominator in eq. 4.

The funny way (eq. 4) that velocities add is the key to understanding a number of the counterintuitive effects of relativity, including the invariance of the speed of light c and the impossibility of accelerating any object to a speed greater than c. The crucial factor is the denominator, which modifies the simple intuitive addition represented by the numerator.

Let's consider the character and meaning of this formula. It's often called the formula for the addition of velocities, which is slightly misleading because it is clearly not addition in the normal sense. The velocities v_0 and v' are measured by different observers, and it's inappropriate to add them. The phrase *composition of velocities* is also widely used and is more accurate.

When trying to understand a new formula, it's a good idea to look at special cases and see if the formula makes sense. Let's subject eq. 4 to this test.

(i) $c = \infty$

If c is infinite the denominator in eq. 4 becomes unity, and the formula reduces to the common-sense relation that I discussed at the beginning of this section:

$$v = v_0 + v' \tag{5}$$

This is because $c = \infty$ implies instantaneous communication, which allows perfect synchronization of clocks and instantaneous comparison of clocks and rulers. Everyone agrees on simultaneity if $c = \infty$, and moving clocks and rulers behave no differently than stationary ones. *Einsteinian relativity is equivalent to Newtonian relativity when $c = \infty$.* We also see in this way that if c is very large, so that the denominator is very close to unity, then the common-sense formula given as eq. 5 is an excellent approximation; that is, it is very nearly right.

(ii) $v' = 0$

The formula reduces to $v = v_0$, the obvious fact that an object that is stationary in (S')'s frame has velocity v_0 relative to S.

(iii) $v = 0$

The formula requires $v' = -v_0$, which expresses the fact that an object at rest in S's frame is moving with velocity $-v_0$ relative to S'.

(iv) $v' = c$

This is the interesting case of a light signal. The result is not quite obvious from the formula, so be sure to work it out and see that you get the proper relativistic result $v = c$. That is, the light signal has the same apparent speed in every reference frame, exactly as required by the principle of relativity.

(v) $v = (1 - \varepsilon')c$, where $\varepsilon' \ll 1$

This is the case where the ball is traveling very close to the speed of light. By judicious use of the approximation formula (eq. 1.4.12) you should be able to obtain (prob. 6.)

$$v = (1 - \varepsilon)c \qquad (6)$$

with ε given by

$$\varepsilon = \frac{\varepsilon'(c - v_0)}{c + v_0} \qquad (7)$$

As a final check on the reasonableness of eq. 4, we need to invert it—that is, to solve for v' in terms of v. The result (you should check this—see prob. 7) is

$$v' = \frac{v - v_0}{1 - v_0 v/c^2} \qquad (8)$$

which is exactly symmetrical to eq. 4, with v_0 replaced by $-v_0$. Note, though, that the flavor of this form is a bit different—it gives most naturally the relative velocity of two objects whose velocities relative to S are both given. It can also be thought of as the basic transformation law for velocities under the Lorentz transformation, analogous to the Lorentz transformation equations 3.1.1–3.1.4 for space and time coordinates. The form that I worked out to start with (eq. 3.1.4) is the inverse transformation for velocities, analogous to eqs. 3.1.10 and 3.1.11, the inverse Lorentz transformation.

PROBLEMS

1. Rewrite the first paragraph of this section, especially the last sentence, from the standpoint that it is the train that's really at rest and the ground is moving by at 88 ft/s in the opposite direction.

2. Try to get a velocity v greater than c by combining two velocities v_0 and v' that are both close to c. For example, try $v_0 = v' = 0.75c$, $v_0 = v' = 0.9c$, and so on. Extra challenge: Construct a theorem that says v is smaller than c for *any* v_0 and v' smaller than c.

3. Find the relative error (the error as a fraction of the result) in the approximate equation $v = v_0 + v'$ (eq. 5) for some common terrestrial speeds and some common astronomical speeds (at least two of each). (Astronomical data are given in Appendix C, §5.)

4. Apply the addition formula (eq. 4) to the case $v' = -c$, and explain your result.

5. Two spaceships, A and B, are approaching each other on a collision course with velocities $0.8c$ and $-0.9c$, respectively. Find the velocity of A as measured by B.

6. (a) Derive the result represented in eqs. 6 and 7. Some further hints about approximations, in the spirit of eq. 1.4.12:

$$(A + \varepsilon)^n = A^n \left(1 + \frac{\varepsilon}{A} \right)^n$$

$$(1 + \varepsilon)(1 + \varepsilon') \approx 1 + \varepsilon + \varepsilon'$$

   (b) Does eq. 7 make sense when $v_0 = 0$? when $v_0 = c$? Explain.

   (c) Taking $\varepsilon' = 0.001$, find ε for each of the following cases: (i) $v_0 = 300$ m/s, (ii) $v_0 = 0.5c$, (iii) $v_0 = 0.999c$

7. Derive eq. 8, which expresses v' in terms of v. Obtain v' for each of the following cases: (a) $v = 0.9c$, $v_0 = 0.8c$; (b) $v = 0.999c$, $v_0 = 0.998c$; (c) $v = 0.9c$, $v_0 = -0.9c$; (d) $v = 0$, v_0 arbitrary.

8. (a) The rule for combining velocities (eq. 4) can be converted to a product rule: $(c + v)/(c - v) = [(c + v_0)/(c - v_0)] \times [(c + v')/(c - v')]$, which has the nice features that the right side is the simple product of a factor involving v_0 and a factor involving v', and that each of these factors has just the same form as the left side involving v. Derive this result.

   (b) Check this form of the rule by putting in the numbers from probs. 2, 5, and 7.

   (c) A spaceship starts from rest on the earth and accelerates by making repeated boosts of $c/4$ relative, for each boost, to its velocity just prior to the boost. Use the rule given in part (a) to find its velocity relative to the earth's frame after each of the first 10 boosts. Prove that the velocity can never reach or exceed c after any number of boosts.

3.3 PROPER TIME AND THE ULTIMATE SPACE TRAVELER

Your biological time as a traveler is no different than the time shown by your clock; you and the clock are both going at the same speed at any moment, and the same physical laws govern the biological processes of a human and the ticking of a clock. How then do you describe the behavior of a clock—or a person—that isn't moving at constant speed? At any given moment you have a definite speed relative to the earth, say, and you have the usual time dilation factor $\gamma = 1/\sqrt{1 - v^2/c^2}$, so in a small time dt (by earth time and earth rules for synchronizing clocks) your clock (and your age) advance by dt/γ. [The expression dt is not a product; it's like Δt, simply a symbolic way of expressing a time interval. You use Δt when the interval is finite and dt when the interval is infinitesimal—or very tiny (we'll look at this case in a moment).] The reading of your clock is your proper time (see §2.6) and is represented by the symbol τ. The advance of your clock during the earth time dt, then, is called $d\tau$, and the relation just described is written

$$d\tau = \frac{dt}{\gamma} \tag{1}$$

Alternatively, you can reverse this and say that when your clock advances by $d\tau$, the amount of earth time that elapses is given by

$$dt = \gamma d\tau \tag{2}$$

Eq. 2 is the appropriate form to use when you want to start with the proper time τ and then work out expressions for earth time and distance traveled as functions of τ. The reason for not using t' to represent the proper time is that you're not moving at constant velocity, and the symbol t' is always associated with a definite uniformly moving reference frame S'. The proper time τ for an observer who doesn't move at constant speed isn't associated with any one uniformly moving reference frame.

To find out the total time that elapses on earth, you have to add up all these small amounts $\gamma d\tau$, starting at the beginning of your trip. That is to say, you break up your trip from the beginning (which I'll call point A) to the present time (which I'll call P) into a very large number of very small time intervals $d\tau$, find out how much earth time elapses during each little interval, and add these up to get the net earth time. You may recognize this as describing an *integration,* which is really nothing but summing a lot of very small parts in just this sort of way. How small do the little intervals have to be? Small enough that your speed can be treated as essentially constant during each interval. For example, if you take the intervals $d\tau$ as being 1 s each, you'll get an excellent answer for the problem we're talking about. You'll get an even better answer if you take intervals of 0.01 s. Mathematicians talk about taking the limit in which dt is infinitely small and they prove theorems about the validity of that limiting process, but it is really equivalent to say simply that *dt must be sufficiently small that it doesn't make any difference if you make it any smaller.* It is common to call dt an *infinitesimal;* the letter d used in this way represents a very small change in whatever quantity it goes with—t in this case.

The sum of these little times, then, is the total earth time t. You can write it as an explicit sum:

$$t = dt_1 + dt_2 + dt_3 + \cdots + dt_N \tag{3}$$

where N is the total number of little pieces, and you can substitute for each of these little times the equivalent expression from eq. 2 in terms of the proper times:

$$t = \gamma_1 d\tau_1 + \gamma_2 d\tau_2 + \gamma_3 d\tau_3 + \cdots + \gamma_N \, d\tau_N \tag{4}$$

We have a compact way of writing expressions like these, using the symbol $\int$, the *integral sign.* Thus for eq. 3 we write

$$t = \int_A^P dt \tag{5}$$

where A and P are the beginning and ending points of the trip. The expression on the right is called an *integral.* As I've tried to explain, the value of the sum is approximately the same regardless of how small you take the little pieces, provided they are small enough, so the value of N is unimportant and doesn't appear in the compact form. In similar fashion, the sum shown in eq. 4 is written in the concise form

$$t = \int_A^P \gamma(\tau)\, d\tau \tag{6}$$

where $\gamma(\tau)$ is the value of the factor γ at the point where the proper time has reached the value τ; it's determined by the speed v, which is the inclination of the world line at that particular point.

The integral sign $\int$ is just an old-fashioned S, for "sum," because the integral in eq. 5 or 6 simply represents the sum of all the little contributions in eq. 3 or 4. The beginning and ending points, A and P, are called the *limits of integration*. (See Appendix A, §5, for more on integrals and integration.)

It's important to remember that the proper time τ is the *natural* time for an observer moving in an arbitrary nonuniform way—it's the time as measured by that observer's wristwatch and by all her own natural processes, such as breathing, heartbeat, and aging. It is not defined with respect to any one uniformly moving frame of reference.

Eq. 6 is meant to express earth time in terms of proper time, because that's the question we're looking at right now; a more common problem in fact is to find the proper time for a traveler whose path is given in terms of the earth time t. In that case you write $d\tau$ as dt/γ, as in eq. 1, and sum up the little $d\tau$'s in the form of an integral:

$$\tau = \int_A^P \frac{dt}{\gamma(t)} \tag{7}$$

You can see that τ will always be less than t (eq. 5) because $1/\gamma$ is always less than 1. The traveler always ages less than the stay-at-home, regardless of the route she takes.

Let's return now to the question I raised in §2.6: How far into the future could you travel in one lifetime, through time dilation, if you were to take a long journey and then return to the earth? The limiting factors, as I mentioned, are time, acceleration, and fuel. I'll put off discussing the fuel needs as not posing a difficulty in principle, though in practical terms the fuel constraints are far more limiting than the biological ones. The interesting factors, I think, are the physiological ones: You have only one lifetime to play with, and you can tolerate only a certain amount of acceleration, so there is a natural limit to the speeds you can reach. I want you to be comfortable, so let's limit your apparent acceleration to g, the normal acceleration of a freely falling object on earth. This means that your apparent weight due to acceleration will be exactly the same as your normal weight on earth. Now g has the approximate value 9.8 m/s², or 9.8 meters per second per second. (In other words, the speed of a freely falling object increases by 9.8 m/s each second.) This turns out (see prob. 2.6.2) to be just a bit more than c per year:

$$g \approx 1.03 c/\text{yr} \tag{8}$$

That is, if the freely falling object could keep falling at constant acceleration for a full year, its speed would be $1.03c$. This ignores relativity, of course, which won't let the object go faster than c, but it would still get to speeds comparable to c in that time. And you in your spacecraft, accelerating by means of rockets or what have you, will also reach speeds close to c in 1 year. In just a few years of your own biological time, you'll be pushing extremely close to c and getting a really big time dilation factor.

The result of this calculation, which is too complicated to include here, is a formula for the amount of earth time t that elapses in the earth's reference frame while you age by an amount τ—that is, while your proper time increases by τ. It is

$$t = \frac{c}{2g} \left(e^{g\tau/c} - e^{-g\tau/c} \right) \tag{9}$$

Here e is a number that plays a very special role in many areas of mathematics; it's equal to 2.71828 . . . , and is often identified as the base of natural logarithms (see prob. 3).

Eq. 9 tells you what time it is on earth (using earth's standard of simultaneity) when you, the traveler, have aged by an amount τ. We can make $c = 1$ by measuring t and τ in years and distances in light-years, which gives g the simple numerical value 1.03 (eq. 8). You can get a good idea of what eq. 9 means physically by looking at large values of τ; I'm assuming that you're prepared to spend a number of years on your trip. For large τ the first term on the right of eq. 9 becomes very large and the second term becomes very small, so it's quite accurate to drop the second term. With $c = 1$, then, you get

$$t \approx \frac{e^{g\tau}}{2g} \tag{10}$$

The same expression gives the distance traveled, in light-years, because most of the trip is at a speed very close to c. It says, for large τ, that both t and x increase *exponentially* with τ; this means that t and x increase a lot for comparatively small increases in τ. That is, the time and distance measured from the earth increase hugely while you age by only a modest amount. The function $e^{g\tau}$ increases by a factor 2 whenever $g\tau$ increases by ℓn 2 (the natural log of 2), which equals $0.693\ldots$. With $g = 1.03$, this means that after the first few years the earth time t will *double* for every additional 8 months (0.693 yr/1.03, approximately) that you spend on your journey. Table 1 gives some cases, with t and τ measured in years and x in light-years.

These numbers show you how huge the effect is. When you're just 10 years older than when you started, the time elapsed on earth is over 14,000 years, and your speed is less than the speed of light by about 3 parts per billion. After 3 more years of your trip, owing to the huge time dilation factor, the earth is over 300,000 years older!

So far we've looked at just a one-way trip. If you want to return to the earth and see what things are like after so long an absence, you'll have to turn around, which is not so easy at this speed. You mustn't slow down any faster than you were accelerating, because that would be just as painful as accelerating too fast. If you want to come back to earth in about 50 years, by your time, you could start by accelerating for 13 years, say, and then decelerate for another 13 years (same magnitude of decel-

.

Table 1   Time dilation at uniform acceleration

τ	v/c	γ	t	x
1	0.774	1.58	1.18	0.56
2	0.968	4.00	3.75	2.91
5	0.999933	86.2	83.7	82.7
10	$1 - 3 \times 10^{-9}$	14,900	14,433	14,432
13	$1 - 5 \times 10^{-12}$	3.3×10^{5}	3.2×10^{5}	3.2×10^{5}
25	$1 - 9 \times 10^{-23}$	7.6×10^{10}	7.4×10^{10}	7.4×10^{10}

Fig. 1 Round trip, with acceleration ±g.

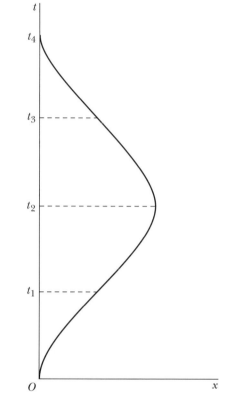

eration), which will bring you to rest someplace far out in space. You then repeat the procedure for the return trip, taking another 26 years to do so. The spacetime diagram of fig. 1 shows your acceleration up to time t_1, your deceleration until t_2, your acceleration homeward until t_3, and finally your slowing down again in order to be at rest when you get home to earth. You are then 52 years older than when you started, and (from the table, using $\tau = 13$ years) the earth is $4 \times (3.2 \times 10^5 \, \text{yr})$, or about 1,300,000 years older. Would you volunteer for the trip?

If you don't care about coming back but just want to travel as far as possible, you can accelerate for 25 years and then slow down for another 25, thereby traveling a distance of $2 \times (7.4 \times 10^{10} \, \text{lt-yr})$, which is something like the present size of the visible universe. *The human body is physiologically capable of exploring the furthest reaches of the universe and times in the future comparable with the present age of the universe.*

Unfortunately, fuel is something of a problem. We haven't discussed energy yet, but you'll find in Chapter 5 that the same factor γ is involved in calculating energies. The amount of fuel you'll need to take along will be something comparable to the entire mass of the sun, even supposing total conversion of mass to energy (by $E = mc^2$), which we are also very far from achieving. But that's another story.

PROBLEMS

1. A clock for use in a space ship consists of a light signal reflected back and forth between two mirrors in a long evacuated tube. The tube is arranged along the direction of travel of the ship, and the time is obtained by counting the number of reflections. If the clock is carefully set up to give accurate time when the ship is on the ground, what does the clock measure as the ship accelerates to relativistic speeds? Justify your answer carefully.

2. (a) Check the figures given for t in table 1 for $\tau = 1$, 10, and 25 yr.
   (b) Use the approximate formula of eq. 10 to answer the following questions.
      (i) If you travel out and return as in the example, but go only far enough to come back 1000 years later by earth time, by how much will you have aged?
      (ii) If you could handle a steady acceleration of 1.5 g_0, where g_0 is the acceleration of gravity at the earth's surface, how far could you travel in one direction in 10 years of your own proper time?
      Note: If $y = e^x$, then $x = \ln y \equiv \log_e y$. If your calculator does not have these functions, you can use $e^x = 10^{x \log e} =$ antilog $(0.4343x)$ and $\ln y = \log y / \log e = \log y / 0.4343$, where $\log y \equiv \log_{10} y$.

3. The number e (eq. 9) can be defined by the property

$$e^\varepsilon \approx 1 + \varepsilon, \quad \text{for } \varepsilon \ll 1 \qquad (11)$$

(Note that for small ε, any exponential A^ε will differ from 1 by a term proportional to ε; the distinctive thing about the case $A = e$ is that the constant of proportionality is equal to 1.)
   (a) Use this property of e (eq. 11) in the form obtained by taking the log of both sides

$$\varepsilon \log e \approx \log (1 + \varepsilon)$$

to calculate e, using ordinary base-10 logs and antilogs and very small values of ε. Make a table showing how the value of e that you get becomes better as you make ε smaller and smaller.

(b) Show that this way of calculating e is equivalent to

$$e = \lim_{N \to \infty} \left(1 + \frac{1}{N} \right)^N \tag{12}$$

a standard expression for e.

(c) Use the property of the exponential given in eq. 11 to show that $\ln (1 + \varepsilon) \approx \varepsilon$ for $\varepsilon \ll 1$.

3.4   VECTORS IN SPACETIME

One of the standard ways to describe how you get around in space is with things known as vectors. A *vector* is defined most simply as something that has both magnitude (length) and direction. When you describe a trip as being 750 miles in a southwest direction, you're specifying a *displacement vector*. When you say that the speed of an airplane is 550 mi/h in a direction 15° east of north, you're specifying the *velocity vector*. We have force vectors, acceleration vectors, and all sorts of others.

Vectors in space

To get an idea of how vectors work, look first at a point P in ordinary Euclidean space, along with a reference frame—an origin O, an x axis, and a y axis (see fig. 1). We use an arrow from O to P, called a *position vector*, to indicate the location of P relative to O, and we represent this vector by the underlined letter $\underline{r}$. The magnitude of this position vector is its length r (*not* underlined—you have to be careful), and the direction is specified by the angle θ that the vector $\underline{r}$ makes with the x axis. The values of r and θ are referred to as the *polar coordinates* of the point P. The vector can also by specified by the coordinates x and y of the point P (shown in the figure), which are called the *components* of $\underline{r}$. The Pythagorean theorem and standard trigonometry tell you that r and θ are related to x and y by the formulas

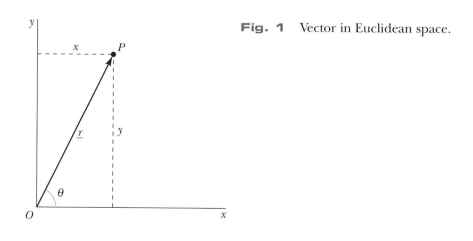

Fig. 1 Vector in Euclidean space.

$$r = \sqrt{x^2 + y^2} \tag{1}$$

$$\cos\theta = \frac{x}{r}; \quad \sin\theta = \frac{y}{r} \tag{2}$$

Let's look now at how vectors are used to describe the separation between points in space. Suppose P_1 and P_2 are two points in space, represented by vectors $\underline{r}_1$ and $\underline{r}_2$ that show the positions relative to the origin O (see fig. 2). The vector $\Delta\underline{r}$, which is a displacement vector, tells you how far and in what direction you'd have to go to get from P_1 to P_2. It can be written as the *difference* of the two vectors $\underline{r}_1$ and $\underline{r}_2$:

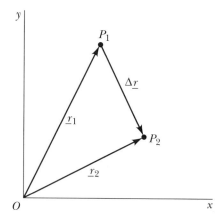

Fig. 2 The displacement vector $\Delta\underline{r}$, representing the separation of two points in space.

$$\Delta \underline{r} = \underline{r}_2 - \underline{r}_1 \tag{3}$$

One way to see that it's appropriate to write $\Delta \underline{r}$ this way is look at the displacement—the trip from P_1 to P_2—in terms of components. Clearly

$$\Delta x = x_2 - x_1 \tag{4}$$

and

$$\Delta y = y_2 - y_1 \tag{5}$$

If P_1 and P_2 are two cities, and we take the x direction as east and the y direction as north, then Δx is the distance you'd have travel eastward (positive in the example shown) and Δy is the distance you'd have to travel northward (negative in the example, so you'd actually be going southward).

Another way of writing the relationship of eq. 3 is to convert the subtraction to a sum in the same way we do with numbers:

$$\underline{r}_2 = \underline{r}_1 + \Delta \underline{r} \tag{6}$$

with the obvious equivalent statements for the components x and y:

$$x_2 = x_1 + \Delta x \tag{7}$$

$$y_2 = y_1 + \Delta y \tag{8}$$

Pictorially, the sum of two vectors is obtained by putting the tail of the second vector ($\Delta \underline{r}$, here) to the head of the first vector ($\underline{r}_1$), and the sum is the vector drawn from the tail of the first vector to the head of the second, as $\underline{r}_2$ is drawn in fig. 2. (You're always allowed to move a vector around, as long as you don't change its length and the direction it points.) Note that the direction of the difference $\Delta \underline{r}$ has nothing to do with the directions of the original vectors $\underline{r}_1$ and $\underline{r}_2$, and it doesn't much matter which of the original vectors is bigger. You'll find more information about vectors of this ordinary kind in Appendix A, §2.

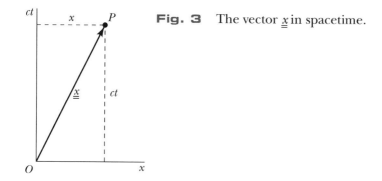

Fig. 3 The vector $\underline{\underline{x}}$ in spacetime.

Vectors in spacetime

I now want to show you how the idea of a vector works in spacetime. Just as in Euclidean space (fig. 1), the location of a spacetime point P relative to the origin O is represented by a vector $\underline{\underline{x}}$, as in fig. 3, with components x and ct. I've introduced a factor c in the time variable so that both components keep the same dimensions (see the discussion and footnote following fig. 2.2.5). The vector $\underline{\underline{x}}$ is commonly referred to as a *four-vector* (4-vector for short) because it's a vector in our four-dimensional spacetime and in general has four components: ct, x, y, and z. I'll use the term 4-vector even when only two dimensions are shown, as in fig. 3, because it's such a standard term. (I'm using a double underline for 4-vectors for the reader's convenience. There's no standard convention.)

Length of a four-vector

What happens to the ideas of magnitude and direction in the case of 4-vectors? The length and angle that you saw in fig. 1 for a Euclidean vector have their natural analogs for 4-vectors, but they look quite strange. I'll show you something of how the magnitude works, because it's useful in what follows and is widely used in physics, but I'll omit the generalized angles because they're a lot more complicated and not especially useful.

It's important that the length, or magnitude, be defined so that it's invariant under Lorentz transformations—that is, so that it has the same value in any reference frame. This means that you can't just use a Euclidean length—what you would measure with a ruler on the spacetime diagram—because that gets distorted when you go to a different reference frame (look at figs. 2.5.2 and 2.5.5). The Euclidean length r is determined by the expression $x^2 + y^2$ (eq. 1), the sum of the squares of the components, a quantity that's invariant under ordinary rotations. The corresponding quantity for the spacetime vector of fig. 3—the quantity that's invariant under Lorentz transformations—turns out to be the *difference* of the squares of the components,

$$\underline{x} \cdot \underline{x} = c^2 t^2 - x^2 \tag{9}$$

I use the symbol $\underline{x} \cdot \underline{x}$ (it's called the scalar product of $\underline{x}$ with itself, actually, but you don't need to know anything about scalar products) for this combination, because I don't have any other handy notation and I need to refer to it from time to time. Eq. 9, then, is the combination that we have to use if we want to define a length for our 4-vector. How do you get a *length* out of an expression like this, which may be either positive or negative, or even zero, in contrast to Euclidean vectors where you find the length by taking the square root (eq. 1)? This lack of a definite sign makes it more complicated to define the length and has a number of interesting consequences. You can see first of all that we have *three* different types of vector, corresponding to these three possibilities: a vector is called *timelike* when $\underline{x} \cdot \underline{x} > 0$, *spacelike* when $\underline{x} \cdot \underline{x} < 0$, and *lightlike* when $\underline{x} \cdot \underline{x} = 0$. (See prob. 1.) For all of these we'll use the square root of $|\underline{x} \cdot \underline{x}|$ to define a length, but the physical interpretation is different in each case.

1.   *Timelike vectors:* For the case $\underline{x} \cdot \underline{x} > 0$ you can see that $|x| < ct$, which means that the points O and P could both lie on the world line of a real particle moving with speed $v = x/t$. The invariant quantity $c^2 t^2 - x^2$ then can be rewritten as $c^2 t^2 - v^2 t^2$, or $c^2 t^2 / \gamma^2$. But t/γ is just the proper time τ for the particle, so we can write this result as

$$c\tau = \sqrt{\underline{x} \cdot \underline{x}} = \sqrt{c^2 t^2 - x^2} \qquad (t > 0) \tag{10}$$

and say that the proper time is the invariant length of the 4-vector $\underline{x}$, apart from the factor c. I've assumed here that the event-point P lies in the future compared to O ($t > 0$); in this case we call the vector $\underline{x}$ *forward timelike*. If the point P lies at an earlier time than O, on the other hand, we call the 4-vector $\underline{x}$ *backward timelike*, and it's appropriate to define the invariant length as the negative square root:

$$c\tau = -\sqrt{\underline{\underline{x}} \cdot \underline{\underline{x}}} \quad (t < 0) \tag{11}$$

2. *Spacelike vectors:* When $\underline{\underline{x}} \cdot \underline{\underline{x}} < 0$ the reasonable definition of an invariant length is

$$s = \sqrt{-\underline{\underline{x}} \cdot \underline{\underline{x}}} = \sqrt{x^2 - c^2 t^2} \tag{12}$$

In this case $|x| > ct$, and the two points O and P could not lie on a particle world line. It *is* possible, though, to choose a reference frame in which the two events O and P are simultaneous (see eq. 2.4.10), and in that frame you have $s = |x|$. That is, the invariant length of a spacelike 4-vector is simply the spatial distance between the two events O and P in that frame in which those events are simultaneous. It's appropriate to call this distance the *proper distance* between the two events O and P.

3. *Lightlike vectors:* A particularly important case is the lightlike vector, for which $c^2 t^2 - x^2$ is equal to zero. In this case we have

$$x = \pm\, ct \tag{13}$$

which is the equation for a light signal traveling in either the $+x$ or the $-x$ direction. More accurately, this means that the point P (see fig. 3) lies on the world line of a light signal sent out from the point O. We have to say that the length of such a vector is zero, even though the vector itself is not zero. For lightlike vectors, just as for timelike vectors, we distinguish between forward and backward directions by whether t is positive or negative.

You should note that the invariance of the quantity $\underline{\underline{x}} \cdot \underline{\underline{x}}$ ensures that the length of a lightlike vector is also zero in any other reference frame, which means that eq. 13 holds also, and hence the velocity of light is invariant, as it's supposed to be. The invariance of $\underline{\underline{x}} \cdot \underline{\underline{x}}$ says it all.

The generalization of the definition of $\underline{\underline{x}} \cdot \underline{\underline{x}}$ (eq. 9) to our full four dimensional spacetime is straightforward (but not very illuminating):

$$\underline{\underline{x}} \cdot \underline{\underline{x}} = c^2 t^2 - x^2 - y^2 - z^2 = c^2 t^2 - r^2 \tag{14}$$

where

$$r = \sqrt{x^2 + y^2 + z^2} \tag{15}$$

the spatial distance between the events O and P. Again, the case where $\underline{\underline{x}} \cdot \underline{\underline{x}} = 0$ corresponds to a point on the world line of a light signal, just as before.

Separation between events in spacetime

I've discussed in some detail the vector $\underline{\underline{x}}$ that represents the location of an event P in spacetime relative to some given origin O. An important generalization of this, which doesn't depend on the choice of origin, is the *separation* between two given events P_1 and P_2, with locations in spacetime represented by vectors $\underline{\underline{x}}_1$ and $\underline{\underline{x}}_2$, as shown in fig. 4, in close analogy to fig. 2, the separation of points in space. We define the separation vector $\Delta\underline{\underline{x}}$ as the vector extending from P_1 to P_2 and write it as a difference:

$$\Delta\underline{\underline{x}} = \underline{\underline{x}}_2 - \underline{\underline{x}}_1 \tag{16}$$

where the four components of $\Delta\underline{\underline{x}}$ are just the differences of the components of $\underline{\underline{x}}_2$ and $\underline{\underline{x}}_1$, namely $ct_2 - ct_1$, $x_2 - x_1$, $y_2 - y_1$, and $z_2 - z_1$. The two vectors $\underline{\underline{x}}_1$ and $\underline{\underline{x}}_2$ depend on the choice of origin, but the difference $\Delta\underline{\underline{x}}$

Fig. 4   Four-vector $\Delta\underline{\underline{x}}$ representing the separation of two points in spacetime.

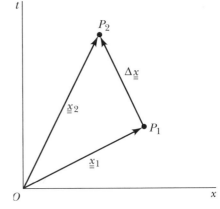

doesn't. As before, the separation $\Delta \underline{x}$ can be timelike, spacelike, or lightlike, and the quantity

$$\Delta \underline{x} \cdot \Delta \underline{x} = c^2 \Delta t^2 - \Delta x^2 - \Delta y^2 - \Delta z^2 \qquad (17)$$

is an invariant measure of how far apart the events are in spacetime. Whether $\Delta \underline{x}$ is spacelike or timelike has almost nothing to do with the character of $\underline{x}_1$ and $\underline{x}_2$ separately.

The displacement 4-vector $\Delta \underline{x}$ tells you all about the possibility of communication between P_1 and P_2. If $\Delta \underline{x}$ is positive timelike, then a traveler at P_1 can herself get to the point P_2, and if $\Delta \underline{x}$ is negative timelike, then P_2 lies in the past relative to P_1 and a traveler at P_2 can get to P_1. If $\Delta \underline{x}$ is positive lightlike, then the person at P_1 can send a message by light signal or radio that will reach a receiver at P_2, while if $\Delta \underline{x}$ is spacelike, there's no possibility at all of communication between the two points, and they'll appear simultaneous to some observers.

PROBLEMS

1. (a) Take $\underline{x}$ as a point on the world line of S', as in fig. 2.4.4, so that the components x and ct satisfy $x = vt$. Show that the 4-vector $\underline{x}$ is timelike and that its invariant length is equal to ct', where t' is also the proper time for observer S'.

   (b) Take $\underline{x}$ as a point on the x' axis in fig. 2.4.4, so that the components satisfy $t = vx/c^2$. Show that $\underline{x}$ is spacelike and that its invariant length is equal to x'.

2. Construct two timelike vectors $\underline{x}_1$ and $\underline{x}_2$ such that the separation $\Delta \underline{x}$ is (a) spacelike, (b) forward timelike, (c) backward lightlike.

3. Find two lightlike vectors whose separation $\Delta \underline{x}$ is forward lightlike.

CHAPTER 4

Dynamics: Momentum and Force

4.1　DYNAMICS

So far we've been talking just about the structure of space and time, and the way in which we describe events and motions, without regard to their *causes* —without any concern, that is, with forces and the way objects respond to forces. The description of motion (Chapters 1–3) goes by the name of *kinematics*, and the subject I want to deal with now, the study of what actually happens—of forces and response to forces—is called *dynamics*. This is where you'll encounter *laws of motion*, the laws that predict future behavior, that tell what will happen next, given the way things are at any given time. Newton's laws, describing particle motion, and Maxwell's equations, describing the EM field, were the classical examples of laws of motion. Although Maxwell's equations survived the relativity revolution completely unchanged, Newton's laws had to be replaced by Einsteinian counterparts; this is because Maxwell's equations are Lorentz invariant and Newton's laws are not.

I believe that the best focus for a discussion of dynamics is the pair of physical quantities *momentum* and *energy*. These are the basic conserved quantities of nature whose changes and interchanges represent the essence of physics and motivate the definitions of the other quantities, like force and work, that play key roles in the subject of dynamics.

You may remember that in the Introduction I advertised a very important relationship between symmetries and conservation laws that runs through all of nature. This was Noether's theorem, which I suggested promoting to Noether's principle because of its apparently very fundamental role in governing the structure of physics. Recall that we're using the word *symmetry* here to refer to the invariance of the laws of physics themselves under some operation or transformation. Every symmetry is associated with a conservation law: the constancy in time of the total amount of some physical quantity.

Momentum and energy are the conserved quantities related by Noether's principle to the most basic symmetries of space and time: invariance under space and time displacements. *Time displacement symmetry* refers to the invariance of physical laws from one time to another. The physics experiment you perform today can be performed again tomorrow with the same results. The athletic skills you learn this year will still be effective next year, because your body and the sports equipment will still operate by the same rules they obey now. Likewise, *space displacement symmetry* refers to the invariance of physical laws from one spatial location to another: The laws of physics work the same way in Tashkent and on Mars as they do in New York and Columbus.

I want to start with the idea of momentum. Although it's a less familiar concept than energy, momentum plays a more immediate role in determining the relationship of force to motion, and thereby in helping us understand the concept of force itself.

It's the invariance under spatial displacements that leads to the conservation of what we call momentum. There are three spatial dimensions, corresponding to our x, y, and z coordinates, so momentum must have three components also, which makes it a *vector,* like the position vector $\underline{r}$ introduced in §3.4. I'll use the symbol $\underline{p}$ to represent the momentum vector.

Let's start by looking at momentum from a classical viewpoint. It was identified by Newton, back in the seventeenth century, as the "quantity of motion" carried by a moving particle. It was therefore reasonable to take momentum as proportional to the velocity of the particle and as proportional also to the mass, because the more stuff there is moving, the more motion there is. In symbols,

$$\underline{p} = m\underline{v} \qquad (1)$$

We've got the desired factors m and v, and we've got the direction of the vector $\underline{p}$ given by the direction of the velocity $\underline{v}$. I haven't tried to define *mass*—it has to be some measure of how much matter there is in the object. In the classical view, mass is an intrinsic property of an object that determines its weight—the force due to gravity—and describes its inertia, which is its resistance to changes in its motion. Mass is measured in kilograms; in the earth's gravitational field 1 kilogram weighs about 2.2 pounds, but you should remember that an object's mass doesn't change when there's no gravity—only its weight changes. Bear in mind also that this Newtonian understanding of mass gets considerably modified in the relativistic picture. The unit for momentum is just kg · m/s; curiously, no special unit was ever introduced. (How about a "mom": 1 M = 1 kg · m/s?)

Newton understood that the momentum of an object must be constant—the "quantity of motion" must stay the same—unless the object is acted on by some external influence. Any change in $\underline{p}$, therefore, must be a measure of the strength of that influence, which we call a *force*. *The force on an object is defined to be just equal to the rate of change of momentum.*

Before we can study this idea of force in more depth, you need to know about the mathematical way of describing rates of change. This is done by means of derivatives, which I'll discuss in §4.2.

PROBLEM

1. Find the momentum of each of the following, using reasonable values for mass and velocity where needed.
   (a) An electron in a hydrogen atom. The speed is around $0.01\,c$ and the mass can be found in Appendix C, §4.
   (b) A thrown baseball
   (c) A moving automobile
   (d) The moon in its orbital motion around the earth (for data, see Appendix C, §5)

 DERIVATIVES

It's time to take a math break and learn what derivatives are and how we use them to describe rates of change. The simplest example is a one-dimensional velocity, simply a rate of travel, which can also be thought of as a *rate of change of position*. If you're traveling at a constant rate, then the velocity is the distance traveled, Δx, divided by the time it takes to travel that distance, Δt:

$$v = \frac{\Delta x}{\Delta t} \qquad (1)$$

Clearly, if you travel for one tenth as long a time, you'll travel one tenth as far, and the quotient will be the same. If you're not traveling at constant speed, then the quotient of eq. 1 gives the *average* speed. If the speed doesn't deviate much, then the average speed is a good measure for the whole interval Δt. Now, how do you describe the *instantaneous speed*? How do you get away from the fact that time intervals like Δt are finite and cover a whole range of different instants? Well, velocities vary smoothly, and it's only necessary to pick a time Δt small enough that your velocity doesn't change to speak of during that time. Such a very small time I'll call dt instead of Δt, and the distance traveled in that time I'll call dx. The distance dx will then be comparably small, and the quotient will give your approximate velocity at that moment. We write

$$v = \frac{dx}{dt} \qquad (2)$$

and treat this as an *exact* expression for your instantaneous velocity, with the understanding that dt is taken sufficiently small to *make* it accurate. (There's a preliminary discussion of infinitesimals like dx and dt in §3.3.) Expression 2 is in fact a *derivative*, and I suggest that you think of it just as a simple quotient of two very small quantities or, more physically, as the rate of change of the quantity x at some chosen moment of time. (See Appendix A, §4, for a more careful introduction to derivatives.) Physicists use a standard shorthand notation, an overhead dot, to indicate a time rate of change—that is, a time derivative—and I'll use it regularly. Thus we'd write

$$v = \dot{x} \qquad (3)$$

and mean exactly the same thing as eq. 2. When you see the dot, think of a rate of change.*

It is often very useful to be able to multiply both sides of eq. 2 by *dt:*

$$dx = v \, dt \qquad (4)$$

where *dt* now represents some specific small time interval and *dx* the distance moved—that is, the change in the value of *x* in that time—this is simply a case of distance = rate × time.

The idea of the velocity as a derivative was hinted at in Chapter 2, in our discussion of spacetime diagrams and world lines. A particle with constant velocity has a world line that is a straight line, and the inclination (inverse slope) of the line, $\Delta x/\Delta t$, is just the velocity. You can see that where the world line is curved, the inclination keeps changing, but that at any given point on the world line, the inclination has a definite value: A tiny segment of the world line is nearly straight and has a well-defined inclination, which is just the derivative dx/dt. This is shown in fig. 1, with the world line of the space traveler from fig. 3.3.1.

We also have to know how to play the derivative game with vectors. In our three-dimensional world we have position vectors, velocity vectors, and momentum vectors, and physics is about the way they change with time. The position of a particle at time *t* can be described by the vector $\underline{r}(t)$, which takes different values as *t* varies. In the small time *dt*, its displacement will be described by the small vector $d\underline{r}$, illustrated in fig. 2. It's the difference between the two vectors $\underline{r}(t+dt)$ and $\underline{r}(t)$, defined (just as in §3.4) as a vector running from the point $\underline{r}(t)$ to the point $\underline{r}(t+dt)$, as in the figure. The vector $d\underline{r}$ represents the actual path that the particle is following during this small time *dt*. Note that $d\underline{r}$ can point in a different direction from $\underline{r}(t)$ and that $\underline{r}(t+dt)$ can be longer or shorter than $\underline{r}(t)$ or even exactly the same length. The velocity vector $\underline{v}$ is the displacement $d\underline{r}$ divided by the time *dt* and can be thought of as the rate of change of the vector $\underline{r}$:

$$\underline{v} = \frac{d\underline{r}}{dt} \qquad (5)$$

$$= \dot{\underline{r}} \qquad (6)$$

---

*You say "*x*-dot" for the right side of eq. 3.

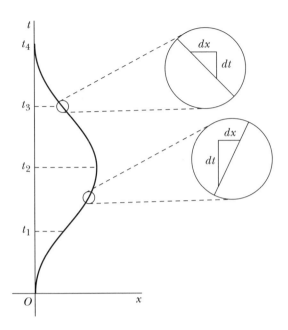

Fig. 1   The inclination at different points of a world line.

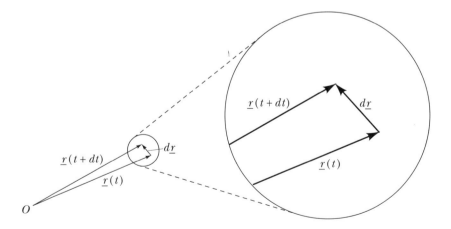

Fig. 2   Infinitesimal change in the vector $\underline{r}$.

The velocity vector $\underline{v}$ points in the same direction as the displacement vector $d\underline{r}$, and its length, rescaled by the factor $1/dt$, is finite—that is, not infinitesimal. As in one dimension, the value of $\underline{v}$ is independent of the value of dt provided that dt is taken to be small enough.

This discussion can be restated in terms of components, and each component of $\underline{v}$ is the derivative of the corresponding component of $\underline{r}$:

$$v_x = \dot{x} = \frac{dx}{dt} \tag{7}$$

$$v_y = \dot{y} = \frac{dy}{dt} \tag{8}$$

$$v_z = \dot{z} = \frac{dz}{dt} \tag{9}$$

You can also differentiate all sorts of other physical quantities; the rate of change of any quantity is indicated by the overhead dot, or by the use of a quotient, as in the above examples. An important case is *acceleration*, which is the rate of change of velocity (review the description in §3.3, in connection with formula 3.3.8, of the acceleration of gravity g). Letting a represent acceleration in general, you write

$$a = \dot{v} \tag{10}$$

$$= \frac{dv}{dt} \tag{11}$$

where eq. 11 tells you that you can calculate a by finding how much v changes (dv) in the small time dt. The two characteristics of a derivative that you'll need to remember in what follows are (1) that it's used to describe the time rate of change of some quantity and (2) that it's all right for our purposes to think of it as a simple quotient of two small quantities: the change in the quantity divided by the time in which that change takes place.

A function $x(t)$ doesn't have to have a simple form for its derivative to be meaningful, but simple functions do usually have simple derivatives also. Let's work out a couple of simple examples here (see also Appendix A, §4).

Example (i)

$$x(t) = At + B \tag{12}$$

where A and B are constants. For a given dt,

$$dx = x(t + dt) - x(t) \qquad (13)$$

$$= [A \cdot (t + dt) + B] - [At + B] \qquad (14)$$

$$= A \, dt \qquad (15)$$

and the derivative is the quotient:

$$\frac{dx}{dt} = \frac{A \, dt}{dt} = A \qquad (16)$$

Example (ii)

$$x(t) = t^n \qquad (17)$$

$$dx = x(t + dt) - x(t) \qquad (18)$$

$$= (t + dt)^n - t^n \qquad (19)$$

$$= t^n \left[\left(1 + \frac{dt}{t} \right)^n - 1 \right] \quad \text{(see prob. 3.2.6)} \qquad (20)$$

$$\approx t^n \left[\left(1 + \frac{n \, dt}{t} \right) - 1 \right] \quad \text{(eq. 1.4.12)} \qquad (21)$$

$$= n \, t^{n-1} \, dt \qquad (22)$$

so the derivative is

$$\frac{dx}{dt} = \frac{d}{dt} t^n = n t^{n-1} \qquad (23)$$

Note that in each case you work out $x(t)$ for both t and $t + dt$ and calculate the difference, the change in x. In the second example, at eq. 21 I used the approximation given by eq. 1.4.12, and we're allowed to take dt sufficiently small that the approximation becomes exact. Note too that in each case you end up by dividing by dt and that the final value of the derivative doesn't depend on the value of dt, except that it has to be very small.

PROBLEMS

1. Write the three-dimensional vector analog of eq. 4, $dx = v\,dt$, and then rewrite this as three separate equations for the components of $d\underline{r}$.

2. Work out the derivative of each of the following functions:
   (a) $x(t) = 4t^2 + 3t + 7$
   (b) $x(t) = 1/t$
   (c) $x(t) = 3t^3 - 2/t^3$
   You may use the result in example (ii) in the text, in the form of either eq. 22 or eq. 23.

3. A boat sails from a point 20 mi east and 5 mi south of Boston to a point 25 mi east and 15 mi north of Boston, in a time of 4.5 h, sailing on a steady course at constant speed.
   (a) Find the initial and final position vectors $\underline{r}_1$ and $\underline{r}_2$ relative to Boston as the origin.
   (b) Find the net displacement vector $\Delta\underline{r}$.
   (c) Find the velocity vector $\underline{v}$. Each vector can be specified in terms of its components, with the x direction due east and the y direction due north.

4. A car accelerates from rest to a speed of 5 m/s in a time of 8 s.
   (a) At the end of the 8 s, how far does the car travel in 0.01 s?
   (b) Find the average acceleration in m/s² and in km/h · s [≡ km/(h · s)] over the 8-s interval. Explain the unit km/h · s.
   (c) If the car continues to accelerate at the same rate, what will its speed be, in km/h, after another 10 s?
   (d) At this same rate of acceleration, what is the change in the car's speed, in m/s, in 0.01 s?

5. A boy is shelling peanuts. Initially he shells them at a rate $u = 10$ pn/min (10 peanuts per minute), but he gradually slacks off, and after 5 h his rate is 6 pn/min.
   (a) Initially, how many peanuts does the boy shell on the average in 0.1 s?
   (b) Define an "acceleration" α (Greek lower-case alpha) that describes the rate of change of u, and give its units.
   (c) Calculate the average value of α over the 5-h period.
   (d) If α is constant, will the boy's shelling rate ever reach zero and, if so, when?

4.3   FORCE AND THE LAW OF MOTION

The law of motion

Let's return now to the idea of *force* as the external influence that causes a change in the momentum of a particle. The rate of change of the momentum p is its time derivative:

$$\dot{p} = \frac{dp}{dt} \tag{1}$$

where dp is the change in p in the small time dt. To be consistent with the way I've introduced the idea of force, then, you have to define the force F so that

$$\dot{p} = F \tag{2}$$

This is a very famous statement, known as Newton's second law of motion, or simply Newton's second law. It's famous because it answers the most fundamental question of dynamics: How does an object move when you apply a force?

In America the most familiar unit of force is the *pound*. Physicists almost invariably use metric units, however, and the standard metric unit of force is the *newton* —named, not unreasonably, after Newton:

$$1 \text{ N} = 0.22481 \text{ lb} \tag{3}$$

Thus, a newton is a little less than a quarter pound—about 3.6 oz. How did they come up with this? It can be related to more standard units by means of eqs. 2 and 4.1.1. Momentum, by eq. 4.1.1, has the unit kg · m/s, and force, which is equal to rate of change of momentum (eq. 2), has therefore the dimension of momentum/time. The appropriate unit of force is (kg · m/s)/s, or kg · m/s²; this combination is what was christened the newton:

$$1 \text{ N} = 1 \text{ kg} \cdot \text{m/s}^2 \tag{4}$$

This is just the force needed to give a 1 kg mass an acceleration of 1 m/s²
(as you'll see when you get to eq. 9).

There's a subtlety in eq. 2 that I want you to catch. I've said that force
is *equal* to the rate of change of momentum, but I've carefully avoided
defining the force to *be* the rate of change of momentum, because that's
not what it is. Force is the external influence that *causes* momentum to
change. This definition makes sense only because the momentum remains constant in the absence of external influences—that is, because
it's conserved. The law and the definition go hand in hand. The force
can be gravitational, electromagnetic, or something else, but there has to
be *something* producing the force and a rule to tell you what the force is
in the circumstances. Then, when you know what the force is, eq. 2 tells
you what effect it produces.

In a time *dt* the force $\underline{F}$ produces a change *dp* in the momentum
given by

$$dp = \underline{F}\,dt \tag{5}$$

(compare eq. 4.2.4). If the force is constant, then the time doesn't have
to be small, and you can write

$$\Delta p = \underline{F}\,\Delta t \tag{6}$$

In words: The change in momentum is equal to the strength of the force
multiplied by the time it acts. The direction of the *change* in p is the same
as the direction of the force and may be different from the direction of p
itself (as with r and dr in fig. 4.2.2).

How is this discussion related to the intuitive picture of force as
pushing or pulling on some object? For example, gravity pulls on me
with a force of around 195 pounds, so if I step off a chair gravity immediately pulls me toward the floor. My momentum increases until I hit
the floor, and then it decreases again because of the upward force of the
floor. If I remain standing on the chair, the chair exerts a force on
me—also 195 pounds—that counters the force of gravity and keeps me
from falling: The net force on me is zero, so my momentum, which is
zero, doesn't change. If my car stalls on the highway and I get out to
push, I have to exert a force on it in order to make it roll to the side of
the road. It's hardest getting it started (because I'm causing the momentum to increase), but once it's rolling it's easy to keep it going if the road

is level (because the momentum isn't changing). In fact, I may have trouble getting it to stop (again a change in momentum) unless there's someone in the car to step on the brakes. If there's no frictional force to reduce its momentum to zero again, a tree may do it for me instead.

When you weigh yourself, you're determining the force that gravity exerts on you. What you're really interested in is probably your *mass* (how much matter there is in your body), but because the pull of gravity is proportional to your mass, your weight is a good measure of your mass. If you go to the moon, though, your weight becomes much less while your mass doesn't change at all. You have to multiply your weight by 6.03 to find out what you'd weigh back on earth, because that's how much stronger the gravitational field is on the earth than on the moon.

Let's look more closely at dp, the change in momentum, and keep to one-dimensional motion for now. Recall that momentum is mv, mass times velocity. In the familiar nonrelativistic world, the mass of a particle is constant, so the change in the momentum depends simply on the change in the velocity:

$$dp = mdv \tag{7}$$

or, dividing by dt to get rates of change,

$$\dot{p} = m\dot{v} = ma \tag{8}$$

by eqs. 4.2.10 and 4.2.11. That is, the rate of change of momentum depends simply on the acceleration. This gives rise, by eq. 2, to a very familiar form of Newton's second law:

$$F = ma \tag{9}$$

In words: The force needed to give an object a certain acceleration is equal to the mass of the object times that acceleration. In three dimensions Newton's second law takes a vector form:

$$\underline{F} = m\underline{a} \tag{10}$$

which is often most usefully given in terms of components:

$$F_x = ma_x \qquad (11)$$

$$F_y = ma_y \qquad (12)$$

$$F_z = ma_z \qquad (13)$$

The advantage of the forms of Newton's law in eqs. 9–13 is that they bypass the less familiar concept of momentum and express directly the relationship between force and motion.

I have a preference myself for turning these equations around so that they tell you directly what the acceleration is for a given force. For the one-dimensional case, I rewrite eq. 9 as

$$a = \frac{F}{m} \qquad (14)$$

which has the virtue of expressing the *effect*—the acceleration—in terms of the *cause*—the force. Notice, too, that for a given force the acceleration varies inversely with the mass: The greater the mass, the less the acceleration. This shows why mass is often identified as a measure of *inertia*—that is to say, resistance to acceleration.

The advantages of the original form of Newton's second law involving momentum (eq. 2) are its more fundamental character (it's directly related to the conservation of momentum) and its more general validity (it doesn't depend on the nonrelativistic assumption of the constancy of mass).*

The forces of nature

I've mentioned that the forces that produce the acceleration of material objects are real forces described by the fundamental force laws of physics. I want to show you now some of the standard examples of force, which we run into most commonly in physics courses and in nature.

---

* *Neither* eq. 2 nor eq. 14 can be used for a system whose mass changes because of emission or accretion of actual matter, such as a rocket that accelerates by emitting a propellant or a snowball rolling down a snowy slope and getting bigger as it goes.

Uniform gravitational field near earth's surface

$$F = -mg \qquad (15)$$

where the minus sign is supposed to indicate, given the convention that "up" is positive, that the force is vertically downward. The force is proportional to the mass m of the object and to the gravitational field g, which is also referred to as the acceleration of gravity. This case is discussed in detail in 4.4. (The field is only approximately uniform because the earth isn't flat, but this is a good approximation over small distances like the size of a baseball field.)

General gravitational field $\underline{g}$

$$\underline{F} = m\underline{g}(\underline{r}) \qquad (16)$$

Here the force is again proportional to the mass of the object, but the strength and direction of the force now depend on the gravitational field at that point.

Gravitational force between two masses (see §6.2) at $\underline{r}_1$ and $\underline{r}_2$

$$F = -\frac{Gm_1 m_2}{r^2} \qquad (17)$$

where

$$r = |\underline{r}_1 - \underline{r}_2| \qquad (18)$$

is the distance between the two masses, and the minus sign indicates that the force is always attractive. This is Newton's law of universal gravitation, which applies with remarkable accuracy to objects on earth and to the sun, planets, and moons within the solar system. It represents an approximation to Einstein's theory of gravity, which is known as general relativity. The constant G, called the gravitational constant, is given in Appendix C, §4. The force on each mass is caused by the gravitational field $\underline{g}$ that is due to the other mass (eq. 16).

Uniform electric field $\underline{E}$

$$\underline{F} = Q\,\underline{E} \tag{19}$$

The force acts on electric charge Q just as the gravitational field acts on mass, and again the strength and direction of the force depend on the electric field vector-$\underline{E}$. Electric charges and electric fields are explained in more detail in §11.5.

Electrical force between two charged particles at $\underline{r}_1$ and $\underline{r}_2$ (see eq. 18)

$$F = \frac{kQ_1 Q_2}{r^2} \tag{20}$$

where Q_1 and Q_2 are the strengths of the electric charges on the two particles. Equation 20 is known as Coulomb's law; notice how similar it is to the gravitational force law, eq. 17. The force is repulsive if the charges have the same sign (like two protons) and attractive if the charges are opposite in sign, like the proton and electron that make up a hydrogen atom. The constant k represents the strength of the electrical force, and its value depends on the units used (see Appendix C, §4). As with the gravitational force (eq. 17), the force on each particle is caused by the electric field $\underline{E}$ that is due to the other particle (eq. 19).

PROBLEMS

1. Use eq. 6, together with the definition for the vector momentum (eq. 4.1.1), to find the final velocity, including direction, for each of the following examples.
   (a) A 0.5-kg ball is moving due east at 20 m/s; a force of 5 N, directed due north, acts on it for 2 s.
   (b) Substitute for the ball in part (a) a pebble of mass 5 g (1 g = 1 gram = 0.001 kg).
   (c) Same as part (a), but with initial velocity due north.
   (d) Same as part (a), but with initial velocity due south. Draw a careful vector diagram of the initial and final momenta for each case, using graph paper and axes marked to show momentum units. Neglect gravity.

2. Experiment with eq. 9 to get a feeling for what it says.
    (a) First check the consistency of eq. 4, the definition of the unit
         of force, the newton.
    (b) What force does it take to accelerate a 1000-kg car to 60 mi/h
         in 60 s? (Use 45 mi/h = 20 m/s.) Get the answer in newtons
         and then convert to pounds. How does this force compare
         with the weight of the car? (1 kg weighs about 9.8 N, or 2.2 lb.)
    (c) What force does it take to give a 1-kg mass the acceleration of
         free fall, 9.8 m/s^2? This is exactly the force of gravity on it—
         that is, its weight.
    (d) What force does it take to give a 5-g ping-pong ball a speed of
         20 m/s in 1/50 of a second? (These are my crude estimates of
         the mass of the ball and of how long the paddle is in contact
         with the ball.)

 **4.4**   THE FORCE OF GRAVITY

Now I want to introduce you briefly to the gravitational force as we
experience it at the earth's surface. The subject is treated in much more
depth in Chapter 6, where I'll describe the transition from Newton's law
of gravity to Einstein's complete theory of gravity, which goes by the
name of general relativity. Locally, the gravitational force is experienced
as weight, the downward pull experienced by every material object. The
simple—and very significant—fact is that this gravitational pull is exactly
proportional to the mass of the object:

$$W = mg \qquad (1)$$

where W is the weight, viewed as a force acting on the object, and g is a
constant of proportionality called the *gravitational field*. There's no obvious reason why the force of gravity should be exactly proportional to the
mass m; mass after all is identified with inertia—resistance to acceleration (see comments on eq. 4.3.14)—a concept with no evident relationship to gravity.*

---

*As you'll see in Chapter 6, this equivalence of inertial mass and gravitational
mass lies at the heart of the general theory of relativity. It was in fact the starting
point of the logic that Einstein followed in developing that theory.

Now, how does an object *move* under the force of gravity (eq. 1)? If you let go of the object it will fall, and its acceleration is given by Newton's second law, eq. 4.3.14:

$$a = \frac{F}{m} \tag{2}$$

(I'm treating the problem as one-dimensional.) If you plug in for F the expression for the gravitational force (eq. 1), you see that

$$a = \frac{mg}{m} = g \tag{3}$$

That is, the acceleration a is exactly equal to the gravitational field g. This looks trivial but is in fact very significant. Make sure you understand the logical steps that lead from eq. 1 to eq. 3. In particular, note that eq. 1 is *not* an example of Newton's second law. It tells you only what the force is, and then eq. 2, which *is* Newton's second law, tells you what the acceleration is. The field g can be a positive or a negative quantity here, depending on the direction of the gravitational force relative to the direction chosen as positive.

From cq. 1 we see that g is properly given in N/kg (because the field is the force per unit mass, and the newton is the unit of force), so you may ask whether the units of a and g agree in eq. 3. In fact, the relation (eq. 4.3.4), 1 N = 1 kg · m/s² (implied in Newton's second law), shows that g can be reexpressed in m/s², which is the right unit for an acceleration. As a consequence, g is commonly referred to as the *acceleration of gravity*, as discussed in §3.3 and displayed in the logic of eq. 3. However, I much prefer to call g the gravitational field, because that term reflects more accurately its basic meaning at the level of our present discussion (pre-Einstein).

The gravitational field at the earth's surface is approximately constant and equal to $-g_0$ (if *up* is taken as positive), where $g_0 = 9.8$ N/kg $\equiv$ 9.8 m/s². Thus motion near the earth's surfacc is always at constant acceleration to a good approximation. For one-dimensional motion in the vertical direction, with y the vertical coordinate, you can figure out the details of the motion, as I'll now show you.

The logic of cause and effect works backwards, you might say: The force determines the acceleration, the acceleration determines how the velocity behaves, and finally the velocity determines the location at any given time. At each step you're given the *derivative* of the quantity you want, and you have to work out the quantity itself. In the case of constant gravitational force, we know that the acceleration is constant, $-g_0$, and this means that the velocity v *changes* at a constant rate:

$$\frac{dv}{dt} = -g_0 = \text{constant} \tag{4}$$

In this case the net change in v over a time t is simply rate × time:

$$\Delta v \equiv v(t) - v_0 = -g_0 t \tag{5}$$

or

$$v(t) = v_0 - g_0 t \tag{6}$$

where v_0 is the initial velocity, at $t = 0$. Thus in a uniform gravitational field, the velocity changes in a simple way with time. If you throw a ball up with initial velocity $v_0 = +30\text{m/s}$, for example, then because of the constant force of gravity, the velocity keeps decreasing at a steady rate, reaches zero when the ball attains its greatest height, and goes negative as the ball descends. You can use eq. 6 to figure out when the ball reaches its greatest height by requiring that $v(t) = 0$. Thus $v_0 - g_0 t = 0$, or

$$t = \frac{v_0}{g_0} = \frac{30 \text{ m/s}}{9.8 \text{ m/s}^2} = 3.1 \text{ s} \tag{7}$$

(Note carefully how the units in the answer combine by the ordinary rules of arithmetic to give the correct result.)

Next you can use the fact that $v(t)$ is itself the derivative of the height y—that is, it's the rate of change of y—to figure out the position of the ball at an arbitrary time t. You have

$$\frac{dy}{dt} = v(t) = v_0 - g_0 t \tag{8}$$

and you need to find a function $y(t)$ that has this derivative. Because the derivative of the function t is 1 and the derivative of the function t^2 is $2t$ (eq. 4.2.22), you can see that the function $v_0 t - \frac{1}{2} g_0 t^2$ has the right derivative—that is, the right velocity at any time t. You can add an arbitrary constant to this, because the derivative of a constant is zero, and get the general result

$$y(t) = y_0 + v_0 t - \frac{1}{2} g_0 t^2 \qquad (9)$$

The arbitrary constant y that we added turns out to be the initial position, because if you look at the case $t = 0$, two of the terms vanish and you see that $y(0) = y_0$. Notice how the formula reflects your intuitive knowledge: The height from which you throw the ball and the initial velocity you give it are yours to choose—they're arbitrary—but the subsequent motion is entirely determined by the gravitational force acting on the ball after you let go of it.

The graph of $y(t)$ from eq. 9, for some arbitrary values of y_0 and v_0, might look like fig. 1. This graph shows the object rising to a maximum height (where $v = 0$) and then falling with ever-increasing speed (v is the slope at any given point) until it hits the ground at $y = 0$. Note that this is not a picture of the *path* of the ball, but a graph. The ball is moving straight up and then down; the horizontal axis of the graph represents the time, not a spatial displacement.

Eqs. 6 and 9 enable you to solve a lot of simple problems involving objects that are falling freely or otherwise moving under the action of gravity alone. Even objects moving horizontally as well as vertically can be dealt with easily, because the horizontal component of velocity is simply constant and the vertical motion is described by these equations regardless of what the horizontal velocity may be. Eq. 6 says that the vertical velocity is constantly changing; the initial velocity v_0 may be positive or negative (corresponding to throwing the object upward or downward), but eventually, as t gets bigger, the second term assures that v will become negative: What goes up comes down. If g_0 were zero—if

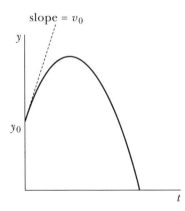

Fig. 1 Graph of $y(t)$, the height of a ball thrown straight up with initial height y_0 and initial velocity v_0.

there were no gravitational force—the ball would keep going up forever at the same constant velocity that you gave it initially.

Eq. 9 describes the position of the object at time t. If the initial velocity v_0 is positive, corresponding to throwing the object up, then at first the quadratic term $- \frac{1}{2} g_0 t^2$ is much smaller than the linear term $v_0 t$ and the value of y increases. Eventually the quadratic term dominates, and y starts decreasing and finally becomes negative, regardless of the initial height y_0 and the initial velocity v_0. If the object hits the ground, then the problem terminates at $y = 0$ (because additional forces enter the picture), but if there is a deep hole in the ground, you can follow the solution down to negative values of y until the object reaches the bottom of the hole.

PROBLEMS

Neglect air resistance in all of these problems.
1. (a) A ball bounces vertically on the ground. Its initial speed as it leaves the ground is $v_0 = 7$ m/s. Find the height of the bounce and the time the ball takes to reach the ground again.
    (b) Show carefully, using the equations for uniformly accelerated motion, that the speed with which the ball hits the ground is also equal to v_0.
    (c) Each time the ball bounces, it leaves the ground with a speed just half the speed with which it hit. Find the time of the nth bounce *after* the initial bounce described in part (a) and the height the ball reaches after that bounce. You'll need the mathematical fact that

$$1 + \frac{1}{2} + \frac{1}{4} + \cdots + \frac{1}{2^{n-1}} = 2 - \frac{1}{2^{n-1}}$$

    (d) Does the ball go on bouncing forever (assuming perfect conditions)? If not, when and why does it stop?
2. (a) A ball is thrown at a vertical wall a distance $\ell = 15$ m away. The initial speed of the ball is $v_0 = 21$ m/s and the angle of elevation is $\theta = 45°$. Obtain formulas for each of the following in terms of ℓ, v_0, and θ, and then evaluate for the given data.
        (i)   The x component (horizontal) of the initial velocity, v_{0x}
        (ii)   The time it takes for the ball to reach the wall
        (iii) The height of the ball when it hits the wall, relative to its initial height

(b) *(For the ambitious)* Find the value of θ, in terms of ℓ and v_0, that maximizes the height at which the ball hits the wall, and obtain a formula for that maximum height. Evaluate θ and h_{max} for the given values of ℓ and v_0.

3. An elevator, starting from rest at a height h_0, accelerates with acceleration a_0, so that its height above the ground after time t is given by

$$y_{elev}(t) = h_0 + \frac{1}{2} a_0 t^2$$

The acceleration is upward if a_0 is positive and downward if a_0 is negative. A passenger on the elevator throws a ball in the air. If the actual height of the ball above the ground at a given moment is $y(t)$, then its height above the floor of the elevator, $y'(t)$, is given by

$$y'(t) = y(t) - y_{elev}(t)$$

(a) Show that the laws of behavior as seen by the passenger are just the same as Newton's laws, but with a different apparent value for g. Derive a formula for g_{app}, the apparent value of g.

(b) What must the elevator's acceleration be to give an apparent g equal to one tenth the normal value? equal to twice the normal value? Be clear about how you relate the sign of a_0 to the direction of acceleration.

(c) What value of a_0, if any, would be needed to reverse the direction of g_{app} but leave its magnitude equal to g_0?

4. This is the three-dimensional version of prob. 3.

(a) A vehicle has constant acceleration $\underline{a}_0$, and the actual gravitational field is $\underline{g}$. By following the same kind of logic as in part (a) of prob. 3, show that the apparent gravitational field for a passenger in the vehicle is $\underline{g}_{app} = \underline{g} - \underline{a}_0$.

(b) Find the apparent field for the case of a racing car accelerating eastward on horizontal ground at 4.0 m/s². Give either the magnitude and direction of $\underline{g}_{app}$ or its components.

5. *(3 significant figures)* **(a)** A ball is thrown straight up with an initial speed of 20 m/s. Find its height, relative to its initial height, at times 1 s, 2 s, 5 s, and 15 min. You are standing on the edge of an abyss, so y can become negative. (Neglect air resistance and changes of g with altitude.)

(b) At what time is the velocity zero, and what is the height at that moment?

   (c)   At what time does the velocity equal –20 m/s, and what is the
            height then?
   (d)   At what time does the height reach –1 km?
6.   A typical air molecule at room temperature has a speed of around
       700 m/s.
   (a)   If its initial motion is horizontal, how far does it drop in cross-
            ing a room 10 m wide?
   (b)   How high does it go if its initial motion is vertically upward?

4.5   RELATIVISTIC MOMENTUM

We now need to discover what the momentum of a relativistic particle is,
bearing in mind that momentum expresses somehow the *quantity of motion* that an object has. You should expect it to agree with the Newtonian
momentum $m\underline{v}$ when the speed is much less than c and to become
infinite as the speed approaches c consistent with the impossibility of
reaching that speed, because any finite change in momentum can be
produced by applying a finite force for a certain finite length of time.

Finding a formula for momentum

Look again at eq. 4.1.1, which expresses the momentum as the product
$m\underline{v}$. The only direction the momentum can point is in the same direction
as $\underline{v}$, so we keep the vector factor $\underline{v}$ and see what we can do with the scale
factor m. Assume, then, that for a given particle with velocity $\underline{v}$ the momentum can be written as

$$\underline{p} = M(v)\,\underline{v} \qquad\qquad (1)$$

The masslike factor $M(v)$ may depend on the magnitude of $\underline{v}$, but not on
its direction, because of the assumed rotational symmetry of the laws of
physics: For a given speed, the magnitude of a particle's momentum
mustn't depend on the direction of its motion. Could this scale constant
simply be a constant? That wouldn't do, because the momentum would
still be finite at speed c, which would make it too easy to reach that speed.
The natural thing to do, as I've said, is to fix it so that p becomes infinitely large as $c \to \infty$, so that when you apply a steady force and p keeps
getting bigger and bigger, the object still never exceeds the speed limit.

Thus we want $M(c) = \infty$. Clearly, too, at nonrelativistic speeds we want $M(v)$ to be m, the nonrelativistic mass:

$$M(0) = m \tag{2}$$

recognizing that, since nonrelativistic speeds are extremely tiny compared to c, the function $M(v)$ won't deviate noticeably from its value at $v = 0$.

Suppose we take out a factor m equal to $M(0)$, the value of the mass function when $v = 0$, and look at the ratio $M(v)/m$, a dimensionless function of v that is unity for $v = 0$ and infinite for $v = c$. These are just the properties of our friend γ, the relativistic factor that governs time dilation and the Lorentz contraction, and it turns out that that's exactly the factor needed. The correct relativistic formula for the momentum of a particle is therefore

$$\underline{p} = m\gamma\underline{v} \tag{3}$$

To derive this result, you have to show that it's required by Lorentz invariance—specifically, the requirement that if the energy and momentum as measured by one observer are conserved, then the values measured by another observer must also be conserved. I'll omit the argument, which is a bit complicated, but I'll show that this definition does satisfy Lorentz invariance. In particular, I'll show in §5.7 that the momentum defined in this way and an appropriately defined energy together constitute a 4-vector, transforming in exactly the same way (eqs. 3.1.1–3.1.4) as the spacetime 4-vector $\underline{x}$. This transformation property makes it very straightforward to show that relationships such as conservation laws that are true in one reference frame are also true in any other.

The mass function

$$M(v) = m\gamma \tag{4}$$

can be thought of in some ways as the apparent mass of a relativistic particle, in the sense that mass is inertia—resistance to acceleration. As an object gets close to speed c, it becomes very hard to make it go faster, because a small increase in speed corresponds to a big increase in momentum.

In teaching, $M(v)$ is sometimes referred to as the relativistic mass and written, for example, m_{rel}; in practice among physicists, the term is very rarely used. As you'll see when I discuss energy, the relativistic mass of a particle is exactly equivalent to its energy, differing only by the constant factor c^2. For this reason the term *energy* is usually enough. What we do need to refer to is the parameter m, which is characteristic of the given particle regardless of its speed. To avoid any ambiguity, m is therefore often referred to as the *rest mass*, though as a practical matter it's most common to hear it referred to simply as the *mass*. Thus, for example, if you were asked for the mass of the electron (no reference to its speed), you should say that it is 9.11×10^{-31} kg. If someone asked you for the mass of an electron whose speed is $0.7c$, on the other hand, you should answer that the question is ambiguous: Does the questioner want the rest mass or the relativistic mass? In the little book that elementary particle physicists carry around in their pockets* there are tables of masses of hundreds of different particles—clearly the rest masses.

Massless particles

There are some things in nature that *do* travel at the speed of light. Light is one of them, of course, and gravity waves almost surely do also. The funny thing is that according to quantum theory (see Part II), electromagnetic waves come in little particlelike bits, called *photons*, and as the waves travel at speed c, so must the corresponding particles, the photons. Gravitational waves undoubtedly obey the same rules; the particlelike bits are called *gravitons* (though we're a long way from detecting them), and the gravitons too must travel at speed c. We've observed particles called *neutrinos* that may also travel at speed c, but we're not sure.

How can this be? After all I've told you about the impossibility of particles' traveling at the speed of light, how can nature get away with it? The question you need to ask is "What's the rest mass?" If $v = c$, then $\gamma = \infty$, and if the momentum is to be finite, then, by eq. 3, m must be equal to zero. A particle that travels at speed c must have zero rest mass.

What happens to the physics when m is equal to zero? Eq. 3 isn't much use because it contains the meaningless product $0 \times \infty$. We don't make use of γ in the case of massless particles but rather describe the

---

* *Particle Properties Data Booklet*, North Holland Amsterdam. Obtainable at no charge from Technical Information Department, MS 90-2125, Lawrence Berkeley Laboratory, Berkeley CA 94720, USA.

particle in terms of its momentum and energy, which are both finite. Massless particles *always* travel at speed c, which is, of course, consistent with what we know about photons; if a massless particle had speed less than c then its momentum would be zero by eq. 3, and then the tiniest conceivable force would immediately give it a very small but nonzero momentum—and speed c.

If you can no longer use eq. 3 to calculate the momentum, what can you do? We got to eq. 3 by assuming that the momentum was determined by the speed, and that turns out to be true only if m isn't zero and the speed isn't c. For massless particles the momentum isn't determined by the speed, because the speed is always c and the momentum can be anything. So what you should do is turn the question around and ask for the speed of a particle in terms of its momentum; this is actually a reasonable way to put it, because momentum is such a basic physical quantity. In prob. 2(a) you're asked to do just this, and the answer, which you're asked to derive, turns out to be

$$v = \frac{pc}{\sqrt{p^2 + m^2 c^2}} \tag{5}$$

an equation, incidentally, that doesn't involve γ. If m is not zero, this gives a meaningful dependence of v on p; it's approximately equal to p/m for small momenta, the correct nonrelativistic form, and approaches the speed of light c as p becomes infinitely large. If m is zero, on the other hand, the equation is still perfectly reasonable but simply reduces to $v = c$ for all nonzero values of the momentum p. Massless particles never have zero momentum, as I just explained.

It is perfectly reasonable to talk about the relativistic mass of such a particle—it's just (see eq. 1)

$$m_{\text{rel}} = \frac{p}{v} = \frac{p}{c} \tag{6}$$

This relativistic mass can be as small as you wish, unlike the relativistic mass of an ordinary particle with a rest mass, for which m_{rel} is always greater than m (eq. 4).

PROBLEMS

1. A constant force F is applied to a particle starting from rest until it reaches relativistic velocities.

(a) Use eq. 4.3.2 to obtain the momentum as a function of time, and then solve eq. 3 for the speed $v(t)$ as a function of the time. (Treat this as a one-dimensional problem.)

(b) Show that the quantity $t_0 = mc/F$ is an appropriate unit for measuring t for this problem, and draw a graph of $v(t)$ with the time scale in units of t_0.

2. Derive formulas expressing the following quantities as functions of the momentum $\underline{p}$ or its magnitude p.

   (a) speed v
   (b) velocity $\underline{v}$
   (c) relativistic mass m_{rel}
   (d) relativistic factor γ

3. Suppose you try to give a relativistic particle a constant acceleration a_0 so that $v = a_0 t$. Find the force needed as a function of time, and describe what happens as v approaches c.

4.6   CONSERVATION OF MOMENTUM AND
          NEWTON'S THIRD LAW

In looking for a definition of momentum, we started with the fact that it's conserved and used this fact in defining the concept of force as the outside influence that causes the momentum of a particle to change. The conservation of momentum has much richer consequences than this, however, both conceptual and practical, and I'd like to explore these with you.

An important consequence that I want to introduce here is Newton's third law, the famous statement that "action equals reaction." This phrasing is too vague, as you might guess; we'll have to look at it more carefully and come up with a more exact statement of it. I also want to tell you why even the more carefully worded statement is not sufficiently general to cover interactions involving force fields, though it works very well for a large number of common nonrelativistic situations.

To look at the mutual forces between two objects, you follow a typical physicist's strategy and simplify the situation: You imagine that the two objects are completely isolated from the rest of the universe and interact only with each other, like two astronauts floating in space and pushing on each other [see fig. 1(a)]. Let $\underline{F}_{12}$ be the force on object 1 due to object 2 [fig. 1(b)], and let $\underline{F}_{21}$ be the force on object 2 that is due to object 1 [fig. 1(c)]. Note carefully that the same contact is responsible for both forces; you should be able to see intuitively that neither force

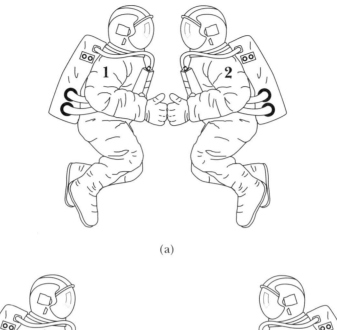

(a)

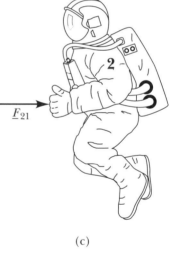

(b)   (c)

Fig. 1 Forces of interaction between two astronauts. $\underline{F}_{12}$ is the force exerted on 1 by 2, and $\underline{F}_{21}$ is the force exerted on 2 by 1.

could exist without the other. Because the objects are isolated from the rest of the universe—no other forces are acting on either of them—the total momentum of the two objects must be conserved; that is, it must not change with time:

$$\underline{p}_1 + \underline{p}_2 = P = \text{constant} \qquad (1)$$

If you take the time derivative of the two sides of this equation, you get a statement about forces. Because the rate of change of a constant is zero, you get

$$\underline{\dot{p}}_1 + \underline{\dot{p}}_2 = \underline{\dot{P}} = 0 \qquad (2)$$

or

$$\underline{F}_{12} + \underline{F}_{21} = 0 \qquad (3)$$

by Newton's second law, eq. 4.3.2. Eq. 3 is Newton's third law: *The forces that two objects exert on each other are equal in magnitude and opposite in direction.* That is,

$$\underline{F}_{21} = -\underline{F}_{12} \qquad (4)$$

What does this mean in practice? The floor you're standing on is exerting an upward force on you exactly equal in magnitude to the downward force that you're exerting on the floor. When you throw a ball, the ball exerts a backward force on your hand exactly equal to the forward force that your hand exerts on the ball. Note that the floor and the ball are not making any effort—they're not doing any work—but they are exerting forces all the same.

We worked out the result by ignoring other forces (eq. 4). Even when other forces are acting, it's generally true that the forces that two objects exert on each other are unaffected and therefore still satisfy Newton's third law.

Those other forces can be confusing. In the example where you're standing on the floor, there's also the downward force of gravity on you, which is equal and opposite to the force the floor exerts on you, as shown in fig. 2. *This is not an example of Newton's third law.* In the first place, both of these forces are acting on the same object, namely you, and the action–reaction pairs of Newton's third law *always,* by the way the forces are defined, act on two different objects. In the second place, the force of the floor on you would be different from the force of gravity if you were

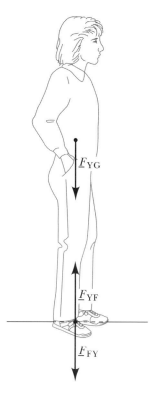

Fig. 2 Reaction forces for a person standing on the floor. F_{YG} is the force on you due to gravity, F_{YF} is the force on you due to the floor, and F_{FY} is the force on the floor due to you. Only F_{YF} and F_{FY} are related by Newton's third law.

accelerating—greater if you were accelerating upward, as in an elevator starting to go up, and less if you were accelerating downward. This is in fact a case of Newton's *second* law, $F = ma$ (eq. 4.3.10), where the net force, the sum of the two forces acting on you, is zero only if your acceleration is zero.

In some cases, such as a collision between billiard balls, it's hard to examine the forces in detail but very easy to apply conservation of momentum directly. However complicated the forces between two objects, their total momentum must be the same after the interaction as before if there are no other forces acting on them from outside.

Where this picture of equal and opposite forces can break down is where fields are involved. Magnetic forces, which I'll tell you about in Chapter 11, are notorious violators of Newton's third law because of the odd way in which they depend on the velocities of the charged particles involved. What happens is that the electromagnetic field has momentum too, so the total momentum is not just the momentum of the two particles and the argument based on eq. 1 just doesn't apply. In the case of simpler forces, like the everyday gravitational force, the same thing is true, but only to an extremely tiny extent. The only way it shows up in the gravitational case is when the objects are far enough apart for the delay time of the gravitational interaction to be important. The point in this case is that the interaction isn't instantaneous, because no signal can travel faster than c. The gravitational force on the moon at a given moment corresponds to where the earth was about a second earlier, since that's how long it takes gravitational signals (like light signals) to travel that far. Thus the momentum being lost by the earth at any given moment isn't gained by the moon until a second later, and the sum of their two momenta doesn't stay constant. As with the magnetic forces I mentioned just now, the failure of eq. 1 is due to the momentum carried by the gravitational field itself. The point to remember is that Newton's third law is less fundamental than the conservation of momentum, and the conservation of momentum is valid in every case.

PROBLEMS

1. You and a friend are on roller skates; you weigh 100 lb, which means your mass is 45 kg, and your friend weighs 200 lb.
    (a) You come up behind your friend and give him a push, so that he moves forward at 1.5 m/s. What happens to you? You're both initially at rest, and your skates roll without friction.
    (b) You are again at rest. Your friend comes up behind you, traveling 2 m/s, and holds on to you so that you both go forward at the same speed. How fast are you going?
2. A brick sits on the ground. Which of the following pairs of forces are examples of Newton's third law?
    (a) The force that the ground exerts on the brick and the force of gravity on the brick
    (b) The force of the brick on the ground and the force of gravity on the brick
    (c) The force of the ground on the brick and the force of the brick on the ground
    (d) The force of gravity on the brick and the gravitational force that the brick exerts on the earth

(e) The gravitational force the brick exerts on the earth and the force the ground exerts on the brick

3. You are floating at rest in the center of a large gravity-free space station, and you throw a piece of chalk in order to recoil and move toward the wall 20 m away. The chalk has a mass of 20 g and reaches the wall in 1 s. How long does it take you to reach the wall? Your mass is 50 kg.

CHAPTER 5

Energy

 5.1   ENERGY AND WORK

I believe that the best definition of energy is the one provided by
Noether's principle:

> *Energy is the quantity that's conserved because of time displacement*
> *symmetry.*

"Time displacement symmetry" refers to the constancy in time of the
laws of physics. This is a very abstract definition, however, and it's important for understanding energy to look at the way it's seen in ordinary use.

The classic definition is this: *Energy is the ability to do work.* You measure work and energy in the same units and define the energy of a system

(an object or a device, for example) as the amount of work it's able to do. Fine. What is *work?* The classic answer:

$$\text{work} = \text{force} \times \text{distance moved} \tag{1}$$

That is, system S (you, for example) loses energy by doing work on some other system S′ (a box you're lifting from the floor, say), and it does work on S′ by exerting a force on it through some distance. Note that S′ has to put up some resistance, which can be due to inertia, to friction, or to other forces (gravity, in the example) that are also acting on it.

If the force is variable, you can use a differential form; for one-dimensional motion through a very small distance dx, the amount of work dW is also very small, because it's proportional to the distance:

$$dW = F dx \tag{2}$$

You can then integrate (see §3.3) to add up the little contributions and get the total work done:

$$W = \int dW = \int F\,dx \tag{3}$$

You know what force is because I told you about it in Chapter 4: It's what produces changes in momentum. For now we'll stick to the intuitive picture of force as the amount of push or pull exerted by one object on another.

In three dimensions, where force and displacement are vectors, the situation is a bit more complicated, because the amount of work the force does depends on the relative directions of the force vector $\underline{F}$ and the displacement vector $d\underline{r}$. The work done is most concisely expressed in terms of what's called the *scalar product* of the two vectors:

$$dW = \underline{F} \cdot d\underline{r} \equiv F\,dr\cos\theta \tag{4}$$

where θ is the angle between the two vectors, as shown in fig. 1. The scalar product of two vectors is the product of their magnitudes, multiplied by the cosine of the angle between them. The point is this: If the motion is along the same direction as the force, then $\cos\theta$ is equal to 1 and the work is just $F\,dr$, as in eq. 2. If the motion is at right angles to the

Fig. 1 Force and displacement vectors. The work done by $\underline{F}$ is proportional to cos θ.

force, then cosθ is zero and no work is done, while if the motion is in the opposite direction to the force, the work done is counted as negative, as I'll explain below. Another way of looking at it is to notice that the quantity $F \cos\theta$ is the component of the force along the direction of motion and that it's only this component that does any work. See Appendix A, §2, for more on scalar products.

If you use the formula of eq. 2 with the force expressed in newtons and the distance in meters, then the work W comes out in joules. The *joule** is a standard unit of energy in the metric system. To relate the joule to your everyday experience, note that the familiar unit of power, the watt (W), is a rate of doing work equal to 1 joule (J) per second. A 100-watt bulb, for example, consumes 100 joules of energy every second.

In some cases work is done *by* the force; in others, it's not. When I'm falling, the gravitational force is doing work because I'm moving in the direction of the force. When I'm just standing on the floor, neither the floor nor gravity is doing work, because I'm not moving. I do work to get an object moving, as when I throw a ball, but when the ball is stopped by my friend's hand, the work done by his hand is negative, because the force that his hand is exerting is in the opposite direction to the motion of the ball. In fact, energy is supplied *to* his hand and body, which get warmer from the impact; this is exactly what we mean when we say that his hand does negative work. The important things to remember here are that work is involved only when a force acts over some distance and

---

*Usually pronounced "jool," even though Sir James Joule, the amateur physicist after whom it is named, pronounced his name "jowl."

that with our technical definition of work, the work done may be either positive or negative.

This approach, relating energy to the ability to do work, is quite satisfactory, and is still very standard, but I don't believe it gets to the heart of the matter. It turns out, even classically, that this way of defining energy hinges on the fact that energy is conserved, as prescribed by Noether's principle: The total amount of energy in the universe can never increase or decrease. It is this fact that's responsible for the very idea of work. We can define work the way we do because the amount of work you can get out of a system doesn't depend on how you get it out; because the work you put into a system is work that you can get out of it again later—no more, no less; and because the work done *by* one system is always work done *on* some other system—and all of these facts follow from the conservation of energy. Here, as elsewhere in physics, the law and the definition are mutually dependent: *You can define energy only because it's conserved.* Instead of defining energy in terms of work, you should turn it around, take energy as the fundamental quantity, and define work as transfer of energy.

In our universe anyway, it appears that energy *is* conserved. If one kind of energy is depleted (such as the chemical energy in fossil fuels), then some other kind increases in amount (such as heat energy). If you burn gasoline to run your car, the chemical energy in the gasoline doesn't vanish but is converted, some of it directly into heat, and some to kinetic energy, the energy of motion of the car. In the end it all goes into heat because of frictional forces, including the final braking to bring the car to rest.

Once energy is in the form of heat, it is more difficult to convert it back to more useful forms. This statement lies at the root of concerns about our energy supplies and is the essence of what's called the *Second Law of Thermodynamics,* a curious law that isn't a law, which we'll come back to in Chapters 8 and 9.

An odd thing about energy is that it too is relative, like the idea of relative velocity that I discussed earlier: Important though it is, the energy of a system is not absolute but depends on your frame of reference. Let me explain. As you can imagine, and as you'll see in what follows, the energy of a moving object depends on its velocity, but to an observer moving with the same velocity, the object appears to be at rest and has no energy of motion at all. The idea of work also depends on your point of view. From the definition of work as (force) × (distance moved) we can see two things. First, the thing or person that does the pushing is doing work and hence is losing energy, while the thing being pushed is having work done on it and is gaining the same amount of energy—the two amounts are exactly equal, by eq. 2 and Newton's third law, eq. 4.6.4. Second, since the distance moved depends on the frame of reference of the observer and is even zero in a comoving frame, it follows that the work done also depends on the frame and that it can be positive, negative, or zero, depending simply on your point of view. We find in fact that energy is one component of a relativistic 4-vector $\underline{p}$, the other three being the components of the familiar momentum vector. This 4-vector is closely analogous to the position 4-vector $\underline{\underline{x}}$ (§3.4), which has both time and space coordinates as its components. When you change your reference frame, the four components of $\underline{p}$ are recombined by the Lorentz transformation, just like the space and time coordinates of a point in spacetime (eqs. 3.1.1–3.1.4). The length of the 4-vector, defined as in §3.4, doesn't change under the tranformation, but the new values of energy and momentum correspond to the different point of view. If energy and momentum are conserved in some process according to one reference frame, then they are conserved according to any other reference frame also, as I'll explain more carefully in §5.7.

PROBLEMS

1. **(a)** Use the definition of eq. 4 to show that the work you do in lifting a weight W straight up to height h is the same as the work you do in lifting it to the same height along a diagonal path as shown in fig. 2. (Neglect forces needed to accelerate and decelerate the weight, as well as forces due to friction.)
   (b) Show that the work you do in lowering the weight again is the exact negative of the work you do in raising it.
   (c) Why do you get winded in running up stairs and not in running down stairs? How might you come down from an upper floor of a building in such a way as to do useful work or store the energy gained for future use?
   (d) In a game of tug of war, which side (the winners or losers) is doing positive work and which is doing negative work?
2. Find the work in joules (ignoring friction) needed
   (a) for a 50-kg person to climb to a height of 50 m.
   (b) for a 10-ton truck to drive a distance of 10 mi down a 1° incline at constant speed. Explain your answer.
   (c) to lift a 10-g thimble onto a book 2 cm thick.

Fig. 2　(see prob. 1)

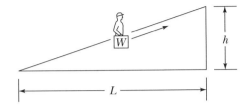

(d)　to push a 50-kg box at constant speed across a distance of 10 m on perfectly smooth ice.

Use scientific prefixes where appropriate (Appendix C, §3).

3.　(a)　A constant force F acts on a relativistic massless particle (§4.5) as it moves along a straight line at constant speed c. Show that the work done by the force is just equal to $c\Delta p$, where Δp is the increase in momentum.

   (b)　Show that the same result holds if the force varies with time in an arbitrary way: $F = F(t)$.

   (c)　The result given in part (a) is related to the fact that the energy of a massless particle is given by $E = pc$. Explain, in terms of energy conservation.

4.　You are standing on the observation platform at the rear of a train moving at 30 m/s, looking back along the tracks as you lean against the wall behind you with a force of 80 N. From the point of view of an observer standing beside the track, figure out the rate at which you are doing work on the wall behind you, in watts (J/s). Is the train also doing positive work on you, and if so, how and at what rate?

5.2　KINETIC ENERGY

I won't in this book be able to deal with all the many different forms of energy. The three kinds that are relevant to our discussion here are *kinetic energy*, the energy due to the motion of an object (the word *kinetic* always refers to motion); *potential energy*, energy due to the configuration of a system or to the location of objects; and *rest energy*, the energy that even an object at rest has on account of its mass. Potential energy is less fundamental in character than the other two, as I'll explain in the next section, but it is extremely useful as a practical concept in most nonrelativistic situations. Rest energy, on the other hand, is a new and totally unexpected concept of fundamental significance, predicted by Einstein and very well confirmed in our experience with nuclear energy, for example, and with processes in which particles are created or destroyed.

What is the kinetic energy K—the energy associated with its motion—of an object moving with speed v? Expressed in Newtonian terms, it's the amount of work needed to accelerate the object to that speed. If we keep to one dimension for simplicity, the increase in K as the object moves through a small distance dx is the work you do in moving it that distance. By eq. 5.1.1, this is then

$$dK = dW = F\,dx \tag{1}$$

Since the motion is nonrelativistic, you can use $F = ma$ (eq. 4.3.9), so

$$dK = ma\,dx \tag{2}$$

$$= m\left(\frac{dv}{dt}\right)dx \quad \text{(eq. 4.2.11)} \tag{3}$$

$$= m\left(\frac{dx}{dt}\right)dv \tag{4}$$

$$= mv\,dv \tag{5}$$

$$= \frac{1}{2}md(v^2) \tag{6}$$

$$= d\left(\frac{1}{2}mv^2\right) \tag{7}$$

In (4) I rearranged the factors using the fact that we can treat the quantities dv, dx, and dt as small ordinary numbers, all related to the same small time interval dt [equivalently, you can write $a\,dx = a(v\,dt) = v(a\,dt) = v\,dv$. In (6) I used the rule for differentiating a simple power function (eq. 4.2.22), taking $n = 2$ and replacing t by v.

Eq. 7 says that the change in K is exactly the same as the change in the quantity $\frac{1}{2}\,mv^2$. To get the total kinetic energy, you start with the object at rest, where you set $K = 0$ because K is defined as the energy of motion, and then add up all the little increases dK (that's an integration) to get the final value. Since the quantity $\frac{1}{2}\,mv^2$ is also equal to zero when the object is at rest, and since by eq. 7 the changes in $\frac{1}{2}\,mv^2$ are equal to the changes in K, it follows that the final values are the same also—that is, that

$$K = \frac{1}{2}mv^2 \qquad (8)$$

To restate this result in the language of integration, we write the sum of the changes dK as $\int dK$ (see Appendix A, §5, for more on integration). The net change in K is then

$$\Delta K = \int dK = \int d\left(\frac{1}{2}mv^2\right) = \Delta\left(\frac{1}{2}mv^2\right) \qquad (9)$$

where the third term, $\int d(\frac{1}{2}\,mv^2)$, means simply the sum of all the little changes in the quantity $\frac{1}{2}\,mv^2$, which must add up to the net change in that quantity, written as $\Delta(\frac{1}{2}\,mv^2)$. Eq. 9 is equivalent to the statement that

$$K - K_0 = \left(\frac{1}{2}mv^2\right) - \left(\frac{1}{2}mv^2\right)_0 \qquad (10)$$

The initial values K and $(\frac{1}{2}\,mv^2)_0$ are the values when the object is at rest, and they are both zero, so you get the result (8), which expresses the kinetic energy in terms of the velocity v.

Eqs. 9 and 10 can be applied to more general situations—for example, to the case where the particle is initially moving and is brought to rest. In this case ΔK is negative, and the negative value of W represents the fact that work is done *by* the particle in the process of being brought to rest; this is consistent with the definition of energy as the ability to do work, as required by energy conservation. In this example, K and v, the final values, are zero, and eq. 10 then states the obvious fact that $K_0 = \frac{1}{2}\,mv_0{}^2$.

Notice that in this analysis we never assumed the force was constant; the result is just the same no matter how the force F may vary over the time in question.

The unit of energy is the joule, the same as the unit of work. The formula for kinetic energy (eq. 8) gives the correct number of joules if the mass is expressed in kg and the velocity in m/s. The energy of motion of a 1-kg mass (which weighs about 2.2 lb) with a speed of 1 m/s is 0.5 J.

PROBLEMS

1. Work out the kinetic energy in joules of various objects—baseballs, automobiles, and so on—using reasonable estimates of masses and speeds. (It's useful to show that 45 mi/h is about the same as 20 m/s.)

2. (a) The *potential energy* of an object raised to height y is the work needed to get it there, or $U =$ force × distance $= mg_0y$. Use Newton's law of motion to show that the total energy for motion in a vertical direction only, given by

    $$E = K + U = \frac{1}{2}mv^2 + mg_0y$$

    is constant—that is, that its time derivative is zero.

    (b) Generalize this result to three-dimensional motion, using potential energy mg_0z, where the coordinates of the object are (x,y,z). (Potential energy is discussed in more detail in §5.3.)

    (c) If a car rolls without friction down a hill of height 10 m (33 ft), starting from rest, how fast will it be going when it reaches the bottom? Give the answer in m/s and in mi/h. Explain why it's not necessary to know the mass of the car or the steepness of the hill.

5.3 POTENTIAL ENERGY

When you do work against some force, the primary result may not be any increase in kinetic energy because the object you're pushing on may not speed up at all. When you carry a heavy suitcase up three flights of stairs (see prob. 5.1.1), you do a lot of work no matter how slowly you perform the task. The question, then, is "Where did the energy you expended go?" Well, you know the energy must be somewhere, because energy is conserved. Furthermore, if you now drop the suitcase out the window, it will acquire a large amount of kinetic energy as it falls to the ground—an amount exactly equal, in fact, to the work you did in carrying it up in the idealized situation in which no heat was generated (no energy was wasted) in the process.

The standard answer of the undergraduate physics course is that the suitcase acquires *potential energy* when it is raised through a gravitational field. The work you did in raising it to a height h is mgh (see prob. 5.1.1), so we say that the potential energy, called U (or, often, V), has increased by that amount:

$$\Delta U = mgh \tag{1}$$

Another way of saying the same thing is by means of a formula for $U(y)$, the potential energy of the object when its height above the ground is y:

$$U(y) = U_0 + mgy \tag{2}$$

where U_0 is some arbitrary potential energy for the object at ground level.

A similar kind of argument can be used for many other kinds of force: nonuniform gravitational fields, electric fields, elastic forces, and so on. In each case the change in potential energy is the work done against the force to alter the configuration of the system—to raise the suitcase, to stretch the spring and so on. The change in U as you move through a little distance $d\underline{r}$ is the work that you, the agent, do (eq. 5.1.3):

$$dU = dW = \underline{F}_{\text{agent}} \cdot d\underline{r} \tag{3}$$

Recall (§5.1) that the scalar product $\underline{F} \cdot d\underline{r}$ may be positive or negative, depending on whether the vectors $\underline{F}$ and $d\underline{r}$ point in more or less the same or opposite directions. The force you have to exert is equal and opposite to the force $\underline{F}$ that is due to the force field (Newton's third law, §4.6), so in terms of *that* force $\underline{F}$, the change in U is

$$dU = -\underline{F} \cdot d\underline{r} \tag{4}$$

The expression $\underline{F} \cdot d\underline{r}$ represents the work done by the force field, and when that work is positive the system loses energy—hence the minus sign in eq. 4. To get the potential energy function $U(\underline{r})$ you have to integrate—add up the little increments dU as you go from the reference position r_0 to the position r that you're interested in:

$$U(r) = U(r_0) + \int dU \tag{5}$$

$$= U(r_0) - \int \underline{F} \cdot d\underline{r} \tag{6}$$

where the integral is along some path from r_0 to r. (The concept of potential energy makes sense only if the work done by the force field— the value of the integral—is independent of the path you choose; this is an important matter that I won't be able to pursue here.) As an alternative to identifying a particular reference point r_0, you can simply replace the first term on the right side of eq. 6 by an arbitrary constant called a *constant of integration*. It doesn't matter much what its value is, because only *differences* in potential energy have direct physical meaning.

In §4.3 I listed some of the forces commonly encountered in nature; in each case there is a corresponding rule for the potential energy, calculated according to eq. 6, which I'll list below. In each case I include an arbitrary constant U_0, which can be set equal to zero without changing any of the physics.

Uniform gravitational field near earth's surface (eq. 4.3.15):

$$U(r) = mgz + U_0 \tag{7}$$

where z is the vertical component of r. The field is only approximately uniform since the earth isn't flat, but this is a good approximation over small distances like the size of a baseball field.

General gravitational field $\underline{g}(r)$ (eq. 4.3.16):

$$U(r) = -m\int \underline{g} \cdot d\underline{r} + U_0 \tag{8}$$

Gravitational force between two masses at r and r_2 (eq. 4.3.17):

$$U(r_1, r_2) = \frac{-Gm_1 m_2}{r} + U_0 \tag{9}$$

where r is the distance between the two masses:

$$r = |\underline{r}_1 - \underline{r}_2| \tag{10}$$

Uniform electric field $\underline{E}$ (eq. 4.3.19):

$$U(\underline{r}) = -Q\underline{E} \cdot \underline{r} + U_0 \tag{11}$$

Electrical force between two charged particles at $\underline{r}_1$ and $\underline{r}_2$ (Coulomb's law, eq. 4.3.20):

$$U(\underline{r}_1, \underline{r}_2) = \frac{kQ_1Q_2}{r} + U_0 \tag{12}$$

The minus signs in eqs. 8 and 11 reflect the more general fact (eq. 4) that the potential energy decreases when the force does a positive amount of work. The minus sign in eq. 9 is due specifically to the fact that the gravitational force is attractive; the potential energy must decrease as the separation of the objects decreases, because the forces are doing positive work. The expression (12) for the potential energy of two charged particles is positive if the charges have the same sign, since the force is then repulsive, and negative if the charges have opposite signs, since the force is then attractive (like the gravitational force).

The idea of potential energy is extremely useful in nonrelativistic physics, where it allows us to use conservation of energy in a powerful and flexible way in a wide variety of situations. You have to recognize, though, that the idea of potential energy is not truly fundamental and that it breaks down in the relativistic world where the dynamics of the force fields cannot be neglected, and where interactions between particles cannot be thought of as acting instantaneously over the distances that separate them in space. The energy that we call potential energy is in fact energy stored in the force fields—gravitational, electric, and so on; it is not simply a function of where the particles are but depends in an intricate way on the entire configuration of the fields in question. This field energy travels through space if there is a wave motion of the force field (gravitational or electromagnetic waves, for example) and cannot of course get from one place to another faster than the speed of light c.

PROBLEMS

1. **(a)** Show that the expression (eq. 8) for the potential energy of a particle in a general gravitational field follows from the idea expressed in eq. 6 that the potential energy is the work that must be done to bring the particle to that location.
   (b) Derive eq. 7 from eq. 8 by using the value of the vector field $\underline{g}$ appropriate to the coordinates used in eq. 7.
2. For a ball thrown into the air, the total energy—kinetic plus potential—is given by

$$E = K + U = \frac{1}{2}mv^2 + mgz$$

Use the constancy of E to determine the following:
   (a) The speed with which you must throw a ball vertically upward to reach a maximum height h relative to the point at which it leaves your hand.
   (b) The maximum height reached by the ball if it is thrown with initial speed v at an angle θ above the horizontal.
   (c) The speed of the ball at a moment when its height relative to the thrower is h if its initial speed is v_0. Do you need to know the angle θ at which the ball is thrown? Explain.

5.4 RELATIVISTIC ENERGY

Now let's see what the kinetic energy of a particle is when we do it right—that is, relativistically. It's not so easy as the nonrelativistic case, but you can follow the same logic as in §5.1 and §5.2. Doing it again in one dimension for simplicity, you start off with:

$$dK = dW = F\,dx \tag{1}$$

$$= \left(\frac{dp}{dt}\right)dx \quad \text{(eq. 4.3.2)} \tag{2}$$

$$= \left(\frac{dx}{dt}\right)dp \tag{3}$$

$$= v\,dp \tag{4}$$

After some further tricky steps, you can show that this is equivalent* to

$$dK = d(mc^2\gamma) \tag{5}$$

Just as in the nonrelativistic case, we've converted the original expression *F dx* into what's called an exact differential, which means that we've expressed it as the infinitesimal change in some expression, *d*(something). This enables us to integrate it (to add up the little increments to get the net change), just as we did with the nonrelativistic case (eqs. 5.2.7–5.2.10):

$$\Delta K = \int dK = \int d(mc^2\gamma) \tag{6}$$

$$= \Delta(mc^2\gamma) \tag{7}$$

Finally, to get the kinetic energy explicitly, you use the fact that for v = 0, the kinetic energy is zero and $\gamma = 1$, so

$$K = mc^2(\gamma - 1) \tag{8}$$

Is this related to the classical kinetic energy (eq. 5.2.8)? You can see the relationship easily by looking at nonrelativistic speeds. For $v \ll c$, we found previously (eq. 1.4.10) that θ has the approximate value

$$\gamma \approx 1 + \frac{v^2}{2c^2} \tag{9}$$

so, for nonrelativistic velocities,

$$K \approx (mc^2) \cdot \left(\frac{v^2}{2c^2}\right) \tag{10}$$

$$= \frac{1}{2}mv^2 \tag{11}$$

---

*The steps, for those with some familiarity with calculus, go like this: from $p = m\gamma v$ (eq. 4.5.3), you square both sides and eliminate v^2 in terms of γ^2, which gives $p^2 = m^2c^2(\gamma^2 - 1)$. Taking the differential of both sides (eq. 4.2.22, with $n = 2$) gives $p dp = (m\gamma v) dp = m^2c^2\gamma d\gamma$, or $v dp = mc2 d\gamma$.

This is just the Newtonian value of the kinetic energy (eq. 5.2.8), with the rest mass m as the nonrelativistic mass, which is consistent with the discussion in §4.5.

You can now see that the Newtonian formula was only approximately correct, though until this century it was impossible to detect the error because the speeds of material objects were always so very small compared to the speed of light. Now, in the big particle accelerators, we routinely accelerate particles to speeds so close to c that we have values of γ as large as several hundred thousand. The Newtonian expression of eq. 5.2.8 is totally inaccurate at these energies, and the relativistic form (8) is confirmed to a very high degree of precision.

As $v \to c$ the kinetic energy becomes infinite, telling us that it takes an infinite amount of work to accelerate an object up to the speed of light, which is consistent with what you learned in Chapter 4. This is an extremely important confirmation of the fact that it's impossible for a material object to reach the speed of light.

As I described in §4.5, there are particles in nature that travel at the speed of light and manage to do so by having exactly zero rest mass. What is the energy of such a particle? You won't get it as a function of the velocity, since the velocity is always c, which is why eq. 8 becomes meaningless, just as eq. 4.5.3 did in the case of momentum. What you *can* look for is a relationship between energy and momentum, since both are finite and meaningful. You can get such a relationship directly from eq. 4, using the fact that $v = c$:

$$K = \int v dp \qquad (12)$$

$$= c \int dp \qquad (13)$$

$$= pc \qquad (14)$$

This tells you that a massless particle can have finite momentum and energy and that the two are simply proportional. See also prob. 5.2.3 and the discussion on massless particles in §4.5.

PROBLEMS

1. Check the units of eq. 14, $K = pc$, for consistency.
2. For an electron the quantity mc^2 (called the *rest energy* of the elec-

tron) has the value 0.511 MeV, where $1 \text{ eV} = 1.6 \times 10^{-19} \text{J}$ ("rest energy" and the unit eV, the *electron-volt*, are explained in §5.5). Find the speed (as a multiple of c) at which an electron's kinetic energy is

(a) 1 eV

(b) 1 J. This gives a speed extremely close to c, so you can give a meaningful answer only by finding the *difference* $c - v$ directly. Show <u>first</u> that if $\varepsilon = 1 - v/c$, then γ is given approximately by $\gamma \approx 1/\sqrt{2\varepsilon}$, and then obtain ε from the appropriate value of γ.

3. (a) At what speed (relative to c) does the error in the non-relativistic formula for the kinetic energy (eq. 11) become 1%? You should not try to get a formula for v, which is pretty nasty, but rather do it by trial and error, using your calculator. It's easier if you first eliminate m from the problem and then express the problem in terms of the quantity $x = v/c$.

(b) At what speed is the kinetic energy of a particle equal to 1% of the rest energy, defined in prob. 2?

(c) At what speed is the kinetic energy equal to the rest energy?

(d) At what speed is the kinetic energy equal to 100 times the rest energy? (Be sure to keep enough figures in your answer to make it meaningful.)

 **REST ENERGY**

The second term in the relativistic kinetic energy formula (eq. 5.4.8) turns out to be quite significant. Notice first that the formula would be much more elegant without it; it's just a constant, and only changes in energy are physically important, so why not just omit it? Suppose we focus on the first term in eq. 5.4.8 and call it the total energy of the particle:

$$E = mc^2\gamma \tag{1}$$

$$= mc^2 + mc^2(\gamma - 1) \tag{2}$$

$$= E_0 + K \tag{3}$$

Eq. 3 expresses the total energy as equal to the kinetic energy plus a constant.

I want you to notice that our new "total energy" is simply proportional to the relativistic mass, m_{rel}, that is defined in connection with the relativistic momentum (§4.5):

$$E = M(v)c^2 \equiv m_{rel}c^2 \tag{4}$$

It's just this equivalence that makes the relativistic mass a somewhat redundant concept, energy being the more commonly useful—and the more fundamental—physical quantity.

The constant in eq. 3, which I've written E_0, is called the *rest energy*, because it's the value of the total energy when the particle is at rest:

$$E_0 = mc^2 \tag{5}$$

In natural units where $c = 1$, so that space and time are measured in the same units, this becomes

$$F_0 = m \tag{6}$$

showing that in essence the rest energy *is* the rest mass. According to eq. 1, when the particle is moving this energy is multiplied by γ, *just like the time on a moving clock*, so energy defined in this way transforms in the most natural possible way under a Lorentz transformation. The rest energy mc^2 is typically very big by everyday standards, because c is very big, as I'll discuss in more detail later on.

Does the rest energy have any physical meaning? Is it a real energy, capable of doing work? Is energy always equivalent to mass? The answer to the first two questions is yes, and the answer to the third question is a qualified yes—it does depend on what you mean by *mass*.

The way the rest energy reveals itself as a real energy is in processes where particles are created or destroyed—something unknown to Newtonian physics but now very familiar. For example, the π^0, the neutral pi meson, or pion, is an unstable particle that once it is created itself, survives for only about 10^{-16} s on the average and then self-destructs; it disappears completely, producing in the process two energetic photons that fly off in opposite directions relative to the original pion. We describe the process by writing

$$\pi^0 \rightarrow \gamma + \gamma \tag{7}$$

where the symbol γ is used for a photon because high-energy photons are often called gamma rays. (This has nothing to do with the relativistic factor γ.)

So how much energy do the photons have? It's always the same, for a pion initially at rest, and it's always exactly equal to the rest energy of the pion, mc^2. The entire mass of the pion, about 2.41×10^{-28} kg, is converted into electromagnetic energy:

$$mc^2 = 2.41 \times 10^{-28} \text{ kg} \times (3 \times 10^8 \text{ m/s})^2 \tag{8}$$

$$= 2.17 \times 10^{-11} \text{ J} \tag{9}$$

In the energy units physicists use for atomic and subatomic processes, this is equal to 135 MeV.* We've become so familiar with the equivalence of mass and energy that we routinely specify the mass of a particle in terms of its rest energy: The mass of the pion is 135 MeV/c^2 (because $mc^2 = 135$ MeV. We say "M E V over c squared.") We often are more casual and just say the mass is 135 MeV, simply equating mass and energy in the spirit of eq. 6.

There are all sorts of examples from particle physics in which one or more particles are destroyed in a collision or some other process, and others appear in their place. If the total rest mass of the particles involved decreases, then the mass that is lost exactly corresponds to the overall increase in kinetic energy, and if the total rest mass increases, then the difference must correspond to a loss of kinetic energy. There must be enough kinetic energy to start with for such a process to take place at all.

PROBLEMS

1. (a) Find the momentum of each of the photons in the π^0 decay, eq. 7, if the pion is initially at rest.
    (b) Determine which of the following decay modes are energeti-

---

* 135 million electron-volts: the energy an electron would acquire in being accelerated by a voltage difference of 135 million volts: $1 \text{ eV} = 1.6 \times 10^{-19} \text{ J}$.

cally possible, and for the modes that are possible, give the total kinetic energy in the final state, assuming the initial particle is at rest.

(i) $\mu^+ \rightarrow e^+ + \nu_e + \bar{\nu}_\mu$

(ii) $p \rightarrow n + \pi^+$

(iii) $K^- \rightarrow \pi^+ + \pi^- + \pi^-$

(iv) $\pi^+ \rightarrow p + \bar{n}$

(v) $e^- \rightarrow e^- + \gamma$

The particle names and masses (in MeV/c^2) are

μ = muon		
$\mu^\pm$ = "mu-plus" or "mu-minus"	106	
e^- = electron	0.511	
e^+ = positron	0.511	
ν_e = electron neutrino	0	
ν_μ = muon neutrino	0	
$\bar{\nu}_\mu$ = muon antineutrino	0	
p = proton	938	
n = neutron	940	
$\bar{n}$ = antineutron	940	
π = pion		
$\pi^\pm$ = "pi-plus" or "pi-minus"	140	
K = kaon (pronounced "KAYon")		
$K^\pm$ = "kay-plus" or "kay-minus"	494	
γ = photon, or "gamma"	0	

2. A relativistic particle of mass m_1 and velocity v_0 in the x direction bounces off a particle at rest of mass m_2. In the final state, the first particle has velocity v_1 at an angle θ above the x axis, and the second particle has velocity v_2 at an angle φ below the x axis, as shown in fig. 1: θ and φ are both positive angles. Write down the equations

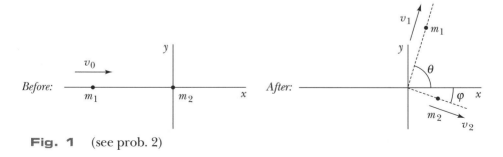

Fig. 1   (see prob. 2)

for conservation of relativistic energy and momentum (both the x and the y components) for this process. Do not try to solve these equations.

3. A charged pion usually decays into a muon and a massless neutrino: $\pi^+ \rightarrow \mu^+ + \nu_\mu$. The rest energy of the pion is 139.6 MeV, and if the pion decays from rest, then the muon's kinetic energy is 4.2 MeV and the neutrino's kinetic energy is 30.1 MeV.
   (a) What is the total energy (rest energy + kinetic energy) of the muon?
   (b) What is the rest energy of the muon?
   (c) What is the relativistic factor γ for the muon?

5.6   EQUIVALENCE OF MASS AND ENERGY

I said in §5.5 that in some sense energy is always equivalent to mass and that it depends on what you mean by "mass." There's an indication of this in the distinction between the rest mass m and the mass function $M(v) = m\gamma$, also referred to as m_{rel}, the relativistic mass, which I introduced in connection with the relativistic form of the particle momentum. However, as you'll see, both ways of defining mass are related in some direct way to energy, and you can regard the equivalence as complete.

Complete equivalence? Yes

The total particle energy E turns out, as I said before, to be simply proportional to the relativistic mass m_{rel} (eq. 5.5.4), and I deemphasized the idea of relativistic mass for just this reason—that it was simply equivalent to the energy. Nevertheless, the point is not trivial. A moving particle does have greater inertia than a particle at rest; it shows greater resistance to acceleration, and it weighs more. Even a photon, whose rest mass is zero, experiences a finite pull of gravity attributable to its nonzero relativistic mass, which is equal to E/c^2, where E is its energy. Historically, inertia and weight are defining characteristics of mass, so identifying m_{rel} as a kind of mass is quite appropriate even if it's not standard usage.

Where the equivalence of mass and energy becomes more compelling is in the case of a composite system—an atomic nucleus, for example—consisting of a number of particles bound together by internal forces. The total energy of the nucleus now includes the relativistic energy (rest energy plus kinetic energy) of the constituent particles (protons and neutrons) as they move around inside the nucleus, as well as the potential energy associated with the force fields that hold these particles together. The significant thing is that the observed mass of the composite object is exactly related to this total energy by the Einstein relation

$$E = Mc^2 \qquad (1)$$

If the nucleus is at rest, then M represents its observed rest mass. If the nucleus loses energy (by emitting radiation, for example, and shifting to a state of lower potential energy), then its mass is reduced by the corresponding amount. This is the classic situation to which the formula is applied, where the mass in question is the standard and directly measurable mass—the rest mass—of the composite object.

You can see, then, how it's valid to apply the same relation to the relativistic energy of a free particle, as in the velocity-dependent mass m_{rel} discussed above, because the relation clearly applies to the relativistic energy of the particles that make up a nucleus, and it would therefore be inconsistent not to apply it to the case of a single particle also. In both cases the total energy E is related to the observed mass of the object, whether simple or composite, in just the same way.

I've mentioned that the mass is related to both inertia and weight. These two effects are seemingly unrelated, because inertia refers to resistance to acceleration while weight refers to the force of gravity. There is no obvious reason why they should be related until you move on to the subject of general relativity, Einstein's theory of gravity as related to the curvature of spacetime. The observed equality of gravitational and inertial mass, already known but not understood, was one of the things that was puzzling Einstein. It was in fact a guide to him in developing general relativity; in the finished theory, this equivalence comes out as a completely natural consequence (see Chapter 6). Anyhow, when we say that mass and energy are equivalent, we mean that *the total energy of a system determines both its inertia and its gravitational weight.*

Now let's look more closely at the masslike properties of energy. Consider a deuteron, which consists of a neutron and a proton bound together by nuclear forces. The energy of a deuteron is a little bit less than the sum of the rest energies of the neutron and the proton. The difference is due to the potential energy of their interaction, which is negative in this case. The magnitude of this potential energy, a positive quantity, is called the *binding energy* and is equal to the amount of energy you'd have to supply to pull them apart. The mass of the deuteron, whether measured by its weight or by its inertia, is equal to that *reduced* energy divided by c^2. That is, it is less than $m_n + m_p$. In the same way, a hydrogen atom, which consists of a proton and an electron bound together by electrical forces, weighs less (by a very tiny fraction) than a proton and an electron separately. A sample of hot gas, on the other hand, weighs more than when it's cool, on account of the greater kinetic energy of the molecules. A moving electron weighs more than an electron at rest, and its mass, measured in this way, is just E/c^2 ($= m\gamma$), the relativistic mass. A photon has a weight proportional to its energy, and so on. Standard ways of producing energy, like burning coal, utilize the extra energy that becomes available when the fuel is oxidized—that is, becomes bound to oxygen atoms. The energy released is equal to the binding energy of the final product, and there's a corresponding decrease in the mass, so that you can say, in conventional language, that in every case mass has been converted to energy. Note that the binding energy of a bound system like a molecule is not real energy available for use: The bound molecule has *less* total energy than the separate atoms. In every case the residue—the ash or other waste product—has greater binding energy, and weighs less, than the original fuel. An increase in binding energy actually corresponds to a *reduction* in the total energy.

The situation is no different when it comes to nuclear energy. You again make use of a change in binding energy, corresponding to a difference in mass, but now the forces involved are thousands of times stronger. Hence the energy released is greater also, and the mass differences are big enough to be measured directly. A uranium or plutonium nucleus, being big and flabby, has a tendency to fall apart into two smaller chunks (this is *nuclear fission*), and because the chunks—nuclei of lighter elements such as barium, krypton, or what have you—are more tightly bound than the original uranium or plutonium nucleus, the net mass is less and energy is released—of the order of 200 MeV, compared with a few electron-volts for chemical processes. This represents a conversion of about 0.1% of the mass, or 1 part in 1000, into energy, compared with around 1 part in 10^8 in chemical processes.

Nuclear fusion, on the other hand, involves protons and neutrons fusing together to form light nuclei and uses the fact that protons and neutrons are very tightly bound in some of the small nuclei, especially that of helium. The "α particle," or helium nucleus, consists of 2 protons and 2 neutrons stuck together with a total binding energy of about 28 MeV. Because the mass of a neutron or proton is about 940 MeV/c^2, the relative reduction in mass due to binding is about (28 MeV)/(4 × 940 MeV) $\approx$ 0.007, or nearly 1%. This is a typical conversion efficiency for fusion processes, though in practice you have to do much more compli-cated things than just bringing protons and neutrons together in order to make it happen. It's this comparatively high conversion efficiency that enables stars like the sun to keep burning so hot for so long and hydro-gen bombs to give such a big bang. Hydrogen is also a lot cheaper and easier to handle than uranium, so it would be nice if we could control it as a power source, but we're not there yet.

The "binding energy" just discussed is, as I said, an example of the general concept of potential energy (§5.3), which is used to describe changes in the total energy of a system that are due to the forces in-volved. The potential energy is typically negative where the forces are attractive—the case of binding energy—because the ability of the system to do work is reduced: Something has to do work *on* the system to pull the particles apart. If the forces are repulsive, on the other hand, the potential energy is usually taken as positive because the system can do useful work as the particles move away from each other. These two effects can be seen in the expressions for gravitational and electrostatic poten-tial energy in §5.3 (eqs. 5.3.9 and 5.3.12), where the gravitational poten-tial energy is always negative (relative to $r = \infty$) and the electrostatic potential energy may be positive or negative, depending on the sign of the two charges.

Total conversion of mass to energy? No

People like to play with the idea of *total* conversion of mass to energy. They used to say that the energy in one train ticket (in olden days people used to travel by train, using little cardboard tickets; they still do in some parts of the world) would be enough to drive the train all the way around the world (see prob. 3). It's that huge factor c^2 in eq. 5.5.4 that does it. A tiny amount of mass is equivalent to a tremendous amount of energy.

As you can see, if we could perform such a conversion it would solve all our energy problems. Why isn't it possible? The answer lies in another conservation law, the conservation of *baryon number*.* The word *baryon* is used to refer to the proton and the neutron and also to a number of similar particles that we know how to manufacture but that aren't found

---

*This conservation law has been called into question in recent years and may in fact be only approximate, but even if so, it's so very nearly exact as to make no difference here.

in nature. In practice, then, this conservation law says that the total number of protons plus neutrons in the universe cannot change. One can turn into the other, as happens in some kinds of radioactivity, but they can't disappear entirely like pions. As you know, ordinary matter is made up of atoms, and atoms are made up of protons, neutrons, and electrons. Almost all of the mass of ordinary matter resides in the baryons, because the neutron and the proton, which both have about the same mass, are nearly 2000 times heavier than the electron. Protons and neutrons can't be destroyed in practice, so the mass of ordinary matter is not available for conversion to energy (a good thing, too, since otherwise, ordinary matter couldn't exist. It would fall apart immediately—and cataclysmically). The energy from fission and fusion comes from tiny differences in the binding energies of different nuclei, not from the destruction of particles, just the same as with ordinary combustion.

Now, there *is* one way around the indestructibility of ordinary matter, which looks fine but in fact has practical difficulties—and a fundamental logical flaw. There are such things as *antiprotons* (symbol $\bar{p}$) and *antineutrons* ($\bar{n}$), which again we are able to manufacture in small numbers but which aren't present in ordinary matter. There are occasional antiprotons found in cosmic rays, but as far as we know, they aren't present in any abundance in the universe at large. Together with anti-electrons—always called *positrons*—they can combine to make *antimatter,* beloved of sci-fi writers. An antiproton or an antineutron counts as (–1) baryons; we say that each has a *baryon number* of –1. This means that if you bring a proton and an antiproton together, they have a net baryon number of zero, and *there is nothing to prevent them from both disappearing,* the combined masses being entirely converted to energy. This kind of event, called *annihilation,* is in fact exploited in some of the big particle accelerators, where a beam of protons going one way around the ring is made to collide with a beam of antiprotons going the other way, producing lots of such annihilations. However, these particles are extremely small, and the energies produced, though large on an atomic scale, are far too tiny to solve any of our energy problems. In order to exploit this on a large scale, you'd have to produce macroscopic amounts of antimatter—not infinitesimal fractions of a gram, but kilograms—which we haven't the remotest idea how to do, and then keep it absolutely and completely out of contact with ordinary matter (what do you keep it *in?*) until you are finally ready to use it. Then you drop it and duck. For a given mass, it would be hundreds of times more powerful than a hydrogen bomb. Nevertheless (and here's the fatal flaw), as a source of energy, antimatter wouldn't help at all, because it would take just as much energy to create it in the first place as the amount of energy you'd get out of it at the end.

PROBLEMS

1. Figure out the mass of fuel that must actually be converted into en-
    ergy and the total amount of fuel needed to run a 100-MW nuclear
    power plant for a year, and then do the same for a fossil fuel plant.
    Use the approximate conversion efficiencies given in this section.

    (1 MW = 1 megawatt = 1,000,000 watts)

2. The mass of the sun is about 2×10^{30} kg, and the radiant power
    reaching the earth from the sun, 1.5×10^{11} m away, is roughly
    1 kW/m^2. Estimate the total radiant power emitted by the sun, and
    estimate how long the sun can keep burning at that rate, assuming
    a 0.7% conversion efficiency. (The surface area of a sphere of ra-
    dius r is $4\pi r^2$, so the fraction of the sun's energy that strikes a sin-
    gle square meter can be calculated by relating that area to the total
    area of a spherical surface centered at the sun.)

3. I mentioned in the text that the mass of one train ticket, if entirely
    converted to energy, might be enough to drive the train all the way
    around the world. Determine whether this is reasonable. (You'll
    need to make some very crude estimates.)

4. (a) What's the maximum number of pions that could be pro-
         duced in a proton–antiproton collision if the proton and anti-
         proton have kinetic energy equal to 6.0 GeV each?

    (b) What is the maximum number of photons that could be pro-
         duced?

    (c) What is the maximum number of neutrons that could be pro-
         duced? Explain the relevance of baryon number conservation
         to this question.

5.7 THE ENERGY–MOMENTUM
         FOUR-VECTOR

In §3.4 you were introduced to vectors in spacetime, like arrows in a
spacetime diagram that point the way from one event to another. A
spacetime vector, or 4-vector, has four components, three that give the
spatial separation of the two events and one that gives their separation in
time. When you look at the same two events from a different frame of
reference, this description of their separation in space and time changes,

and the components in the new frame are given by the Lorentz transformation (eqs. 3.1.1–3.1.4). Thus physicists say that this transformation rule is characteristic of a 4-vector and that any four quantities that transform according to the same rule, the Lorentz transformation of eqs. 3.1.1–3.1.4—can be called the components of a 4-vector. In this section I want to tell you about another important 4-vector in physics, the energy–momentum 4-vector.

Since energy is related to time and momentum is related to space, through Noether's principle, and since time and space variables together form the 4-vector $\underline{x}$ (§3.4), you can reasonably ask whether energy and momentum form a 4-vector also. Momentum and energy are both related to velocity, so let's start by trying to relate velocity to a 4-vector and then work from there.

A 4-vector, as I said, is a combination of four components that transform in the same way as x, y, z, and ct under a change in reference frame—that is, according to the Lorentz tranformation of eqs. 3.1.1–3.1.4. The fact that we're working with ct instead of c alters the form of the equations slightly but doesn't change the logic. To get a velocity 4-vector for a particle, you simply differentiate its position 4-vector $\underline{x}$ with respect to the proper time τ, the point being that the proper time is invariant and doesn't change the transformation rules:

$$\underline{u} = \frac{d\underline{x}}{d\tau} \tag{1}$$

This tells how the four components of $\underline{x}$ vary with τ as you move along the particle world line. The derivative $\underline{u}$ is guaranteed to be a true 4-vector because $d\underline{x}$ (the separation between two events very close together along the world line) is a true 4-vector and $d\tau$ is invariant because the proper time is the same no matter what frame you're in. The timelike part of $\underline{u}$ is given by

$$\frac{d(ct)}{d\tau} = c\frac{dt}{d\tau} \tag{2}$$

$$= c\gamma \tag{3}$$

because $dt/d\tau$ is just the time dilation factor γ, by eq. 3.3.1. The spacelike components of $\underline{u}$, which I'll call $\underline{u}$, are given by

$$u = \frac{dr}{d\tau} \tag{4}$$

$$= \left(\frac{dt}{d\tau}\right)\left(\frac{dr}{dt}\right) \tag{5}$$

$$= \gamma v \tag{6}$$

Here the factors dt cancel between the two quotients in eq. 5 (the chain rule of differential calculus—see Appendix A, table 4.1), and those two quotients are simply equal to γ and v, respectively. The components of the velocity 4-vector can thus be written compactly as

$$u = (c\gamma, \gamma v) \tag{7}$$

It's a timelike vector because dx, which lies along a world line, is timelike, and its length, then, is given by the same rule as eq. 3.4.10:

$$\sqrt{u \cdot u} = \sqrt{(c\gamma)^2 - |\gamma v|^2} = \gamma\sqrt{c^2 - v^2} = c \tag{8}$$

The velocity 4-vector always has the same length. It's the *direction* it points in spacetime (related to the inclination, introduced in §2.2), not its length, that tells you how fast the particle is going. For a particle at rest, u points in the time direction (its spatial components are zero), while for a particle moving at a speed close to c, u points in a direction very close to the the world line of a light signal—and has very large components.

Now we're ready to construct the energy–momentum 4-vector. Because the energy is equal to $mc^2\gamma$ (eq. 5.5.1) and the momentum is equal to $m\gamma v$ (eq. 4.5.3), you merely have to multiply u by the rest mass m and you have it:

$$p = mu = (mc\gamma, m\gamma v) \tag{9}$$

$$= (E/c, p) \tag{10}$$

The time component is $mc\gamma$, or E/c, and the spatial components form the relativistic momentum $m\gamma v$. This deceptively simple relation (eqs. 9 and 10) (or weren't you deceived?) is deeply significant. It tells you first of all that the energy and momentum we came up with in this and the previous chapter are related to the motion of the particle in the simplest way possible, through eq. 9, and it tells you that the four components of $\underline{p}$ (E/c and the three components of the momentum p) transform in a particularly simple way under a change in reference frame, according to the standard Lorentz transformation equations (eqs. 3.1.1–3.1.4). With the appropriate factors of c, the transformation rules for energy and momentum become

$$p'_x = \gamma_0\left(p_x - \frac{v_0 E}{c^2}\right) \tag{11}$$

$$p'_y = p_y \tag{12}$$

$$p'_z = p_z \tag{13}$$

$$E' = \gamma_0(E - v_0 p_x) \tag{14}$$

The same transformation equations apply even when E and $\underline{p}$ are the energy and momentum of a complicated system, with lots of internal degrees of freedom that all look very different in different reference frames.

What is the invariant length of $\underline{p}$? Since $\underline{u}$ is forward timelike, eq. 9 tells us that $\underline{p}$ must be forward timelike also, and since the length of $\underline{u}$ is c (eq. 8), we see from eq. 9 that the length of $\underline{p}$ is just mc.

$$\underline{p} \cdot \underline{p} = m^2 c^2 \tag{15}$$

Apart from the constant factor c, then, the *rest mass* can be interpreted as the invariant length of the energy–momentum vector; all electrons have vectors $\underline{p}$ of length $m_e c$ regardless of their velocity, all protons have vectors $\underline{p}$ of length $m_P c$, and so on.

Equation 15 gives you a very important relation between the energy and the momentum of a particle. It tells you (with an extra factor c^2) that

$$E^2 - p^2 c^2 = m^2 c^4 \tag{16}$$

or

$$E = \sqrt{m^2 c^4 + p^2 c^2} \tag{17}$$

To physicists this is a more important way of writing E than the expression $mc^2\gamma$ (eq. 5.5.1), because, as I've said, a particle's momentum is in many ways a more fundamental characteristic of its motion than its velocity is. Eq. 17 is more practical too, because momentum is a more accurate measure than velocity for speeds very close to c and can vary by a large factor while the speed changes by only a minute amount. The graph of eq. 17 has the form of a hyperbola, shown in fig. 1(a).

Look at the limiting behavior of eq. 17 for small p and for large p. When $p \ll mc$ you can use the approximation given in eq. 1.4.12:

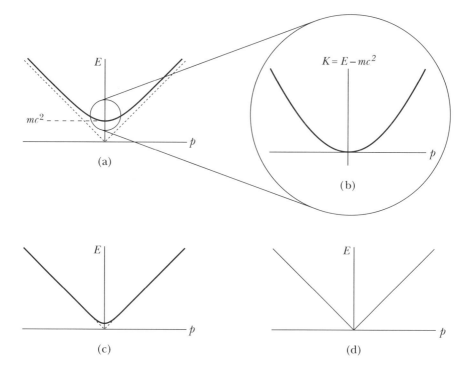

Fig. 1   Relation between particle energy and momentum. (a) Relativistic relation (eq. 17). (b) Kinetic energy in the nonrelativistic limit (eq. 20). The energy scale has been expanded in this graph. (c) The case of very small mass, or of energies very large compared to the rest energy (eq. 21). (d) The limiting behavior for a massless particle (see eq. 5.4.14).

$$E = mc^2 \cdot \sqrt{1 + \frac{p^2}{m^2 c^2}} \tag{18}$$

$$\approx mc^2 \cdot \left(1 + \frac{p^2}{2 m^2 c^2}\right) \tag{19}$$

$$= mc^2 + \frac{p^2}{2m} \tag{20}$$

The term mc^2 is the familiar rest energy, and the second term is then the kinetic energy in the nonrelativistic limit. Notice that the kinetic energy doesn't involve c (Newtonian physics doesn't know about c) and that this expression for the kinetic energy is consistent with our previous non-relativistic expression (eq. 5.2.8) if you replace p by mv (eq. 4.1.1).

For very large p ($p >> mc$), on the other hand, you get a good approximation by neglecting $m^2 c^4$ in eq. 17. You get

$$E \approx pc \tag{21}$$

an expression that becomes exact, in fact, for a massless particle like a photon or a massless neutrino (see eq. 5.4.14). Eq. 21 describes the situation where the rest energy is so small compared to the total energy that it can be neglected. You can get some feel for the expression mc, which appears as a standard of comparison for momenta, by thinking of it as the momentum that a Newtonian particle of mass m would have if it could move at the speed of light. [In fact, of course, the speed is always less than c because of relativity, and when $p = mc$, v takes the value $c/\sqrt{2}$. Compare also prob. 5.4.3(c).]

Let me emphasize one final important consequence of the fact that energy and momentum form a 4-vector:

> *If energy and momentum are conserved in one frame of reference, then they are conserved in every reference frame.*

This is because if two 4-vectors are equal in one frame, then they're equal in any other because they transform the same way. Thus if $\underline{P}$ is the total energy–momentum 4-vector for the system you're looking at and $\underline{P}$ final = $\underline{P}$ initial in one frame, then the same is true in every frame. This shows the consistency of the conservation law with the requirement of Lorentz invariance. Note that energy conservation alone is not consistent, because the energy in a new frame depends, by the Lorentz transformation law (eq. 14), on the *momentum* in the old frame; thus momentum nonconservation in one frame leads to energy nonconservation in the other. The point is that the energy and momentum in the new frame depend only on the energy and momentum in the old frame, *and on nothing else.* If they're both conserved in one frame, then they're both conserved in the other, even though momentum and energy get mixed together by the transformation.

PROBLEMS

1. **(a)** Use eq. 3.4.9 for the square of the length of a timelike 4-vector to show that $d\underline{x} \cdot d\underline{x} = c^2 d\tau^2$ if $d\underline{x}$ is a small displacement along the world line of a particle moving in one spatial dimension. (Hint: Note that $dx = v \, dt$.)
   (b) Use the result of part **(a)** to check the validity of eq. 8 in the form $\underline{u} \cdot \underline{u} = c^2$, using the expression $\underline{u} = d\underline{x}/d\tau$.
2. Explain carefully why the velocity 4-vector $\underline{u}$ (eq. 7) is always *forward* timelike (see §3.4).
3. Check that all four components of the momentum 4-vector $\underline{p} = (E/c, \underline{p})$ (eq. 10) have the same units.
4. **(a)** Obtain a formula for the relativistic factor γ in terms of the momentum p by comparing expression 17 for the energy with the more familiar form $mc^2\gamma$ (eq. 5.5.1).
   (b) Check that the formula for γ from part **(a)** gives correct values for γ in the nonrelativistic limit and in the limit of very large momentum.
   (c) Show that the formula is equivalent to the standard form (eq. 1.4.7).

5.8 SUMMARY OF SPECIAL RELATIVITY

This completes our discussion of special relativity. Let's summarize the key ideas:

1. *The principle of relativity,* which says that the laws of physics are truly the same in *any* uniformly moving reference frame and thereby makes it meaningless even to speak of an absolute state of rest

2. *The odd behavior of moving clocks and rulers* as observed in any one reference frame—specifically, time dilation and the Lorentz contraction—which conspires to make all frames look the same

3.   *The absolute speed limit c for the universe,* the same in every reference frame

4.   *The lack of any absolute standard of simultaneity,* which is due to the impossibility of instantaneous communication and to observer-dependent procedures for synchronizing clocks

5.   *The combining of space and time into a single four-dimensional spacetime*

6.   *The Lorentz transformation,* the analog in spacetime of rotations in Euclidean geometry, showing mathematically the relation between measurements in different reference frames

7.   *The four-vector,* a combination of timelike and spacelike quantities that shows their relationship to each other and their transformation properties. Examples are the position 4-vector $\underline{x}$ for the location of an event in spacetime, the velocity 4-vector $\underline{u}$ for the rate of change of $\underline{x}$ with respect to proper time, and the energy–momentum 4-vector $\underline{p}$ comprising both the energy and the momentum of a particle or system. Four-vectors may be timelike (forward- or backward-pointing), spacelike, or lightlike.

8.   *Invariant quantities,* often recognized as the invariant lengths of 4-vectors. Examples are the *proper time* τ for the invariant measure of time along the trajectory of a particle, the *proper distance s* for the invariant measure of a spacelike separation between events, and the *rest mass m* for the invariant mass of a particle or the invariant length of its energy–momentum 4-vector.

PROBLEMS

1. Write an expanded explanation of each of the key ideas listed in this summary for a reader who is not familiar with special relativity.
2. As an exercise, go through Chapters 4 and 5 and write out all the important equations with c set equal to 1. This is supposed to make the structure more transparent, because it lets you think of space and time on the same footing and makes clearer the symmetry between the space and time dimensions.

CHAPTER 6

Gravity and the Curvature of Spacetime

You may think the moon is going around in circles, but in fact it's doing its best to go in a straight line.

 GENERAL RELATIVITY

Although Newton questioned his own understanding of gravity, his theory was nonetheless so far in advance of his time that it stood solid for over two hundred years and still gives highly precise descriptions of gravity on the earth, the behavior of the moon and planets, the orbits of earth satellites, and so on. When Einstein developed the general theory of relativity, he wasn't trying to explain any experimental discrepancy but to probe more deeply the significance and character of familiar gravitational effects. There were in fact some tiny discrepancies, one already known and two more predicted by Einstein, which provided a direct test of his new theory but were not the prime motivation for Einstein himself. The places nowadays where general relativity actually makes a big difference are at the astrophysical and cosmological level:

stellar collapse to a black hole, for example, and the origin, shape, and fate of the universe. These all involve a fair amount of conjecture, though the evidence keeps coming in, and in all areas the confirmation of Einstein's picture keeps getting stronger.

The commitment of physicists to general relativity has rested not only on the direct evidence, I think, but also on a feeling of *rightness*. The elegant way in which it explains the fact that you can't tell gravitational effects from inertial effects (like centrifugal force and the apparent weight changes due to acceleration) and the beauty of the idea that force-free motion in a curved space can describe all the gravitational effects we observe give us a sense that this is just the sort of form that a proper theory of gravity ought to have.

Einstein's theory of gravity is deeply related to the theme of symmetry that runs through all our discussion, and in particular to the ideas, mentioned in the introduction, that every continuous symmetry must be a *local* symmetry and that the associated conserved quantity should interact in a fundamental way with an associated gauge field. The symmetries here are the displacement symmetries of spacetime, and they are local symmetries because of their invariance under arbitrary, locally contorted changes in the coordinate system. The conserved quantities are energy and momentum, and the gauge field is the gravitational field itself. There are important—and tantalizing—differences between this and other known gauge fields, so there are many unsettled questions. My discussion will follow the more traditional line developed before gravity was identified as a kind of gauge field. This traditional treatment of general relativity still carries a great deal of weight, because it is evidently valid up to the point where quantum effects have to be included, and quantum effects are exceedingly tiny at the astronomical scale where gravity is important. Besides, efforts to treat gravity by the methods of quantum field theory are still very speculative—and so far not very successful.

6.2   THE GRAVITATIONAL FIELD

Isaac Newton thought of gravity in terms of *forces*. His law of gravity (eq. 4.3.17) was that every object exerts an attractive force on every other object, according to the formula

$$F = -\frac{Gm_1 m_2}{r^2} \tag{1}$$

That is, the force is proportional to each of the two masses, m_1 and m_2, and inversely proportional to the square of the distance r between them. In the metric units that we've been using, the constant of proportionality G has the value

$$G = 6.672 \cdot 10^{-11} \text{ Nm}^2/\text{kg}^2 \tag{2}$$

This means that two 1-kg masses ($m_1 = m_2 = 1$ kg) 1 m apart ($r = 1$ m) would exert an attractive force on each other of 6.672×10^{-11} N.* The fact that G (called the *gravitational constant*) is so very small explains why, with the entire earth right under our feet, a 1-kg mass weighs only a couple of pounds. To compare this with electrical forces, for instance, suppose that we separated the positive and negative charges in that 1-kg mass and placed the positive charges (the nuclei) at the center of the earth and the negative charges (the electrons) at the surface. The electrical force between them would be around 5 billion tons (the U.S. billion, equal to 1000 millions). And if these positive and negative charges were brought to 1 m apart, the force would be about 40 trillion† times larger still. This is to be compared, then, with that minute gravitational force of 6.7×10^{-11} N.

The formula doesn't give the direction of the two forces, but that can easily be included by the use of vectors: Let $\underline{r}_1$ and $\underline{r}_2$ be the locations of the two masses m_1 and m_2 relative to some arbitrary fixed origin, and let

$$\underline{r} = \underline{r}_1 - \underline{r}_2 \tag{3}$$

represent the vector separation between them—that is, the location of m_1 relative to m_2, as shown in fig. 1. Then let

$$r = |\underline{r}| \tag{4}$$

---

*Recall (§4.3) that 1 N (one newton) is about 3.6 oz.
† This is because the electrical force is proportional to $1/r^2$, just like the gravitational force (eq. 1), and we'd be reducing the separation r by a factor of 6.4 million—the radius of the earth in meters.

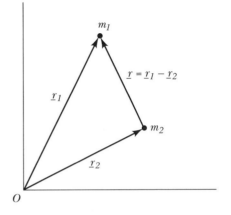

Fig. 1   The vector separation between two particles.

be the magnitude of the distance between them, and introduce a unit vector—that is, a vector of length 1 (no units)—

$$\hat{r} = \underline{r}/r \tag{5}$$

to represent the direction from m_2 to m_1. Finally, if we let $\underline{F}_{12}$ represent the vector force on m_1 that is due to m_2, then the complete statement of the force law is

$$F_{12} = -\frac{Gm_1 m_2}{r^2}\,\hat{r} \tag{6}$$

The magnitude is as given by eq. 1, and the direction is given by the unit vector factor $-\hat{r}$. The minus sign shows that the force is attractive.

This law is completely adequate for terrestrial and planetary calculations, but it isn't perfectly accurate. Apart from the corrections due to general relativity, which I'll tell you about in this chapter, there are several effects due to special relativity that come in if the objects are moving—tiny effects, to be sure, but they have to be allowed for, for example, in high-precision tracking of satellite orbits. These effects are as follows:

1.   The force on m_1, say, depends not on where m_2 is at the given time but on where it was at a certain earlier time, because the information about the mass and location of particle 2 can't travel from 2 to 1 any faster than the speed of light c. This *retardation effect* is shown in fig. 2, a spacetime diagram showing the world lines of m_1 and m_2. The force on

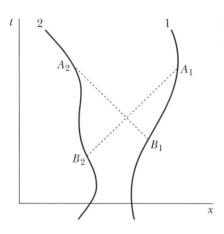

Fig. 2 Retarded forces. B_2A_1 and B_1A_2 are light paths.

m_1 at the spacetime point A_1 depends on the location of m_2 at B_2 (not at A_2); the dotted line from B_2 to A_1 represents a light path. Similarly, the force on m_2 at A_2 depends on the location of m_1 at the earlier time B_1.

2. The gravitational force depends not on the rest mass but on the relativistic mass $m\gamma$ (eq. 4.5.4)—on the energy—of each of the objects, where m is the rest mass (including any reduction due to binding energy if they are composite objects—that is, if they consist of several particles bound together by attractive forces, as explained in §5.5).

3. There are additional forces, closely analogous to magnetic forces (see §11.5), that depend on the velocities of m_1 and m_2. These forces are needed to satisfy the requirement of Lorentz invariance.

I won't go into all these fine points; I mention them for completeness and to make it clear that gravity is much more complicated than Newton's picture, even without general relativity. For these and other reasons we've come to think of gravity in terms of *gravitational fields*, which I now want to introduce.* Concentrate on a point in space a foot

---

*The idea of associating gravitational forces with a "gravitational field" is a logical outgrowth of Newton's understanding of gravity. The idea was apparently not conceived of before Einstein came up with general relativity. Thus, although the concept of a gravitational field is very useful whenever Newtonian gravity provides an adequate picture, the idea was actually overthrown before it was even thought of.

in front of your nose. You know that if an object is put there, it will experience a gravitational force due to the earth and that if you let go of the object, it will not just float there but will accelerate downward—fall—under the influence of that force. Even if there is no object at that point in space, you can say that the state of things at that point in space is such that there *would* be a force on any object placed there. The way we describe that state of affairs at that point in space is to say that there's a certain *gravitational field* at that point. We use the symbol g for the field, or $\underline{g}$ if we want to stress that it's a vector quantity with a definite direction as well as magnitude.

The force on any object placed at that point in space would be proportional to the object's mass, and this fact provides the basis for a definition of the field. We define g as the force per unit mass on such an object. The formula for the force on such an object is then

$$\underline{F} = m\underline{g}(\underline{x}) \tag{7}$$

where $\underline{x}_1 \equiv (\underline{r}, t)$ represents the location of the particle in space and time at the moment we're interested in. That is, the *force* is given by the *mass* times the *force per unit mass*. The *gravitational field* at $\underline{x}$ is $g(\underline{x})$, a vector that thus determines both the magnitude and the direction of the force $\underline{F}_1$ on a particle located at $\underline{x}$. The field g, like the force, might be due to just one other mass, as in eq. 1, or to many masses distributed through space. Note that this description of the gravitational field puts the discussion of §4.4 on a solid footing and identifies the gravitational field at the earth's surface as a special case of a very general situation.

If you now compare eqs. 6 and 7, you find that what I've done in eq. 7 is to divide up the physics represented by eq. 6 into two separate parts. We say that the force on m_1 is caused by a gravitational field g at the point A_1 where the object is, while the field at A is caused by other objects at other locations, such as m_2 at B_2. As I said, we think of this field as existing at A_1 whether or not there's an object there to be acted on by it. At first this way of separating the effect into two stages may seem artificial, but in the end it's essential. The fields, or the curvature effects that take their place in general relativity, lead an independent existence, traveling in waves and carrying energy and momentum across billions of miles of space long after the objects that produced them have gone their separate ways or even vanished. At any range, short or long, it's the dynamical characteristics of the field that determine the nature of the interaction between objects.

Going back to eq. 6 and comparing it again with eq. 7, we see that the field at $\underline{x}_1$ due to the mass m_2 has to be given by

$$\underline{g}(\underline{x}_1) = -\frac{Gm_2}{r^2}\,\hat{r} \tag{8}$$

The vector $\underline{g}$ points directly toward the spatial position $\underline{r}_2$ of the mass m_2 (provided we can neglect the retardation effect).

The discussion of §4.4, eqs. 4.4.1–4.4.3, concerning the motion of an object in a gravitational field applies quite generally. Both the field and the acceleration are vector quantities, and by exactly the same logic, the acceleration is given by

$$\underline{a} = \frac{\underline{F}}{m} = \frac{m\underline{g}}{m} = \underline{g} \tag{9}$$

That is, the acceleration vector $\underline{a}$ of a particle is exactly equal to the gravitational field vector $\underline{g}(\underline{x})$ at the location of the particle.

PROBLEMS

1. (a) Use eq. 8, together with astronomical data from Appendix C, §5, to calculate the gravitational field g_0 at the surface of the earth. (It's legitimate for this purpose to treat a spherical mass as if all its mass were located at the center.) Compare your result with the value of g_0 used in §4.4.

   (b) At what distance above the earth's surface, in miles, does g drop to half its value at the surface?

   (c) What is the field due to the earth, in units of g_0, at the location of the moon?

2. (a) Calculate the magnitude g_{sun} of the gravitational field due to the sun (i) at the location of the earth, (ii) at the location of the planet Mercury, and (iii) at the surface of the sun.

   (b) At what distance from the sun is g_{sun} equal to g_0, the earth's gravitational field at the surface of the earth? Express the result in terms of R_{earth}, the radius of the earth's orbit.

   (c) At what point between the earth and the sun, in terms of R_{earth}, is the net gravitational field equal to zero? (The radius of Mercury's orbit is about 5.8×10^{10} m. See Appendix C, §5, for other astronomical data.)

3. Find the net force due to the earth and the moon, both magnitude and direction, on a 50-kg astronaut floating at a point in space that forms an equilateral triangle with the earth and the moon. Give a careful verbal specification of the direction of the force.

4. Estimate the gravitational force between two 50,000-ton battleships, side by side.

6.3   THE PRINCIPLE OF EQUIVALENCE

As I understand it, what got Einstein thinking about gravity was the great similarity between gravitational forces and inertial effects. What I mean by *inertial effects* are the various forces you feel, or seem to feel, when you are being accelerated in different ways—speeding up, slowing down, or going around a sharp curve. Think about some examples:

1. You're in an elevator that starts to go up. As it gets going you feel heavier, as if the gravitational force had increased. It hasn't, of course, and all you are *really* feeling is the extra force of the floor on your feet that is needed to get you accelerating along with the elevator.

2. When the elevator is rising at uniform speed, you don't feel any heavier than normal; your body is neither speeding up nor slowing down, so no extra force is needed. Then, when the elevator slows down in its upward motion, for a while you feel lighter. The floor is pushing up with less than your actual weight, and the difference—the excess downward gravitational force—is just what's needed to slow down your own motion so that you'll come to rest when the elevator does.

3. The elevator cable breaks and you fall freely (in real life elevators have emergency brakes, so with luck this won't happen to you). For a little while you experience *weightlessness*. Gravity hasn't stopped pulling on you, so what's going on? On the one hand the floor is no longer pushing up on your feet, so you don't *feel* heavy, and on the other hand you don't see yourself moving relative to the elevator cage around you (because it is falling with you), so you don't seem to be falling. You aren't experiencing the normal consequences of weight, so you seem weightless. If you try to drop a penny from your hand, it will appear not to fall either, because in fact you and the cage are falling (*accelerating* downward) as fast as the penny is.

4. The astronauts in a space shuttle are in exactly the same situation. They and the spacecraft are both in free fall. They don't hit the earth's

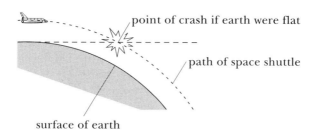

point of crash if earth were flat

path of space shuttle

surface of earth

Fig. 1 Space shuttle in free fall.

surface because the surface, being curved, falls away as fast as the space-craft drops. If the surface were flat, the shuttle *would* hit, as shown in fig. 1 (which is a modern version of a very similar picture that Newton himself drew). Note that the direction of the earth's gravitational field keeps changing in such a way that the spacecraft is continuously chang-ing its direction of fall, which is always tending toward the center of the earth.

5. The driver of the car you're in takes a sharp curve at high speed, and you are thrown against the side of the car. It is just as if the gravita-tional field were pointing for a short time toward the side. What is actu-ally happening, though, is that your body is trying to continue moving in a straight line while the car, because it is on a curved path, inter-feres. The sideways force the car exerts on you is needed to give you your sideways acceleration (here the acceleration is a change in the *di-rection* of $\underline{v}$). The same kind of thing happens on amusement park rides and in all sorts of other situations.

It's important to realize that no matter what is happening, you never feel the gravitational force itself. What you feel are the forces of contact with other objects around you, especially the ground or whatever else is supporting you, and internal stresses in your muscles and joints. It is *these* forces, even when you are simply standing on the ground, that give you the sensation of weight. You feel the ground pushing up on your feet, and you feel the pressure at your knee joints because those joints have to hold up the rest of your body, *but you don't feel the pull of gravity itself.* When the other forces are absent, as in free fall, you don't feel any force at all, even though gravity is still acting. There's nothing mysterious about this; *any* force that acted equally on all the molecules of your body would be similarly impossible to feel, because it wouldn't of itself cause any stress—any tension or pressure—on your body tissues. (If you could devise such a force, you could give people huge accelerations without any stress, making interstellar space travel much more rapid and comfortable, as science fiction enthusiasts are aware.)

In each of these examples we've seen that it feels as if the gravitational field were different than it actually is, and I've tried to explain what is "really" happening. This may remind you of special relativity, where even if you're moving, you can believe you're at rest, and there's no way to tell whether you are or not, but what's "really" happening is that your motion causes your measuring implements to behave in just such a way as to deceive you. Now as you can tell, *I've been cheating in my use of language.* I've been talking in this and the preceding paragraphs as if there were a single, true rest frame, and when I've used words like *actually* and *really,* I've been pretending that you can say absolutely whether you're moving or not and whether you're accelerating or not. By now you should be used to the fact that words like these are inappropriate for uniformly moving observers, though they may be convenient in giving you a point of reference. What's new is that they seem to be inappropriate for accelerated frames also. Einstein concluded that it doesn't in fact mean anything to say that one reference frame is accelerating or rotating and another is not. He put this in the form of what he calls the principle of equivalence.

> The principle of equivalence: *Inertial effects—the effects of acceleration—are completely indistinguishable from gravitational effects.* *

This is a major jump beyond the simpler kinds of equivalence that we discussed in the first three chapters, and it requires a further substantial overhaul of the laws of physics and our concept of the nature of space and time. Einstein had a feeling for how the world ought to behave, and it has turned out that, as far as we know, he was right—or at least as right as he could be without considering quantum effects (difficult, as I've said).

Another way of stating the principle of equivalence, a way that better reflects its name, is to say that *all* reference frames, including accelerated

---

*The converse is true locally, in small regions of space, but is *not* true in general. It's easy to think of a gravitational field arrangement—the nonuniform field that surrounds the earth is a good example—whose effects over large distances cannot be mimicked by accelerations alone.

reference frames, are equivalent, that the laws of physics take the same form in *any* reference frame. What appears as an inertial effect in one frame is seen as a gravitational effect in another. In a freely falling reference frame—the space shuttle or the freely falling elevator—it becomes correct, at least over small distances, to say that there is no gravitational field at all. And it is also correct to say that the Copernican view (with the sun at the center) and the Ptolemaic view (with the earth at the center) are equally valid and equally consistent! In the Ptolemaic view you would have to introduce strange and wonderful gravitational fields in order to account for the complicated motion of the planets and the immensely rapid motion of the stars, but these fields would still be completely consistent with the laws of physics. A new and marvelous set of laws— Einstein's general relativity—has had to be devised in order for this to work, and experiment has confirmed the new laws in more and more detail with the passing years.

PROBLEMS

1. Using the same logic as that of example 2 on page 180, describe a downward trip on the elevator as it starts, continues, and comes to rest again.
2. The apparent weight of a person in an elevator is equal to the magnitude of the force F_f that the floor exerts on her feet, while the *net* force F, which is what produces the acceleration $a = F/m$, is the difference between her weight mg (eq. 4.4.1) and the force of the floor. Taking the downward direction as positive, you have

    $$F = mg_0 - F_f$$

    Calculate the acceleration of the elevator in terms of g_0, the acceleration of gravity, if her apparent weight is **(a)** 5% greater than her actual weight, **(b)** half her actual weight, and **(c)** twice her actual weight. Can you in each case deduce the direction of her motion?
3. Give three additional examples of inertial forces. In each case explain what's happening as an example of $\underline{F} = m\underline{a}$, and identify the direction of an apparent gravitational field.
4. Describe how it would feel to be in a nonuniform gravitational field that exerts different forces on different parts of your body (such forces are often referred to as tidal forces). To be specific, you might suppose that your body is in an orbit with a radius of a few

meters around a "mini-black-hole"—just think of it as a very tiny, very massive planet—and consider different possible orientations of your body. What's your best strategy for avoiding discomfort? (Estimate the mass of the mini-black-hole for the gravitational field strength to be of the order g_0 at a distance of 2 m.)

6.4   THE CURVATURE OF SPACETIME

A key consequence of equivalence

The principle of equivalence (§6.3) says roughly that the apparent forces due to accelerated motion of any kind can, if you prefer, be accurately understood instead as *real* forces due to appropriate gravitational fields. That is, there's no way to tell whether your motion is accelerated or not. This principle has unforeseen consequences that are analogous to the unforeseen consequences of the principle of relativity. The effect I want to focus on here is the remarkable fact that spacetime must appear *curved*. I must explain what this means, of course, and I must also explain that whether spacetime is really curved or not depends (typically!) on how you look at it.

What does "curvature" mean here? You've seen curved lines and curved surfaces, and you can tell they're curved because you're outside and can look at them. If you could stand outside our three-dimensional space, you might similarly be able to tell whether it's curved—but you can't, of course. Can you tell from inside, so to speak? What you can see from inside is that the rules of Euclidean geometry break down, just as they do on a curved two-dimensional surface. Thus a circle of radius r won't appear to have circumference $2\pi r$, and you won't be able to construct a square with equal straight sides and four right angles. I'll describe these odd effects in more detail later on.

As usual there are different ways of looking at the situation. One way is to suppose that space is really curved, and another is to take space as "flat" (Euclidean geometry applies) and conclude that gravitational fields have a distorting effect on your rulers that just makes it *appear* that Euclidean geometry no longer works. And as usual both views are totally consistent, but one works much better in practice. If spacetime is curved (the curvature effects extend into time as well as space), then it turns out that the equivalence principle falls out as a natural consequence, while if you take spacetime as flat, then you have to take the principle of equivalence as a complete coincidence. You have to ask why it is that the distortions of rulers and clocks due to gravitational fields have the effect of exactly concealing the difference between gravitational effects and inertial effects due to accelerated motion.

From a pragmatic point of view you might as well say that spacetime *is* curved, because there's *no experimental way to tell that it's not.* There are no stores where you can buy undistortable rulers and clocks by which you could determine that the geometry really is Euclidean (or, rather, Minkowskian)—that is, flat in the sense of special relativity. Einstein found that he could give a complete account of gravitational effects in terms of this curvature, eliminating entirely the idea of a separate gravitational field (see the footnote on page 177):

> *There are no gravitational forces. Gravitational effects are entirely due to the curvature of spacetime.*

As I mentioned, you may insist that the space is really flat and that it's the measurements that are wrong, but it doesn't get you far to do so. For one thing, it's doubtful whether the idea of a "true" distance has any meaning when there is no way to measure it. For another thing, the curvature can also show itself in other ways, known as topological, the simplest being that space may be finite and closed on itself like the surface of a sphere. That would mean that you could travel only a finite distance before you'd come back to where you started from and that no two points would be more than some maximum distance apart. Such an effect could not easily be blamed on the misbehavior of your rulers.

The metric

What does it mean to say that a space is *curved*? It will be easier for now to think of a surface and to imagine that we are two-dimensional creatures living *within* this two-dimensional world. How do you tell mathematically whether the surface is curved or flat? And if it is curved, how do you describe the curvature? Is it rounded like a ball or curved in opposite directions like a saddle or the bell of a tuba? It's a matter of making measurements to see whether the rules of Euclidean geometry—or some other rules—apply, and our basic tool in this inquiry is what's called the metric.

The first thing you have to do is describe distances. You can't talk about a straight line between two points on a curved surface, but you can measure the distance along an arbitrary path and express that distance somehow in terms of whatever coordinate system you're using. The rule for doing this is called the *metric* of the surface. The math is beyond the scope of this book, but the idea is pretty simple. You start with an infinitesimal displacement from a point P to a neighboring point P', corresponding to an increment (dx, dy) in each of the coordinates, and then you express the distance ds from P to P' in terms of dx and dy. Note that x and y can't be ordinary cartesian coordinates because you can't fit that kind of coordinate system onto a curved space, but it's always possible to lay *some* coordinate system on a surface—like the lines of latitude and longitude on the spherical earth. The rule for the distance ds in terms of the increments dx and dy involves something called the *metric tensor* $\underline{\underline{g}}$ at the point P, which consists of a set of related functions of the coordinates x and y. Once you know the metric tensor, all the properties of the surface, such as the curvature, can be worked out according to standard formulas. The curvature is described by another tensor $\underline{\underline{R}}$, which is again a set of related functions of the coordinates x and y. These functions are all determined by the functions that make up the metric tensor, and they give a complete description of how the surface is curved in the neighborhood of P. If the surface is flat, then the curvature tensor $\underline{\underline{R}}$ is zero everywhere, regardless of what coordinate system you're using.

One simple way to determine whether the surface is flat, as I mentioned at the beginning of this section, is to draw a circle of radius r and see whether its circumference is equal to $2\pi r$. If the surface is a sphere or is rounded like a sphere, then the circumference is less than $2\pi r$, and if it's saddle-shaped, then the circumference is greater than $2\pi r$. Note how important the metric is for this test. To draw a circle you have to find all the points a distance r from the given point P (that's the *shortest* distance between P and the point on the circle), and then you have to measure around the circle to get the circumference (this is the integral, $\int ds$, the sum of all the little distances ds as you go around the circle in tiny steps). The process I've described is an experimental procedure involving rulers, and it corresponds to a mathematical description using the metric $\underline{\underline{g}}$ to get the little distances ds. Why do you need the mathematical description? Because you might be trying to understand *why* the space is shaped the way it is and developing a mathematical theory to predict it. The result of the theory would be a specification of the metric $\underline{\underline{g}}$ over the space, and you'd test the theory by comparing its prediction of the curvature with your measured result.

This discussion applies equally well to a three-dimensional world and even to a four-dimensional spacetime. With a three-dimensional space there's no way for you to stand outside and look at it to see if it's curved, but you can perform internal measurements to see what the geometry is like: Is the surface area of a sphere greater or less than given by the Euclidean formula $4\pi r^2$? What are the corner angles in a cube? And so on. In spacetime, the idea of length is the same as I told you about in §3.4: You can now have both spacelike distances and timelike distances, but you can still detect the curvature by doing geometrical measurements of a similar type, using clocks now as well as rulers.

6.5 THE NEW LAWS OF SPACETIME

You now know something of what it means for spacetime to be curved. What's needed now is a whole new set of rules: rules for the behavior of a particle in a curved spacetime and rules to determine how the curvature itself evolves with time. It's these rules that constitute Einstein's general theory of relativity.

Motion of a particle—geodesics

As to the first question—How does an object move?—there's a very elegant answer that does away entirely with the idea of a gravitational *force:* The world line of a particle follows what's called a *geodesic,* which is simply the nearest thing to a straight line that you can have in a curved space. Think of the case of a curved surface. If you look at a small enough bit of the surface it appears flat, and a geodesic is a line on the surface that looks like a straight line in any such small region. It's the path that a tiny bug must follow if it wants to take the shortest route across the surface to where it's going, without veering to the left or right. Even though the path is curved when you stand back and look at the whole surface, it's as straight as you can get in every local region. The geodesics on the surface of a sphere of radius *r,* for example, are the *great circles*—circles of radius *r* that divide the surface into two equal hemispheres, such as the equator and the lines of longitude. Think of a region near the north pole on a globe, for example; the bug would clearly have to veer to the left if she walked eastward on one of the small circular latitude lines, but walking down a longitude line toward the equator would not involve veering in either direction. (Note that the great circle route is the shortest distance between two points only on the local scale. You might follow the equator eastward to a destination three quarters of the way around the globe, and though you followed a geodesic, it would obviously have been shorter to go west instead of east.)

The story is much the same for spacetime. In the neighborhood of any event point, spacetime is approximately flat, in the sense that you can set up a local reference frame that looks just like a reference frame for special relativity if the neighborhood is small enough, and with respect to which a particle moves in a straight line at constant speed. (We have to assume there are no *other* forces—electric or magnetic, for example—on the particle.) If you think that a freely falling object doesn't move in a straight line, it's because you're using the wrong kind of reference frame! The local frame of reference that makes spacetime look flat in this sense, and in which objects appear to move in straight lines, is the freely falling frame that I spoke of in §6.3, in which there seem to be no gravitational forces. Of course the paths don't look like straight lines to a person who's not falling, because from his point of view there's a gravitational field that makes things accelerate and follow curved paths. This rule, that in the local freely falling reference frame an object moves in a straight line at uniform speed, is just the rule that particles follow geodesics in spacetime. When you look at the particle's path over a larger region, in which the curvature shows up (corresponding to nonuniform gravitational fields), then this geodesic world line no longer looks straight, but even so, as with the bug on the surface, it's as straight as you can get in the curved spacetime. I don't know any simple way to visualize geodesic paths in curved spacetime, but the mathematics is straightforward and works very well.

One curious difference between ordinary space and spacetime is that a geodesic world line in spacetime is the *greatest* distance between two points rather than the least. This fact generalizes the twin paradox described in §2.6, where the twin who ages the most is always the one who stays at home—that is, who doesn't undergo any accelerations. A curved world line—a nongeodesic—is *shorter* than a straight one, in the sense of proper time, because of a generalized version of the time dilation factor.

It's useful here to try to visualize what these spacetime geodesics might look like in terms of the real world. The earth going around the sun (see fig. 2.2.6) is following a geodesic—following its nose across spacetime in as near to a straight line as possible, according to the picture I've tried to draw for you. What makes this reasonable is the idea that the space around the sun is curved, like a big dimple with the sun at the center, and that the natural path around a dimple is curved inward by the shape of the dimple itself. It's as if you drew a straight line on a piece of paper and then rolled the paper into a cone; the straight line becomes a curve that follows around the cone. The picture is more complicated than this, of course, because time is involved in the curvature as well as space, but all I'm trying to convey is the reasonableness of the picture. In a related example, you might throw a ball straight up and watch it rise and then return to the ground. In a conventional spacetime diagram, like fig. 1(a), the path doesn't follow a straight line, returning

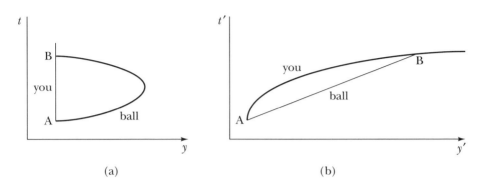

Fig. 1 Spacetime diagram for a ball thrown vertically upward. (a) Reference frame fixed on the ground. (b) Freely falling reference frame (its initial velocity is zero, relative to the earth's surface).

on itself, but follows a curve extending in the forward time direction, because the downward part of the motion occurs later than the upward part. The funny thing is that the curved path is *straighter,* relative to the curved spacetime, than the straight line between the same two points in the conventional diagram. Like the moon, the ball is following its natural path around the dent in spacetime that surrounds the earth. Curiously enough, the ball ages more than you do, like the twin who stayed at home in the twin paradox, because the ball is following its free path through spacetime while you are held to a nongeodesic path—that is, kept from falling—by the external force of the ground on your feet. If you look at the same sequence in a freely falling reference frame, as in fig. 1(b), you appear to be following a curved path and the ball appears to be following a straight one. You are in fact accelerating upward relative to the freely falling frame.

The shape of spacetime—the Einstein equations

It's time to talk about what determines the shape of our curved spacetime. There are lots of possible shapes, but which is the right one? Physics doesn't provide a straightforward answer; it gives you, as we've seen in other contexts, the equations of motion, which means in this context that physics tells you how the universe will evolve *given* its initial configuration, but it doesn't tell you what that initial configuration is. It does this for general relativity by means of the *Einstein field equations,* which can be thought of as the gravitational analog of Maxwell's equations for the electromagnetic field, mentioned in §1.1.

The mathematical machinery, known as Riemannian geometry, for the description of curved spaces was already around, and Einstein had to learn it and adapt it to his purposes. Because no coordinate system is more natural or appropriate than any other, you have to use an arbitrary nonrectangular coordinate system for describing events in space and time. This generalizes the coordinates x, y, z, and t that we used in special relativity and all the older work. You have to describe the curvature in terms of this coordinate system and then look for a basic law for the curvature that's independent of the choice of coordinate system. This is such a tight constraint that Einstein had very little choice in what the law should be, and he came up with what we now call the Einstein field equations. We won't be able to do anything with them here, because they're complicated nonlinear partial differential equations that are way beyond the level of this book. However, these famous equations can be written schematically as a relation between two tensors:

$$\bar{\bar{R}} = -8\pi\kappa\underline{\underline{T}} \tag{1}$$

The symbol $\bar{\bar{R}}$ represents what's called a reduced curvature tensor, which is derived from the full curvature tensor $\underline{R}$ that I told you about in §6.4. Remember that $\bar{\bar{R}}$, like $\underline{R}$, is a collection of a number of different quantities that are related to each other somewhat as the different components of a vector are related. The symbol $\underline{\underline{T}}$ represents the *energy–momentum tensor*, whose components give the distribution of nongravitational energy and momentum present in the universe. This includes all ordinary matter (remember that mass is equivalent to energy) and electromagnetic fields but doesn't include the energy and momentum of the gravitational field itself. There are in fact terms concealed within the left side of eq. 1 that correspond to the energy and momentum of the gravitational field; these could have been included in $\underline{\underline{T}}$, but only at the expense of spoiling the elegance of the equations and undermining some of their special symmetry properties.

The factor κ in eq. 1 is an important factor of proportionality—important because it provides the link between the purely geometrical quantities on the left side of the equation and the components of $\underline{\underline{T}}$ on the right, which have to do with energy and momentum. This fundamental link between geometry and energy is amazing to me—it lies at the heart of Einstein's vision.

The constant κ is essentially the same as the gravitational constant G introduced by Newton (see §6.2):

$$\kappa = \frac{G}{c^3} = 2.4765 \times 10^{-36} \text{ s/kg} \qquad (2)$$

which suggests a specific link between mass and time and hence, if you throw in a factor c, between mass and length:

$$\kappa c \equiv 7.4243 \times 10^{-28} \text{ m/kg} \qquad (3)$$

This means that there's an intrinsic equivalence between mass and length, analogous to the equivalence between length and time that we found in special relativity. Just as 1 s is equivalent to 186,000 mi (the distance light travels in 1 s), so a mass of 1 kg is equivalent to about 2.5 $\times 10^{-36}$ s or, multiplying by c, to about 7.4 $\times 10^{-28}$ m. You find, then, that the mass of the sun is equivalent to a length of 1.5 km, just half of what's called the Schwarzschild* radius of the sun, discussed in §7.1 (see eq. 7.1.1).

The field equations 1, which are very deep, are used to determine the correct curvature tensor $\underline{\underline{R}}$. They don't determine it completely, however, for two reasons: (1) The initial state of the universe is still completely arbitrary and (2) the equations must allow for an arbitrary choice of coordinate system. The field equations are just restrictive enough to determine the actual physical shape of the universe in the future, provided its initial shape is known, but they impose hardly any restriction on the initial shape. If you try to solve the field equations for some given initial state (an impossible calculation in practice except in some very special cases), you find an infinite number of solutions, all corresponding to the same physical shape but with all the infinitude of different possible coordinate systems.

I have very little feeling myself for the physical meaning of the reduced curvature tensor $\underline{\underline{R}}$ that appears on the left side of eq. 1, beyond the fact that it's equal to zero in any matter-free region. As I understand it, the equations were arrived at primarily from the principle of equivalence (§6.3), which leads to the requirement that the equations take exactly the same *form* in any arbitrary coordinate system. That is, the symmetry requirement alone narrowed the choice almost completely and led to field equations of just the right character and of remarkable power and accuracy.

---

*Pronounced "SHVARTZshilt."

Predictions and confirmation

The next step is to apply all this math to familiar cases and see if you get the familiar answers. That is, do the Einstein equations give the right behavior for the moon and the planets—not to mention all the commonplace objects on earth that experience the force of gravity, such as baseballs and human bodies? The case of a planet around the sun is general enough to test the theory, because the path of a thrown baseball, for example, is just like a planetary orbit on a much smaller scale. So you ask the Einstein equations the following questions: What is the shape of spacetime outside a stationary, spherically symmetrical, massive object (the sun)? And how does a particle (one of the planets) move around the object, given that shape of spacetime? For the shape of spacetime around a massive object, the Einstein field equations give a definite answer. This is the *Schwarzschild solution,* which is quite reasonable at large distances from the object but does peculiar things very close to the center if the object is sufficiently small. The sun isn't that small—it would have to have all its mass concentrated within a 3-km radius—so this peculiarity, which is called a *black hole,* doesn't show up in this case. I'll come back to black holes in §7.1.

You now have to ask how a planet moves given this curvature, using the geodesic paths we talked about. The answer is almost exactly the elliptical orbits that we're familiar with from Kepler's laws, Newton's mechanics, and, of course, astronomical observations. Thus Einstein's theory, though it is radically different from Newton's in its picture of the nature of space and time, predicts almost no observable difference in the behavior of the planets and other satellites or in any of the everyday effects of gravity. This is a point in favor of the new theory, because the classical Newtonian description agrees with observation to a very high degree of precision.

According to Einstein, the elliptical orbit of each planet should be shifted by a tiny amount each time it goes around the sun, as illustrated in fig. 2. For just one of the planets, Mercury, such a shift, called the *precession of the perihelion,* had actually been observed and had caused some anxiety among astronomers. Einstein's prediction for the shift, about 43 seconds of arc per century, corresponded nicely to the observed value; it represents one of the important successes of the new theory. Thanks to more precise modern observations, the corresponding precession of the earth's orbit has now also been observed; the shift is larger but is much harder to see because the earth's orbit is more nearly circular. Needless to say, gravitational effects on earth—baseballs, weight watching, and so on—are not detectably different from the Newtonian theory, even though the conceptual picture is so different.

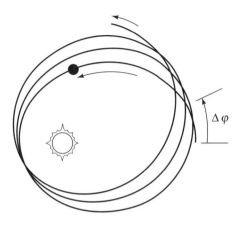

Fig. 2 The precession of the orbit of the planet Mercury, much exaggerated. The shift of each successive orbit, $\Delta\varphi$, is actually an angle of about 0.14 second, or 1/26,000 of a degree.

Two other effects need to be mentioned: Einstein predicted that light as well as material objects would follow a curved path in a gravitational field, as illustrated in fig. 3. The idea was not new—Newton himself, believing that light consisted of particles, questioned whether light, like other particles, might be deflected by gravity. In fact a German astronomer named Johann Georg von Soldner, in 1801, made a prediction for the deflection of light by the sun, based on Newtonian particle mechanics, that was surprisingly close to Einstein's result. What is new is

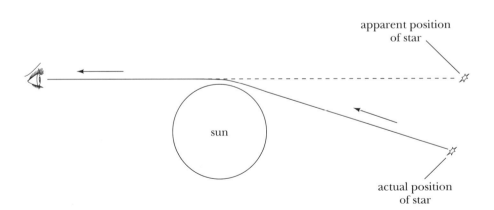

Fig. 3 Bending of light by a gravitational field.

that the light path, like the path of a planet, is a geodesic in the curved spacetime around the sun: The gravitational field has been supplanted by curvature as the cause of the observed deflection.

To observe the deflection (of the order of 1 second, 1/3600 of a degree) people had to wait for a total eclipse of the sun, because only then are stars close to the sun visible. (The moon plays a passive role; it serves to block out the sun but has much too little mass to cause any observable deflection of the light.) Einstein's first attempt to calculate this effect, which was based on special relativity, was off by a factor of 2. There were several attempts to observe the effect before Einstein managed to correct his result, but—fortunately perhaps—such efforts were not successful until 1919, after he had worked out the ideas of general relativity (1915) and corrected his error. The observed effect, though not very precise, agreed nicely with Einstein's corrected prediction, and almost instantaneously Einstein became world-famous. Our understanding of gravity—seemingly built on solid rock since the time of Newton— had been radically altered, and the man who did it, who by pure thought had penetrated so deeply into the mysteries of the universe, was cast by the world (with the help of the newspapers) in the role of hero and saint.

Einstein did his calculation treating light as an electromagnetic wave—a difficult job. It's interesting to note that you get exactly the same result by considering the path of a photon, viewed as a massless particle following a suitable geodesic. This is a much easier calculation, and Einstein certainly knew about photons, having pioneered in their discovery, but he consistently refused to make *use* of quantum theory because he regarded it all his days as an unfinished theory. More on that in Part II.

Einstein's theory also predicts that light emitted in a star's gravitational field, close to the star's surface, will be shifted to longer wavelengths as it moves out of that gravitational field. This effect, which has also been confirmed experimentally, is called the *gravitational red shift*. It is a direct consequence of the more basic result that there is a time dilation associated with the gravitational field: Both atoms and clocks perform their functions more slowly at the star's surface, so the frequency of the light emitted by the atom looks normal to the observer who is right there, but it looks slow to the observer outside, whose clock is not running so slow.

You can understand the gravitational red shift also in terms of photons: Each photon loses energy as it leaves the star (just as a ball slows down when you throw it straight up), and since for a photon the energy is related to the frequency by the quantum relation $E = hf$ (h = Planck's constant, the basic constant of quantum theory), this means that the frequency decreases, and hence the wavelength increases ($\lambda = c/f$). Einstein consistently refused to base any conclusions on quantum theory and photons, as I have mentioned, but his prediction is completely consistent with the photon picture.

PROBLEMS

1. If you start a journey at longitude zero and south latitude 45°, traveling due west initially but following a geodesic, do you eventually cross the equator and, if so, where? If you follow the 45° south latitude line as you travel west, does your route veer to the right from a geodesic, to the left, or neither?

2. (a) Describe the geodesics on a simple circular cylinder with flat ends, recognizing that a geodesic is a straight line on either of the ends or on the unwrapped cylindrical surface and that it goes over the edge from the cylindrical surface to one of the ends by making equal angles with the edge on both sides of it. Show that two geodesics need not intersect each other and that a single geodesic may intersect itself any number of times.

   (b) Find on a cube a closed geodesic that visits all six faces.

CHAPTER 7

The Shape of the Universe

7.1   BLACK HOLES

I want to return now to the issue discussed in §6.5, the shape of space-time in the neighborhood of a massive object. The appropriate solution of Einstein's field equation, the Schwarzschild solution (see §6.5), shows that if a star is dense enough—if it is sufficiently small given its mass—then some kind of singular behavior should occur at a certain radius, the *Schwarzschild radius*. Can stars become that dense? It seems very likely. You need a star a lot bigger than the sun, and as it uses up its fuel* it cools off, the pressure drops, and it starts to collapse. The collapse tends

---

*Hydrogen "burns" by fusion to give helium, and then helium burns to give heavier and more stable nuclei. See the more detailed description of the process in §5.6.

to heat it up again, and it may go through some big explosions (appearing as a supernova) and lose a lot of its matter in the process. In the end, if it was sufficiently big to start with, there will still be enough matter left to make it keep on collapsing under the force of gravity. The smaller the radius r gets, the bigger the gravitational forces become, because the forces are proportional to $1/r^2$ and quadruple every time the radius is halved. If the gravitational forces get stronger than the electrical forces that keep atoms apart in ordinary matter, then the star collapses further and becomes a dwarf star, and if it's sufficiently massive, it may go right on collapsing until it consists only of nuclear matter—neutrons compressed to a density like that of atomic nuclei. The sun, if compressed to this density, would be a sphere of radius 20 km (12 mi), and its density would be around a billion tons per cubic centimeter. Because neutrons are extremely hard to compress, a star may stop collapsing at this point, even though the gravitational forces are now monstrous, and may remain as a *neutron star*, rotating very rapidly—sometimes as fast as 1000 revolutions *per second* (this has actually been seen, in the form of pulsars). If the mass is large enough, the compressional forces can still win out; the star has lost a lot of mass, but if over 3 solar masses are left, the gravitational forces are great enough to squash even the neutrons, and the star keeps on collapsing. There is no longer any force strong enough to stop it. As its radius* approaches the Schwarzschild radius, given by

$$r_{\text{SCH}} = \frac{2MG}{c^2} \tag{1}$$

things start to get very strange. The gravitational field g gets so strong that even light becomes trapped, unable to leave. Matter—or light—that falls in never leaves again and cannot be seen from outside. It is this kind of totally collapsed star that is called a *black hole*. The Schwarzschild radius for the mass of the sun is about 3 km. This means that if the mass of the sun were compressed into that small a radius, the sun would become a black hole.

The distortion of time is so great in the neighborhood of a black hole [remember how clocks slow down in a strong gravitational field (§6.5)]

---

*The concept of a radius gets confusing as a result of the severe distortions of spatial measurements, but there's a technical definition that enables us to speak accurately about it.

that if a person were to fall into it, it would seem to her to take only a finite time to get inside the Schwarzschild radius, although the fall would appear to an outside observer to take literally forever, an infinite length of time. She essentially disappears from sight in a modest length of time, to be sure, but she never quite reaches the Schwarzschild radius according to earth time. From the point of view of the space explorer falling in, nothing special happens as she passes the Schwarzschild radius, or "event horizon," but a very short time after that she reaches a point where the gravitational forces pull so unequally on different parts of the spaceship (tidal forces—see prob. 6.3.4) that it is torn apart. All this takes place infinitely far in the future from the earth's perspective.

In fact, the star's collapse itself takes infinitely long from the outsider's point of view, just like the fall of the space traveler, so that we on earth can never see anything but an incomplete or imperfect black hole as a result of stellar collapse. It takes only a short time, though, for it to get so close to the real thing as to make no observable difference.

PROBLEMS

1. **(a)** Estimate the size of a 1-kg cube of nuclear matter. Where could you keep such an object?
   (b) What would be the radius of the earth and the moon if they were compressed to this density? (The radius of the sun at this density is given in the text. Other astronomical data are given in Appendix C, §5. The volume of a sphere of radius r is $\frac{4}{3}\pi r^3$.)
   (c) Find the Schwarzschild radius for each of the masses in parts **(a)** and **(b)**.
2. **(a)** Using the conventional Newtonian picture (eq. 6.2.8), show that the acceleration of gravity g (that is, the gravitational field) at a distance of the Schwarzschild radius (eq. 1) from a point mass M is given by $g_{SCH} = c^2/2r_{SCH}$.
   (b) Show that this is just the acceleration needed to reach speed c in a distance r_{SCH}, according to nonrelativistic mechanics. You may use information about nonrelativistic motion from Appendix B, §1.
   (c) Evaluate g_{SCH} for the earth's mass and for the sun's mass.
   (d) Find the mass, and the value of r_{SCH}, for which $g_{SCH} = g_0 = 9.8$ m/s^2, the standard value at the surface of the earth.

 THE BIG BANG

If space and time are curved, as we now believe, then in addition to asking about the curvature near a star—the local dimples in spacetime— you can also ask about the global shape of things, which we'll consider in §7.3. When you look at a region containing millions of galaxies and ignore the local dimples, what overall shape do you see? Is the overall curvature positive, like the surface of a sphere (where the circumference of a circle of radius r is always less than $2\pi r$), or negative, like a saddle or a mountain pass (where the circumference of a circle is always greater than $2\pi r$)? Does the universe close on itself like the surface of a sphere, which has a finite area but no edge? And how do its size and shape change with time? In this section I'll talk about the fact that the universe we see is expanding from an initial very tiny size, and in the next section we'll look at the shape of the universe, both in space and in time.

Hubble's law

One of the more striking things we see as we look beyond our own galaxy is that the universe seems to be expanding. More precisely, other galaxies are all moving away from ours, with speeds proportional to their distance away:

$$v = H_0 d \tag{1}$$

where v is the speed and d is the distance from the earth. Edwin Hubble observed this in 1929 and made the first estimate of the proportionality constant H_0, known as the *Hubble constant*. It is now believed to have roughly the value

$$H_0 = 80 \text{ km/s/Mpc} \tag{2}$$

or 80 kilometers per second per megaparsec, where the *megaparsec* is a standard astronomical unit of distance. It's equal to a million *parsecs,* and a parsec is 3.26 lt-yr, or 1.9×10^{13} mi. The parsec (parallax plus second) is defined to be equal to a distance from the sun such that the radius of the earth's orbit would subtend an angle of 1 second of arc, or 1/3600 of a degree, as illustrated in fig. 1.

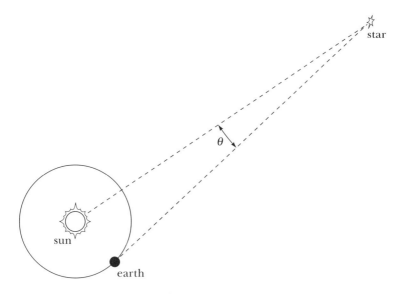

Fig. 1   The definition of a parsec. The star is 1 parsec away from the sun when the angle θ is just 1 second of arc.

The *Local Group* of galaxies (two dozen or so) extend over a region about 1 Mpc in diameter. Hubble's law says that a galaxy 1 Mpc away is moving away from us at 80 km/s, a galaxy twice that far away is moving twice as fast, and so on. The measurements that give us Hubble's constant are not at all precise, and it's felt that H_0 could lie anywhere between 50 and 100 in the units given.

Now, before you can convince yourself of Hubble's law, you have to be able to tell how far away a galaxy is, and that isn't easy. If it's nearby, you can tell its distance by its *parallax*—the shift in its apparent location (relative to more distant objects) as the earth goes around in its orbit, which is the basis for the definition of the parsec. For an object 1 parsec away, for example, the parallax corresponding to the diameter of the earth's orbit is just 2 seconds (see fig. 1). For more distant galaxies the parallax is no longer observable, and you have to rely on their apparent brightness, which decreases the farther away they are. If you assume that all galaxies of a given type are equally bright, then you can estimate how far away a galaxy is by comparing its brightness with nearby galaxies whose distance you know. (I really should say, "how far away it *was* when it emitted the light we now see.")

You also have to be able to tell how *fast* a galaxy is moving, because Hubble's law is a relationship between distance and velocity. For this you look at the "red shift of spectral lines." *Spectral line* is a term we use for one of the definite wavelengths of light emitted by any given kind of atom or molecule (see the more detailed discussion of spectral lines in §12.1), and the *red shift* is the shift of such a line to a longer wavelength (toward the red end of the spectrum) because the atom emitting it is moving away from the observer.* The patterns of spectral lines for the different elements are well and precisely known, so it's possible to identify them even when they are shifted.

The red shift is a particular example of what's called a *Doppler shift*, a decrease in the apparent frequency of any traveling wave—light, sound, or what have you—if the source is moving away, or an increase in apparent frequency if the source is moving toward you. This is easy to understand if you think of the waves as a series of signals sent out at equal intervals by the source and traveling to the observer. The signals could be wave crests in the case of waves on water, pressure maxima in the case of sound waves, or maxima of the electric field strength in the case of electromagnetic waves, including light. If the source is moving away from you, then each successive signal has farther to travel, and takes longer to reach you, than the one before, so the time between successive arrivals is greater than the time intervals at which the signals are sent out by the source. In other words, the frequency of arrival of the signals is less than the frequency with which they are emitted.

The Doppler shift

The formula is easy to derive. If T_0 is the period at the source—that is, the time between successive signals—and v is the speed with which the source is moving away from the observer, then each signal has to travel a distance $v T_0$ greater than the signal before. If the speed of the signals (the waves) is v_0, then the extra time it takes each signal to reach the observer, compared to the previous signal, is distance/velocity = $v T_0 / v_0$, and you have

$$T = T_0 + \frac{v T_0}{v_0} = \left(1 + \frac{v}{v_0}\right) T_0 \tag{3}$$

---

*See §1.3 for a description of waves and wavelengths.

where T is the period observed—that is, the time between the *arrivals* of successive signals. The frequency f of a wave (see §1.3) is just the reciprocal of the period

$$f = \frac{1}{T} \tag{4}$$

so the Doppler-shifted frequency is given by

$$f = \frac{f_0}{1 + v/v_0} \tag{5}$$

For sound waves, the frequency of the waves is directly related to the pitch; the pitch is high for high frequency and low for low frequency. Formula (5), then, with v_0 equal to the speed of sound, gives an accurate description of the drop in pitch of a fire engine siren that you hear as the engine passes you, first approaching and then departing (see probs. 1 and 2).

Eq. 3 is all there is to it if the speed of the source is nonrelativistic—much less than c, the speed of light—but if v is comparable to c, then there is the additional feature of time dilation. We should relate T in that case not simply to T_0, the period of the source as measured by the stationary observer S, but rather to T_0', the period as measured by S', the observer moving with the source. This requires an extra factor γ, because

$$T_0 = \gamma T_0' \tag{6}$$

If we also put in $v_0 = c$ because the waves we are interested in are light waves, we get

$$T = \gamma\left(1 + \frac{v}{c}\right)T_0' = T_0' \sqrt{\frac{c + v}{c - v}} \tag{7}$$

What we're interested in is the shift in wavelength, and we can get that directly because the wavelength is just the distance light travels in one period:

$$\lambda = cT \tag{8}$$

If λ_0 is the wavelength the light would have if the light source were not moving, then by the principle of relativity, it is also the wavelength as observed by S'. That is,

$$\lambda_0 = cT_0' \tag{9}$$

and we get the final result

$$\lambda = \lambda_0 \sqrt{\frac{c+v}{c-v}} \tag{10}$$

which is the shift in wavelength due to the motion of the source.

The Bang

The effect observed by Hubble is just what you would see if there had been an explosion at some definite time in the past, and the stars and galaxies were thrown out with various speeds from the point of the explosion. The faster ones would have traveled farther than the slower ones in just this way. In fact, Hubble's law (eq. 1) can also be written

$$d = vT_{\mathrm{H}} \tag{11}$$

where

$$T_{\mathrm{H}} = \frac{1}{H_0} \tag{12}$$

which says that the distance d is just how far a galaxy would travel in the time T_{H} if it were traveling at speed v. You can see this on the spacetime diagram of fig. 2, where the inclination of each galactic world line gives its velocity (fig. 2.2.4). In this discussion, and in fig. 2, I'm neglecting the time it takes for the light from a galaxy to reach the earth. For nearby stars this time is negligibly small, and for really distant sources, the whole picture is seriously altered because of the curved shape of the early universe. In any case, the distance measurements we're able to make are too crude for the time delay to be a factor.

If we continue the world lines of fig. 2 backward in time, they meet at a point, as shown in fig. 3. The time T_{H}, which I'll call the *Hubble time*, is thus crudely equal to the age of the universe. It turns out to be about 12 billion years when we use the value of H_0 from eq. 2, but it could be

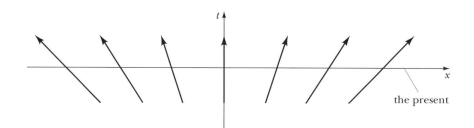

Fig. 2 Galaxies obeying Hubble's law.

anywhere from 10 to 20 billion years. This is the simplest version of the argument for the Big Bang—an initial state for the universe of extremely high temperature and density, with all the matter and energy in the entire universe compressed into a point and then moving explosively outward in all directions. We don't have to be at the center for it to look this way—it will look the same to an observer on any galaxy. When I say "extremely high temperature," I really mean it. Try for a moment to imagine all the energy of the universe—millions of galaxies each con-

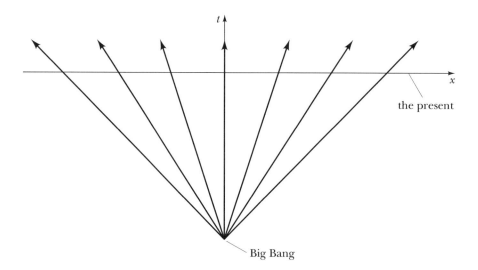

Fig. 3 Hubble's law implies a Big Bang.

taining a trillion stars, more or less—compressed into a volume the size of a pinhead, or even much smaller. The pressure has to be incredibly high, and the work that it would require to compress it so much corresponds in a direct way to the temperature of that compressed state.

Physicists and astronomers in modern times have been very reluctant to suppose that it requires a god to set the universe in motion, and indeed many of them resisted mightily the notion of a Big Bang for just that reason—it seemed much too much like the finger of God. The phrase *Big Bang* was in fact coined by the physicist and cosmologist Fred Hoyle as a term of contempt for a theory that requires spontaneous creation of the universe out of nothing. To my mind, however, a universe of infinite duration in the past and future (such as Hoyle insisted on) points just as much—or as little—to God as does a universe of finite duration. The mystery is why there should be a universe at all. This mystery is one that physics can never solve, because the laws of physics, which are all we have to lean on, are nothing more than generalizations about the universe we see. You've got to have a universe to start with before you can look for the laws governing its behavior, so those laws can't tell you about other possible universes or about the likelihood or unlikelihood of having no universe at all. Physicists have little choice, then, but to accept the universe as given and proceed from there. This cannot and should not prevent them from speculating about the deeper questions, or even from making reasonable guesses about them, but physics will never enable them to get at the truth of the matter.

Actually, you can't conclude from Hubble-type observations that the observed universe started at a single *point,* because the correlation between positions and velocities is much too crudely measured to give more than a very rough approximation to this picture. The arguments that lead to the picture of a single initial point are of three kinds: (1) a study of the equations of general relativity, (2) the search for an understanding of stellar (and planetary) structure and composition—where did the elements come from?—and (3) a consideration of elementary particle physics—how did the (assumed) very symmetrical laws of physics, which we haven't yet identified but are strongly suggested, give rise in the developing universe to the imperfectly symmetrical laws that we see in the universe as it now is?

This third argument, which I'll mention first, has to do with physicists' suspicion that there may be complete symmetry among all (or at least *most*) of the apparently different forces of nature. The strong nuclear force, the weak electromagnetic force, the very weak "weak interaction," and possibly even the gravitational force (the weakest of the lot) are all to be seen as different manifestations of the same force law, and they are all assumed to be equally strong at sufficiently high energies—the extremely high energies that particles would have at Big Bang temperatures. The great differences among these forces are understood as the result of *broken symmetry*—a lack of symmetry in the way the universe cooled down, but not a lack of symmetry in the basic physical laws. Physicists have already seen a strong indication of this in the case of the electromagnetic and the weak interactions, unified now as the *electroweak interaction,* and with this as inspiration they keep coming up with unification schemes to try to unify the other forces of nature as well.

It is also this process of cooling off from an initial fantastically high temperature that has provided the basis for our current understanding, such as it is, of the initial formation of stars and galaxies, the relative abundance of photons and matter in the universe, and the relative abundance of the chemical elements.

The evidence for the initial Bang is circumstantial but very compelling. Physicists try, at least, never to take anything for granted, and we keep reexamining even the most reasonable of our assumptions about physics and the universe. Very few of us, though, have much doubt that the age of the universe is finite and that it started off in a state of unbelievably high density and temperature.

PROBLEMS

1. From the Doppler shift formula (eq. 5), figure out how the apparent frequency of a fire engine's siren changes as the fire engine approaches you (negative v) and then moves away. Take the speed of sound as 1100 ft/s, the speed of the fire engine as 88 ft/s (60 mi/h), and the frequency of the siren as 500 Hz (500 cycles/s). What would happen to the frequency you hear if the fire engine were to approach you at the speed of sound? At slightly less than the speed of sound? Faster than the speed of sound?

2. A semitone shift in musical pitch corresponds to a change in frequency by a factor $2^{1/12} = 1.0595$. Work out a formula for determining the speed of a vehicle in terms of the shift in pitch as it passes you. Remember that there's a Doppler shift up while it's approaching and a Doppler shift down while it's moving away. Use the approximation of eq. 1.4.12 where appropriate.

3. (a) Check the correctness of the final form given in eq. 10 for the Doppler shift for light, using expression 1.4.7 for γ.
   (b) How fast would a galaxy have to be moving to produce a 1-nm shift in a spectral line of wavelength $\lambda = 500$ nm?

(c) How far away must the galaxy be according to Hubble's law?
4. Check the value of 12 billion years given for the Hubble time (eq. 12).

7.3   IS THE UNIVERSE FINITE OR INFINITE?

Returning to the equations of general relativity, we can ask if there are solutions that correspond to the expanding universe suggested by the Big Bang picture, assuming a more or less uniform distribution of matter. It turns out that the Einstein equations do allow a universe that can expand or contract; in fact, they do not allow a static universe. There are found to be two basically different types of solution, one corresponding to *closed universes*, finite in spatial extent and finite in duration, and one to *open universes*, infinite in extent and infinite in duration. There is actually a host of possible universes allowed by the Einstein equations, many of them very exotic; I want to describe here two fairly simple models that physicists take seriously as possibly describing the real state of affairs.

The closed universe model is closed on itself like a three-dimensional analog of the two-dimensional surface of a balloon. It starts from a single point, expands to a maximum size, and then gets smaller again until it collapses to a point and vanishes. The other model that we'll look at is an open universe, spatially infinite and not closed on itself. The extent of the universe at any one time would be infinite, but we'd see only a small, finite part of it because the light from more distant parts would not have had time to reach us. In this picture the universe is always expanding. The expansion is always slowing down on account of gravitational attraction, but the density of matter is not great enough to bring the expansion to a halt, so the universe never collapses. This universe is therefore infinite in time as well as in space; its past is finite (it too started with a big bang), but its future goes on forever.

In both of these models the universe starts off expanding (just as we observe it to do, as in Hubble's law), and the expansion slows down because the masses distributed throughout the universe are always pulling on each other. The difference lies in whether the mass density is great enough to reverse the expansion and cause the universe to collapse again.

The open universe corresponds to a lower average mass density in the universe, at any given stage in its expansion, than the closed universe, so we should in principle be able to tell the difference by direct observation. The visible stars and galaxies have in fact far too little mass to account for a closed universe, but we know also that there must be a large amount of *dark matter* that we're unable to detect—maybe as much as 90% of all the mass. This invisible matter makes itself known by its effect on the motion of stars within their galaxies and of galaxies within their clusters; however, there are large intergalactic regions of the universe where there are no such motions to be observed and where it's therefore impossible to guess at the amount of dark matter. One obvious form of dark matter would certainly be neutrinos, which are virtually impossible to detect because they interact so weakly with light and with other forms of matter. Other possible kinds of dark matter have been suggested, such as dead or dying stars, massive magnetic monopoles, and WIMPs (weakly interacting massive particles), but none of them has been confirmed. Our estimates of the overall mass density in the universe do keep improving, but they always seem to be too close to the borderline between the two cases to enable us to tell whether or not the universe will close. Physicists refer to these puzzles as the dark matter problem.

One popular theory has us precisely on the borderline—another "conspiracy," reflecting perhaps some important fundamental principle that we haven't identified yet. (To be right on the borderline, incidentally, would in fact imply an open universe; there's no way for it to be neither open nor closed.) The reason for such a supposition is simple: Cosmologists have shown that if the universe were even a little way off the borderline at an early stage in its life, its evolution up to the present would have increased the effect tremendously, and it would be very easy to tell which case was the correct one. The fact that we are even moderately close to the borderline now means that we must have been incredibly close in an earlier epoch, which suggests very strongly, because physicists don't believe in coincidences, that there must be some reason for us to be exactly on the borderline. As you can imagine, furious speculation rages about the significance of such a borderline case and about scenarios constructed for the early moments of the universe that would account in some natural way for this apparent exact coincidence.

Now let's look more closely at these two models and try to visualize what it means for the universe to be open or closed. A lot of our understanding as physicists hinges on the metric (§6.4), which enables us to measure distances and times as we move around the universe. As you saw in §3.4, there are different kinds of "distance" in spacetime: timelike, spacelike, and lightlike. The curvature of spacetime doesn't alter this fact, and in any small neighborhood, in fact, the picture of spacetime provided by special relativity is accurate. This is like the fact that a very small portion of a curved surface such as a sphere is approximately flat, which is why Columbus had so much trouble convincing people that the earth is round.

General relativity, then, gives a picture of spacetime that combines these two features: It is curved, and it has an *indefinite metric*, which means simply that small displacements can be timelike, spacelike, or lightlike. Light signals still provide a natural absolute speed limit. The distinction between spacelike and timelike intervals in spacetime still holds, as the following examples illustrate.

Closed universes

A picture of a possible spacetime (that doesn't obey Einstein's equations, though) is shown in fig. 1, where space (labeled by x) is one-dimensional and closed, like a circle, and time (labeled t) continues indefinitely. In this picture the size of the one-dimensional space decreases at first, reaches a minimum at time $t = 0$, and then increases thereafter. The two

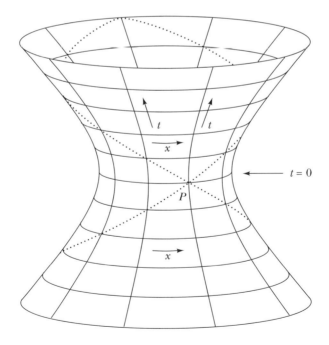

Fig. 1   A curved spacetime.

dotted lines through the point *P* are the world lines of two light signals traveling in opposite directions, first converging to the point *P* and then traveling outward in both directions until they meet again at the opposite side. An interval along one of the dotted lines would be lightlike, an interval in the direction of one of the arrows marked *t* would be positive timelike, and an interval in the direction of one of the arrows marked *x* would be spacelike. In the near vicinity of any point, the picture is just like the spacetime of special relativity (like the various figures in Chapter 2).

Another example is shown in fig. 2, which is a crude picture of the big bang (the closed universe scenario), again with just one spatial dimension and one time dimension. Time progresses outward from the center, starting at the moment of the bang, $t = 0$. Each concentric circle represents all of space at a single moment of time, so again space is finite and closed on itself: If you could follow along the circle, you'd get back to your starting point after moving a distance equal to the circumference. At $t = 0$ space is just a point, and shortly after that it's a very small

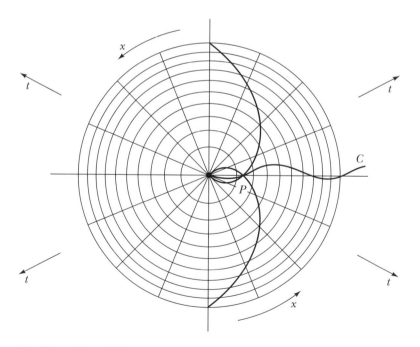

Fig. 2 Spacetime right after a big bang.

circle, getting bigger as time goes on. The circles represent equal time intervals (in the proper time of an observer whose world line is one of the radial lines in the figure), and because the spacing of the circles, which represent equal time intervals, decreases as you go outward, you can see that the pace of expansion decreases with time.

The local spacetime geometry in this picture is suggested by the two light paths through an arbitrarily chosen point *P*, together with a possible particle world line *C*. Time progresses outward in the picture, so the forward, or future, timelike intervals point *outward* from the point *P*, and the backward intervals point in toward the center. The fact that the particle must always travel slower than light means that the world line must always fall *between* the two light paths; it could go radially outward in the picture, for example, but it couldn't follow a spacelike path such as one of the concentric circles. The question of the evidence for this kind of expanding universe was dealt with in §7.2.

These pictures show a universe that is spatially closed and simply continues to expand. According to Einstein's theory, as I've said, if the universe is closed like this it means that the matter in the universe is dense enough for gravitational forces to cause it collapse again after a certain time. The picture then might look more like fig. 3, where the circles that represent space start (at the point *BB*) by expanding, just as in fig. 2, but eventually start getting smaller again and finally fall back

Fig. 3   A closed universe that expands and then collapses. The point *BB* is the Big Bang.

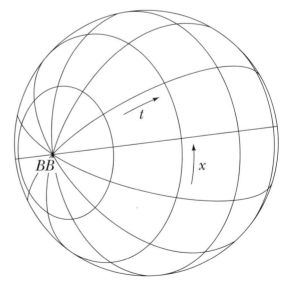

into a single point. There is no such thing as a time "before" the initial bang, and there is no time "after" the final collapse. Remember that space is one-dimensional in this picture: Each circle represents the universe at a single time. I don't believe that in such a picture time would start to go backward when the universe started collapsing, as people sometimes claim. (A more careful discussion on the direction of time appears in Chapter 8.)

I've drawn these pictures for a universe with just one spatial dimension. Try to visualize how it would look for two spatial dimensions. At a given moment of time the spatial universe would be the surface of a sphere, and the radius of the sphere would get bigger and bigger as time progressed. You could think of it as a round balloon, with the stars and galaxies looking like little dots painted all over it, as in fig. 4. The balloon gets bigger with time, and the little dots get farther and farther away from each other. In spatial terms, the volume inside and outside the sphere has no meaning—the whole universe lies on the surface. To make a spacetime diagram of it, you have to imagine a series of concentric spheres representing the different values of the time t, with a single point at the center representing $t = 0$, when the sphere has zero radius. This would be an exact analog of fig. 2.

The final step, to a picture of the real three-dimensional world, requires still another dimension. The universe at any given moment is the

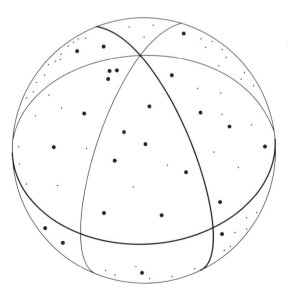

Fig. 4 The expanding universe.

surface of a *hypersphere,* a closed three-dimensional world analogous to the two-dimensional surface of a sphere. Mathematically, a hypersphere is defined as the set of points in a four-dimensional Euclidean space that are all at the same distance r from a given central point, but you must remember that we're not interested in the Euclidean space; only the surface is real. This surface is three-dimensional in character—it takes three parameters to specify a point on it—just like the space we're familiar with. And again the spacetime diagram would consist of a series of concentric hyperspheres corresponding to successive moments of time.

Let's return now to the expanding universe picture of figs. 3 and 4. This corresponds to just one very special choice of initial conditions; if the initial shape were not spherically symmetrical, then the shape at later times would not be spherically symmetrical either. The universe that the astronomers see has some strange large-scale structures—huge gaps of various sizes in the distribution of distant galaxies—but it also has some remarkably uniform features, especially in the distribution of primordial electromagnetic radiation. This radiation, left behind, so to speak, as the universe cooled off from its initial fantastically hot state, has itself cooled off as the universe has expanded until it now corresponds to a temperature of about 3 K—three degrees above absolute zero. It was thought until very recently to be completely uniform in all directions (apart from a Doppler shift due to the relative motion of the earth), but the primordial electromagnetic radiation has now been seen (1992), by observations from the COBE (Cosmic Background Explorer) satellite, to have tiny irregularities that are consistent with the nonuniformities in the distribution of matter that have built up over billions of years.

The distributions that we see suggest some kind of approximate symmetry in the overall shape that's consistent, if the universe is closed, with the kind of spherical shape (hyperspherical, of course) shown in fig. 4, or, if the universe is open, with some correspondingly symmetrical shape. As usual, we make the simplest choice until we have reason to do otherwise. In any case, we have no idea *why* the universe should start off the way it did rather than some other way. All the physical laws we have learned about so far are rules of *behavior,* in the form of equations of motion, which tell what will happen next but, as I described in §4.1, contain no information about how it all gets started. Physicists sometimes dream about finding laws that determine the initial conditions too, but it may very well be that there is no such law.

I want you to try to imagine what it would be like to live in a small *closed* universe, of finite extent like the surface of a sphere, only in three dimensions. It would be like this: You're floating at a point in space (there's no earth in this universe, I'm afraid) and you start constructing spherical surfaces of different radius; for small spheres the area would be very close to the Euclidean result $4\pi r^2$, where r is the radius of the sphere. Then for bigger spheres you find the formula is no longer accurate—the area is less than $4\pi r^2$. The area still increases for increasing r, but not as fast as the formula says. At some point (say, at $r = 5$ miles) the area of these spheres actually stops increasing and starts decreasing, even though you continue to increase the radius r, until finally, for a sphere of radius 10 miles, the area has decreased all the way to zero. You could explore these spheres by sending out agents in all directions, in little flying machines—one going up, one going down, and four more going north, east, south, and west. When they had all gone 10 miles, they would meet each other, all converging on a single point in space. They would appear smaller and smaller to you for the first 5 miles, and then they'd appear to get bigger again as they approached the convergence point 10 miles away. Your universe would act like a big lens, and the effect of this magnification would be that the six little spaceships would all appear to be coming at you backward—that is, you would still be looking at their tails, but they'd be getting bigger and bigger as if they were converging on you. They wouldn't hit you, though, and in fact any nearer objects would block your view of them. If they managed to avoid hitting each other at the convergence point, they would then return to you; you'd see them coming head on, getting smaller until they reached the 5-mile point and then getting bigger again.

This whole picture is very similar to sitting on the north pole and looking at circles of various radius on the earth's surface, with their centers at the pole—the lines of latitude. The circumferences are less than $2\pi r$, as we have seen, and when you get to the equator the circumferences of the circles start decreasing even though the radii—the distances from the north pole measured along the lines of longitude—keep on increasing. The agents I was talking about (there are only four of them now) are traveling down different lines of longitude, and for you, sitting on the north pole, these lines of longitude are your lines of sight. As an agent gets to within a few feet of the south pole, she spans just as many lines of longitude as if she were that close to the north pole, so she appears just as big to you as if she were nearby. When she's sitting *on* the south pole you see her in every direction you look—she appears to surround you!

Return now to your closed three-dimensional universe, where you can't travel more than 10 miles in any direction away from your starting point. The fact that a sphere's area drops to zero at $r = 10$ miles means that at that point you have exhausted the possibilities—you can't make the spheres any larger in radius, which is to say that you can't get any farther away than that from the central point you started with. *The metric of your three-dimensional space determines the structure of the space.* You don't have to conceive of it as being imbedded in some Euclidean space of higher dimension. I think it's only our psychological need for a Euclidean framework that makes it hard to conceive of a space that is truly finite.

Open universes

The other major possibility that I've mentioned is that the universe is *open,* which means both that it is infinite in spatial size and that it will keep on expanding forever. I've yet to understand in terms of physics why these two features, finite spatial size and finite lifetime, are exactly tied together; all I can say is that that's what the equations say! I wonder myself whether more complicated and unsymmetrical solutions to Einstein's equation might allow other combinations.

To visualize an open universe that starts with a big bang, you might imagine an infinite sheet of rubber, again with dots uniformly painted on it to represent the galaxies, and now constantly expanding, as I've tried to show in fig. 5. As you look back close to the initial time, you find the dots very dense on the sheet, but the sheet is still infinite in extent, right back to the Big Bang, the moment when the density was infinitely great. All the galaxies that we can see now would have come from a very tiny region—of great density—at a moment close to the initial moment.

At the present moment we can see only a small, finite part of this infinite universe, because light signals could not have reached us from the more distant parts. Nevertheless, light signals from even the most distant star will eventually reach our earth, despite the fact that the distance between us is increasing at a rate much greater than the speed of light c. The prohibition on sending signals faster than the speed of light isn't exactly violated in general relativity, but it *is* modified almost beyond recognition, because the possible paths of a light signal in curved spacetime depend in nonintuitive ways on the large-scale structure.

Gravitational fields

Now, what has happened to the notion of a *gravitational field* in these complicated pictures of curved universes? It all depends on how you pick your reference frame. In fig. 2, for example, any one of the radial lines could be the world line of an observer S sitting in a galaxy that to him

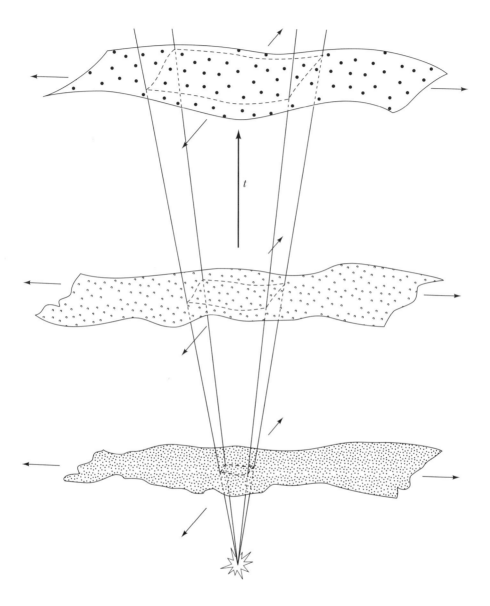

Fig. 5   Spacetime diagram for an open universe. In this picture the universe contains an infinite number of galaxies and stars, but only a finite number are visible to us. The bottom scene, at t_1, represents a moment
immediately after the Big Bang.

seems to be at rest, while all the other galaxies seem to be moving away from him, as in the expanding balloon of fig. 4. The fact that the pace of expansion is decreasing means that the neighboring galaxies appear to be decelerating—the speed with which any given galaxy is moving away is decreasing with time. (This effect is too tiny to have been observed, but it's undoubtedly there because galaxies necessarily exert a gravitational pull on each other.) Our observer attributes this to the gravitational field due to the matter in his own vicinity; from his point of view there are gravitational fields everywhere except right where he's sitting, as shown in fig. 6. An observer S′ on one of the other galaxies, on the other hand, interprets the situation differently. She regards *herself* as being at rest, and she sees S as moving away from her and decelerating as he does so. And she then attributes his deceleration to gravitational fields around *her* position, in just the same way.

You'll recall, in fact, that in any reference frame associated with a freely falling observer, such as either S or S′, there appear to be no gravitational fields at all and the local geometry is just the ordinary spacetime geometry of special relativity. This means that near any point in spacetime there's a choice of coordinates—the coordinates of the freely falling frame—that look like the spacetime coordinates x, y, z, and

Fig. 6   Gravitational fields
as seen by S.

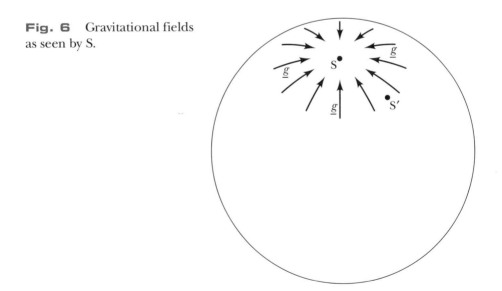

t of an inertial reference frame in special relativity. This is just like saying that an ordinary curved surface is approximately flat near any given point and that there's some coordinate system that looks locally like Cartesian coordinates, even though it must necessarily get distorted as you move away from that point. An example for a more familiar curved surface is the lines of latitude and longitude very close to the equator.

PROBLEMS

1. (a) Show that when a circle of small radius r is drawn on a sphere, the circumference of the circle can be written approximately as $2\pi r(1 - \alpha r^2)$, where α is a constant for a given sphere. Give an expression for α in terms of the radius of curvature r_0 of the sphere. For small angles θ, given in radians, you can use the approximation $\sin\theta \approx \theta - \theta^3/6$ (note that the first term alone is not enough).

   (b) Suppose you land on an unknown planet and find that a circle of radius 1 km has a circumference that is just 1 m less than $2\pi r$. Find the radius of the planet.

2. Obtain a formula for the circumference of a circle of arbitrary radius r (not necessarily small) on the surface of a sphere of radius r_0.

3. In the example of the small closed universe given in this section, describe the events from the point of view of one of the agents, including the scene a few moments after departure, the scene at the 5-mile point, and the scenes immediately before and after the rendezvous at the convergence point. (The 5-mile point is tricky.)

4. (a) It is noted in the text that the different types of displacement from the point P in fig. 2 can be forward or backward timelike, forward or backward lightlike, or spacelike. Make an enlarged copy of fig. 2 and construct examples of each type of displacement. Draw several possible particle world lines.

   (b) If you continue the picture outward, the light paths continue to spiral around, encircling the center any number of times. Draw the light paths that represent the experience of seeing the same point on the opposite side of the universe by looking in either direction. Draw a light path that corresponds to seeing yourself as you were at some previous time.

5. Give a verbal description of a three-dimensional spacetime diagram of a two-dimensional spherical universe that expands and then collapses in analogy to fig. 3.

6. Give a one-paragraph description of each of the following, addressed to a reader who is unfamiliar with these ideas: **(a)** principle of equivalence in general relativity, **(b)** curvature of spacetime, **(c)** geodesic, **(d)** black hole, **(e)** Hubble's law, **(f)** Big Bang, **(g)** open universe, **(h)** closed universe.

CHAPTER 8

The Direction of Time?

 THE PROBLEM OF TIME

In the previous chapters I've had a lot to say about time, but if you look back you'll see that I've stressed the *similarity* of space and time and have not examined the special way in which we experience time. I've talked about spacetime as if it were a single four-dimensional space of some sort and have insisted that ordinary events are to be thought of as located in that four-dimensional space. I've shown that to moving observers, space and time get mixed up with each other just like the different dimensions of space when you look at them from a rotated point of view.

The unique asymmetry of time

The fact is, though, that space and time present themselves to our human perception in very *different* ways. Space is somehow where things *are*, while time has something to do with things *happening*. In this chapter I want to explore the distinctive character of time and see if we can get some insight into the physical reasons for the subjective ways we experience it. An important part of this exercise is to analyze these subjective perceptions—to describe them accurately and formulate questions that probe the difficulties. Following are some crucial questions that I'll try to address in this chapter.

1. Why is there such a big difference between the past, the present, and the future? Why does the present moment, and *only* the present moment, seem immediately real to us, and why is the present moment the point of decision or choice, the place where we can control events (or think we can)? Why are past events known not by direct perception, like the present, but at best only by fading memories? Why do our minds contain memories of the past and not of the future while our present decisions and actions affect the future and not the past, so that the future is controllable while the past is immutable? Is our apparent ability to control the future in conflict with the laws of physics, which are supposed to determine future events entirely in terms of the state of the universe at the beginning of time? Is the lack of symmetry between the future and the past a consequence of physical laws? Is it real or is it merely a consequence of our subjective human view of the world?

2. How is it that we keep moving on from one moment to the next? Why do we have the sense of the *flow* of time—the feeling that time moves inexorably forward and can never be made to stop or go backward?*

3. Why does the rate at which time progresses seem to vary so? Is there some kind of *subjective* time against which you measure the pace of events in ordinary time? Does this subjective time come to a stop when you're unconscious (completely unconscious, as when you're under total anesthesia) even though physical time goes on?

---

* *The Moving Finger writes; and, having writ,*
  Moves on: nor all your Piety nor Wit
    Shall lure it back to cancel half a Line,
  Nor all your Tears wash out a Word of it.
                                        Edward Fitzgerald, with the help of Omar Khayyam

4. Why are the ordinary events of everyday life so completely irreversible? Try to imagine some commonplace events as they would appear if they were to happen in reverse.

a. *The unburning of a match:* The particles of soot and molecules of carbon dioxide and water gather around the burnt remnant and draw warmth out of the air to reverse all the chemical processes and rebuild the unburnt match. The flame and light are drawn back into the match, and at the end of the process the match suddenly becomes completely cool.

b. *The unspeaking of a sentence:* The thermal energy in the walls of the room and in the eardrums and tissues of the hearers refocuses into coherent vibrations that send sound waves through the air back to the mouth of the speaker, which in turn collects them and converts them back into the mental state that preceded the speaking. The hearers are left with no memory of the sound at all and no knowledge of the content of the sentence.

c. *The unspilling of a glass of milk:* The milk that soaked into the carpet and clothes spontaneously works its way out and acquires the energy from its own molecular vibrations and those of the materials to leap up from the floor and lap to the table surface. There it joins the puddles of milk on the table and pours itself back into the glass, whose broken pieces have meanwhile leapt up and reassembled themselves into an undamaged vessel. The sound and light waves that carried information about the event must also reconvene to the scene of the accident and contribute a small part of the energy needed to lift the milk and pieces of glass and reassemble them. All tears (for the spilt milk) return to the eyes and are reabsorbed by the lacrymal glands, and all knowledge of the event is erased from the minds of the observers.

As you'll see in what follows, the key to understanding all these questions is the famous second law of thermodynamics, which focuses all these mysteries and distills them into a single supermystery. The character of the *Second Law* (as it's often called for short) is totally different from that of all the other physical laws, some of which I've discussed in the preceding chapters. The Second Law is a non-law, an extra law added to a set of laws that is already complete, a negative law that tells not what will happen but only what won't happen. It's like a partial conservation law—it says that something called the total *entropy* in the universe cannot decrease but can increase; in fact it *will* increase. And even this half-prediction is weakened by the admission that the entropy *might* decrease, though the likelihood of its doing so is so fantastically tiny that it's a mistake to think of it as possible at all. As you may have guessed, all those reversed events I've just described would involve a decrease in entropy—a violation of the Second Law.

In this chapter and the next we'll explore these matters. I'll try to give you an idea of where the Second Law comes from and why it has the odd, somewhat imprecise, character I've mentioned. And you'll find out what entropy is and what it's good for.

The Second Law—the trend to disorder

I believe, then, that the unsymmetrical nature of time is due to the character of the physical world we inhabit, in particular to the second law of thermodynamics, which says that on the macroscopic scale all processes are in practice irreversible and that there is an inevitable trend from order to disorder. I believe too that this law is itself in turn a consequence of the initial state of the universe at the moment of the Big Bang and, indeed, of some extremely unlikely characteristics (as I'll explain in §8.4) of *our* Big Bang as distinct from all the possible varieties of big bang that might have occurred. In simple terms the argument is this: By any reasonable definition of *orderly*, there are vastly more disorderly states of a system than orderly ones. Thus if the system starts in an orderly state (low entropy) it will quickly find itself in a disorderly state (high entropy), and once it is in a disorderly state it will be exceedingly unlikely to return to an orderly state—so unlikely for complicated systems like our universe, or even a small macroscopic object like a rubber ball, as to be completely impossible for all practical purposes.

To see this, think about dropping the ball. It bounces a few times and comes to rest, its energy all converted to thermal energy. In the initial state all the molecules of the ball (something like a million million million million—that's 10^{24}—of them) are moving in the same downward direction, while in the final state they are all moving thermally in different random directions. There is only one way for all the molecules to be moving in the same direction, but there's a huge number of ways, indistinguishable to you, in which they could be moving randomly. The time you'd have to wait to see the process reverse itself, to see the random thermal motions of the molecules just happen to be all in the same direction at the same time, would be forever and a day, by which I mean ridiculously long compared to the age of the universe. I'll try to give you some feeling for the incredible immensity of such times in §8.2.

PROBLEMS

1. What do *you* feel are the crucial questions regarding the distinctive
    character of time? Formulate them carefully, especially if they're
    different from mine.
2. Construct three or four additional examples of time-reversed
    events, giving in each case as complete a verbal description of the
    reversed sequence as you can. Can you think of any ordinary events
    that would *not* look strange if played backward?

 MAXWELL'S BABY

One example of the Second Law is the way in which a child messes up a
room—not out of malice but just because there are so many *more ways* for
the room to be untidy than for it to be tidy. To make a concrete model
for this, imagine a baby* in a room with 10 toys on a shelf. The baby plays
with a toy for 5 minutes and then tosses it at random. We'll generously
allow the toy a probability of 1% of landing on the shelf, and we'll call
the room *tidy* if all the toys are back on the shelf.

The odds against tidiness

The first question is how long it takes for the room to become completely untidy—for all the toys to be on the floor. (Before you read on,
try to make a careful estimate of the time.) It takes longer than 10 turns,
obviously, because when most of the toys are on the floor the baby is
more likely to pick up one of the ones on the floor, so it takes a while for
him to get to the last few on the shelf. In this dismantling process we can
neglect the chance of a toy's getting back on the shelf—that remote
chance has a very small effect on the answer. The first toy goes on the
floor in one turn—5 minutes. For the second toy it's only 90% likely that

---

*I've called the baby *Maxwell's baby*, in contrast to Maxwell's demon, a hypothetical imp invented by Maxwell who produces order out of chaos without violating
the laws of physics, simply by sorting out molecules and sending them where
they belong. Maxwell not only was the primary inventor of classical electromagnetic theory (described by Maxwell's equations, which are discussed in Chapter
1) but also made major contributions to our understanding of thermal physics.

the baby will get a toy off the shelf in one turn, so on average it takes 10/9 of a turn to get a second toy on the floor. For the third toy the chance is 80%, so it takes 10/8 of a turn on average. To get them all on the floor, then, takes an average of

$$1 + \frac{10}{9} + \frac{10}{8} + \frac{10}{7} + \frac{10}{6} + \frac{10}{5} + \frac{10}{4} + \frac{10}{3} + \frac{10}{2} + \frac{10}{1} = 29.3$$

turns. (For the last toy it takes about 10 tries before the baby picks the one on the shelf.) At 5 minutes a turn, then, it takes 150 minutes, or 2.5 hours, to produce chaos.

Now, how long will it be before the room is tidy again? (Give this some thought and again make an educated guess.) It will take about 100 tries before 1 toy gets on the shelf, or about 8 hours. The baby then has about 9 tries to get a second toy onto the shelf before he spoils it by picking up the one that's already on the shelf and throwing it back on the floor, or a chance of $9 \times 1/100 \approx 1/11$, one in eleven, of getting a second toy on the shelf before he loses the first one. So on average it will take him 11 such chances, or 90 hours of play, before he gets 2 toys on the shelf. What about 3? Once he gets 2 on the shelf, he has only about 4 chances to try for the third one before he spoils it (it takes an average of 5 turns to lose 1 of the 2), or a chance of $4 \times 1/100 = 1/25$ of succeeding. How many times, then, will he have to get 2 on the shelf before he'll succeed in getting a third? Clearly, 25 times, which means that we get an additional factor of 25, and it take 25×90 hours, or about 3 months, to get 3 toys on the shelf. Repeating this logic, we can estimate the length of time to get different numbers of toys back on the shelf as shown in table 1.

You'll need to provide a new baby from time to time. T_H represents the Hubble time—the age of the universe—which I've taken as about 10 billion years (it's not known very well, as I've mentioned), and it will take 100,000 times the age of the universe before the room just happens to get tidy again. During most of these billions and billions of years all the toys are on the floor; occasionally one or two are on the shelf, only to come off again before anything else happens. At the end, completely unpredictably, you would see all 10 toys go on the shelf in rapid succession—in a space of about 2½ hours on average and of course purely by chance.

Notice in table 1 that the total number of tries required to get all the toys on the shelf can be written 100^{10}. This is also equal to the average number of tries it would take if instead of using the baby to move toys

Table 1   Tidying-up times for Maxwell's baby

Toys on shelf	Factor	Number of tries	Time
1	100	100	8 h
2	$100 \times 1/9 \approx 11$	1,100	4 d
3	$100 \times 2/8 = 25$	28,000	3 mo
4	$100 \times 3/7 \approx 43$	1,200,000	11 yr
5	$100 \times 4/6 \approx 67$	8×10^7	750 yr
6	100	8×10^9	75,000 yr
7	150	1.2×10^{12}	100 million yr
8	230	2.8×10^{14}	3 billion yr
9	400	1.1×10^{17}	10^{12} yr $\approx 100\ T_\mathrm{H}$
10	900	1×10^{20}	10^{15} yr $\approx 100,000\ T_\mathrm{H}$

around one at a time, you could just shake the room each time like a dice cup. Then each try would represent a fresh random arrangement with no relation to the previous one. On each try, then, the chance of getting all the toys on the shelf is simply $(1/100)^{10}$—a factor of $1/100$ for each of the 10 toys—and the average number of tries to get it right is just the reciprocal of that, 100^{10}. Over that span of time, the process you choose to use for producing successive random arrangements makes only a very tiny difference.

This is supposed to give you a feel for how unlikely it is for the room to be tidy if only random forces (babies) are at play. As you can imagine, having 100 toys would increase this unlikelihood by a very large factor: The time to get all the toys on the floor increases only to about 40 hours, but the 10^{20} turns that we deduced for getting all the toys back on the shelf would become 100^{100}, or 10^{200}—a *huge* increase.

It's hard to get any feeling for numbers this big. Think for a moment about how many seconds there are in the Hubble time—the age of the universe. Think about the hours, days, years, centuries, thousands of years, millions of years, up to 10 billion years, which is around 10^{18} seconds. Now try to imagine *that many Hubble times*—one Hubble time for every second in the age of our universe. That's 10^{36} seconds and we're still nowhere close. We'll have to keep repeating this process 11 times: Call 10^{36} seconds a *double Hubble,* and put 10^{18} of *those* end to end, one for each second in the age of our universe. That's a triple Hubble, 10^{54} seconds. A quadruple Hubble is 10^{18} triple Hubbles, and so on, till we get to an undecuple (that's elevenfold) Hubble, 10^{198} seconds, which is getting close. [Actually we need 10^{200} *turns,* which means another factor of 300 (number of seconds per turn), but that hardly seems to matter any more; we're only off by a factor of 10^5 or so!] Note that it's not hard to write down this number in conventional decimal form:

$$10^{200} = 100,000,000,000,000,000,000,000,000,000,000,000,$$
$$000,000,000,000,000,000,000,000,000,000,000,000,$$
$$000,000,000,000,000,000,000,000,000,000,000,000,$$
$$000,000,000,000,000,000,000,000,000,000,000,000,$$
$$000,000,000,000,000,000,000,000,000,000,000,000,$$
$$000,000 \tag{1}$$

The reason this is such a huge number is that each zero represents a *multiplication* by 10: The number gets 10 times bigger with every zero.

Now, back to Maxwell's baby and the 10-toy scenario. What makes the process go so quickly one way and so slowly the other? Why does it take only 2 hours for the room to get messed up and 100,000 ages of the universe to tidy it up again? Why the asymmetry in time? It's simple: *The room was tidy to begin with.* Presumably the baby's father or mother had just cleaned it up. If you film the *whole* process up to the time that the room gets tidy again and then play the film backward, it will look just the same as it does forward. (You'll have to make the baby put the toys down carefully each time instead of throwing them, so that it will look OK played backward.) It will take the same 2.5 hours or so for the room to get messy (in real time, that's the time it takes for the final winning streak when all the toys go back on the shelf in rapid succession), and there will be the same long waits to see 3, 4, . . . , up to 10 toys back on the shelf.

Our tidy universe

The description I've just given was for only 10 or 100 toys. If you now replace the toys by the atoms in the universe (something like 10^{100}

atoms), you might get tidying-up times like $10^{10^{100}}$ seconds—or years, or Hubble times—it doesn't matter any more. To write this number in the usual form would require a 1 followed by 10^{100} zeros. If all the matter in the universe were made into paper, there wouldn't be enough paper to write that many zeros. Then you have to realize that each of those zeros represents a *factor* of 10: $10 \times 10 \times 10 \times 10 \times 10 \times 10 \times \ldots -10^{100}$ times. Alternatively, consider the fact that $10^{10^{100}}$ has 10 times more zeros than $10^{10^{99}}$, and that $10^{10^{99}}$ has 10 times more zeros than $10^{10^{98}}$, and so forth, and then think about what it means for one number to have 10 times more zeros than another.

I've tried in the previous paragraph to illustrate what an unimaginably long time you would have to wait before random events would begin to look like a reversal of the ordinary, familiar events of nature. I'm trying to drive home the utter unlikelihood of a chance reversal of the tide of increasing disorder that we call the second law of thermodynamics, and I'm also trying to put into words my own ever-multiplying sense of astonishment that an orderly universe exists at all.

We live in a universe that is almost completely tidy in comparision with the really untidy ones in which all the toys are on the floor—in which all the particles in the universe are randomly distributed in position and velocity. In such a really random universe, everything would be at the same temperature—nothing hot, nothing cold. *There would be no forward direction of time* because the picture would look just the same whether you played it forward or backward. We live in a universe (using the metaphor of the toys) such that every door you open reveals a room with a baby busy playing and 8 or 9 toys on the shelf; everywhere you look in the visible universe there is the same degree of tidiness—hot stars and galaxies happily burning and losing their energy to the surrounding empty space, corresponding to toys being thrown on the floor. The obvious inference is that 10 or 15 minutes ago the toys were all on the shelf in every room, because we can see the babies playing and the disorder increasing everywhere we look. For our universe, that moment 10 or 15 minutes ago corresponds to the Big Bang. The age of our universe, T_{H}, should be thought of as extremely short compared to the time it will take even for the universe to run down—something like 10^{90} seconds, or a quintuple Hubble (see pages 227–228), and enormously short compared to the time required for tidiness to reemerge. The only reason there is a second law of thermodynamics is that the universe was so orderly such a short time ago and therefore is inevitably getting more disorderly. The Second Law is a characteristic of that short period of time when the universe is running down (the first 2 hours in the playroom), and in a totally rundown universe there is no longer a Second Law—*and without the Second Law there is no direction of time.*

The Big Bang, the apparent explosive beginning of our universe, is the key—the very essence of the mystery. It is that point of maximum tidiness that *makes it possible* for the universe to run down and thereby gives us the sense of time as we know it. The Big Bang may give the impression of being very untidy—hot and chaotic—but as I'll argue more carefully in §8.4, it is a *far tidier* state than it appears. The Big Bang is, uniquely, an event with no prior history, the ultimate "initial condition" without which the laws of physics themselves are incomplete.

You saw in Chapter 7 that general relativity gives a description of the Big Bang but that this description is very incomplete: The thinking of the cosmologists still involves a great deal of speculation. One unanswered question is this: Does such an explosive beginning necessarily yield a universe with a Second Law (a direction of time at the level of our local observations): stars, for example, that start hot and give up their energy by radiation to cooler surroundings? This is not at all obvious, as I'll discuss in §8.4. My own belief is that for a locally tidy universe like ours to arise from a big bang is just as improbable as for it to arise spontaneously out of a random, rundown state.

PROBLEMS

1. In the example of Maxwell's baby, the average number of random tries needed to get the room tidy came out to be 100^{10}. Construct the corresponding formula for the average number of turns needed when there are n toys, with probability P of getting a toy on the shelf in one throw.

2. Construct the table corresponding to table 1 for the case of a room with 5 toys; with 20 toys. How many toys would allow the room to get tidy in a single lifetime? Now work out the 5-toy table with several different choices for the odds of a toy landing on the shelf, such as, 1 in 50 and 1 in 200.

3. The tidying-up time for the universe was estimated in the text to be something like $10^{10^{100}}$—in seconds, years, *or* Hubble times. To see that it doesn't matter which units are used, express $10^{10^{100}}$ seconds in terms of Hubble times and describe in words the difference in the two numbers. Use $T_H = 10^{18}$ seconds.

4. Try to describe $10^{10^{100}}$ seconds by using the double-Hubble trick described in the text.

8.3   CONSCIOUSNESS OF TIME

Where does our own sense of the flow of time come from? It comes, the way I understand it, from this same process of increasing disorder taking place in our brains. It is a consequence of the fact that every event that happens to me—to my perceiving brain in particular—leaves a mark that was not there before: a record, a trace, a *memory* in fact. In fig. 1 I've drawn a spacetime diagram of the brain (as in §2.2), looking like a long sausage of small spatial extent but extending for the traditional three score years and ten in the time direction. In fig. 2 I've tried to show the effect of impinging events on the brain: nerve impulses coming in from the senses as well as internal activities involving thoughts and emotions. Because the brain is free of any marks (memories) at the beginning of its life, the effect of these events can only be to leave a trace extending *forward* in time. If there were no initial tidy state, the brain would be a big jumble of random marks both before and after the event, and although

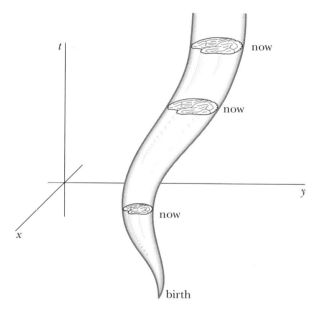

Fig. 1   Spacetime diagram of the brain.

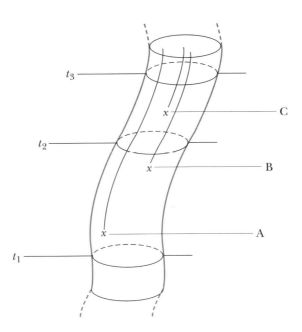

Fig. 2   Memories.

the event would make a tiny difference, it would not correspond to a distinctive mark either before or afterward. If all the toys are on the floor and someone moves one, there's nothing distinctive about the final state to suggest that a toy has been moved, while if they are all on the shelf and someone drops one on the floor, you can tell just what has happened. The memory trace represents a kind of order in that it carries information, but the information is like that of the single dropped toy: It's the deviation from total orderliness that carries the information. When many toys have been dropped—or when the mind has acquired too many memories—the jumble becomes too great and you can no longer get information from it.* It's my guess that loss of memory with advancing age has more to do with this kind of jumble effect, and with the

_____

*Human lifetimes at present don't seem long enough for this to be a problem—though I'm already aware myself that as you get older you see something of this effect. I believe that if large advances are made in increasing human longevity, it could become serious.

consequent difficulty of information retrieval, than with any actual erasing of long-term memory.

There is consciousness at *every* point in time, and to that consciousness that moment of time is a "now," as I've indicated in fig. 1. (For deeper speculations on the meaning of consciousness, see Chapter 15.) At t_1 in fig. 2 there is consciousness but no trace (no memory) of the events A, B, and C. At time t_2 the consciousness includes memory of events A and B but not C. Consciousness at t_2 remembers time t_1 but consciousness at t_1 does not remember time t_2. What is it that gives you the sense of the *flow* of time? The recollection at any given time of immediately preceding events, present feelings of anticipation together with recollections of previous experiences of anticipation and their fulfillment—or disappointment—all these give you the vivid awareness of the relationship of that moment to times before and times after that constitutes the sense of flow. All moments of consciousness are equally real, equally immediate, and equally aware of time's flow. They are *all* now's.

Note the variety of different perspectives that enter into this discussion. Each moment of consciousness has a distinct perspective, different from all the others, in which that moment is real and all the others are classified as past or future. Physics, on the other hand, has a perspective that is outside all those moments. This perspective is illustrated by the spacetime diagram, which treats all times as equally real and isn't committed to considering them in any particular order. In fact, from the point of view of physics, even cause and effect can be viewed in arbitrary sequence: The differential equations of physics enable you to deduce past states from future conditions just as readily as you can deduce future states from past conditions. The equations of motion of physics are very much concerned with time but are completely neutral as to which direction it goes.

The feature that I haven't touched on here is free will: a person's ability to control physical events. Whether this ability is real or only imagined—I'll come back to this in Chapter 15—our normal experience in exercising control is that we can affect the future and not the past. This is completely consistent with the picture of the brain presented here: If you think of the act of initiating muscular activity as an event in the brain, and if the initial state of the brain is required to be blank, then the result of that particular brain event—a decision in this case—must lie in the future, just like the memory traces that I spoke of earlier. The occurrence of that event, if you believe in free will, is not a deterministic consequence of the previous state of the brain. Thus our memory is of the past, and our sphere of influence is the future. The past, from the perspective of any one moment of time, is the domain of the actual, and the future is the domain of the possible.

PROBLEM

1. Consider carefully, and and try to put down on paper, what it is in your mind at any one moment of time that gives you the subjective sense that time is flowing. I've given my own version of this exercise in the text, but it's not intended—or expected—to be the last word.

8.4   THE COLLAPSE OF THE UNIVERSE

If the universe collapses again toward a big bang in reverse (often called the Big Crunch)—the closed-universe option mentioned in §7.3—docs that mean a reversal of time? Does the universe start getting tidier? People argue about this, but I really think not, because the time scale is far shorter than that required for a reversal of time by chance (if that's even meaningful, which I doubt), and indeed there's no rule that says a smaller universe is necessarily tidier.

A reversal of time?

The behavior of the physical universe is governed by equations of motion together with specified initial or final conditions. You cannot specify both initial *and* final conditions, because that overdetermines the problem. In our universe we're in the position of knowing the initial conditions, in the form of the Big Bang and other information that we have about the past, and of not knowing anything about the future except what we can predict using the equations of motion. The fact that the universe may collapse to a final singular state is not a further specification but rather a consequence of its present state, and there's nothing in the present state to make it in the least likely that things will start to get more tidy, any more than if the universe were not to collapse. If the universe reaches maximum size and then slowly starts getting smaller, there is nothing in that to cause events to start looking to the future for their causes, as would have to be the case in a time-reversed universe: Photons would have to start seeking out stars rather than being emitted by them, and thermal energy would have to start focusing itself into more coherent, ordered forms of energy in the various exotic ways described in §8.1. In fact it's inconceivable to me that at the local, terrestrial level or even in the behavior of our own galaxy you could see any direct effects of such a cosmic turnabout. Hubble's constant (§7.2) would change sign, to be sure, but it would take several million years for astronomers to find out about it, because it's the observation of other galaxies than our own that gives us Hubble's law, and even the nearest galaxies are several million light-years away.

Even if the universe collapses, then, I'm convinced that there is a big asymmetry between the initial bang and the final crunch and that this asymmetry is peculiar to our universe. This leads me to a point I made earlier, in §8.2, which I now want to expand on: the argument that big bangs don't necessarily lead to locally tidy universes—universes with a second law of thermodynamics.*

Let's consider just closed universes, which necessarily start with a bang followed by expansion and end with a bang following a final collapse. (There is in fact a theorem, due to Hawking and Penrose, that says that if a universe does collapse it necessarily collapses to a state just as infinitely catastrophic as the initial bang.) Now if a collapsing universe doesn't have to exhibit increasing local order, as I've been arguing, then it follows that an expanding universe doesn't have to exhibit decreasing local order either. That is, a universe doesn't have to have a Second Law at all. If a stagnant universe can collapse without getting tidy, then the time-reversed picture, which must also be consistent with the laws of physics, shows the possibility of an initial bang leading to a universe that is simply stagnant, with no local sense of time other than the single simple fact of expansion. Furthermore, a universe that is stagnant to start with, if it is closed, will eventually collapse again without ever ceasing to be stagnant, because without any requirement that it do otherwise that is by far the most likely thing for it to do.

There's an immensely larger number of possible stagnant universes than tidy ones, and any one of those stagnant universes can collapse without any increase in tidiness other than that of getting small. Each of these scenarios can then be run backward, giving rise to a vast host of possible universes in which an initial big bang leads to a stagnant, though expanding, universe. Thus of all the possible big bangs that might initiate universes, the chance of having one that leads to a Second-

---

*The same conclusion is argued very effectively by Roger Penrose in *The Emperor's New Mind* (New York: Oxford University Press, 1989), Chapter 7, "Cosmology and the Arrow of Time."

Law universe, one with a direction of time like ours, seems to be fantastically tiny—something like 1 chance in $10^{10^{100}}$ or some huge number like that. The initial state of our universe, then, despite its apparent similarity to any other exceedingly hot, dense state, was *far* more orderly, I believe, than most of this vast number of possible alternatives. That is, our own big bang is highly exceptional as big bangs go. The question of why our big bang should be so exceptional is overwhelming, and I think that, as human beings if not as scientists, we are compelled to look for a reason. For myself, I find the situation very suggestive of some sort of divine initiative, though I think most physicists would not agree.

I've been speaking casually about locally stagnant universes, as if a uniform gas of matter at uniform temperature were necessarily the most disorderly state for a universe to be in. This is probably not so. Because gravity is a long-range attractive force, matter tends to cluster, so the picture would not be one of simple uniformity. Astrophysicists argue that this clustering would lead to star and galaxy formation and a natural appearance of the Second Law. However, the kind of argument I've been making in this chapter leads me to think that this picture of star and galaxy formation is like the macroscopic order produced by human civilization (see §9.3): It can take place only in a universe that already has a direction of time associated with ever-increasing microscopic disorder. If the universe were locally stagnant, as I understand things, there could be clustering of matter but no temperature differences; there could be no "unburning" stars and therefore no burning stars. A locally stagnant universe must look the same when you run the picture backward, and therefore it could not lead to any stellar and galactic evolution of the sort we're familiar with in our universe.

Back-to-back universes?

It is conceivable, though it seems very unlikely in terms of our present understanding, that the mechanical motion of the universe might be traced farther back in time than the moment of the Bang. This would mean that space and time were not so distorted as to make such tracing back meaningless.* If so, it seems to me very likely that, in comparison

---

*This would constitute, as I understand it, a violation of Einstein's theory, which requires that all universes begin or end with a completely catastrophic singularity. Remember, though, that Einstein's theory, like *any* theory that physicists come up with, can always be questioned.

with times both before and after the Bang, the moment of the Bang itself would be the moment of maximum tidiness and that the time direction associated with the second law of thermodynamics would therefore be reversed for times before the Bang. Entropy—untidiness—would increase the farther away you get from the Bang in either direction. And if that prior universe showed an evolution at all like ours and allowed the development of some kind of intelligent creatures, then their sense of time would also be reversed from ours. To them as to us, the Bang would be seen as the beginning of time. Neither for them nor for us would there be any possibility whatsoever of deducing the history of the universe prior to the Bang; they could not possibly know about us, and we could not possibly know about them.

PROBLEM

1. It has also been suggested that a closed universe may bounce—that each successive collapse is followed by a new bang, ad infinitum. Write a critique of this idea on the assumption that the time between bounces is very small compared to the time needed for the universe to get tidy spontaneously. There may be different possible scenarios, depending on whether information is wiped out in each successive crunch.

CHAPTER 9

Thermodynamics

9.1 THERMODYNAMICS AND ENTROPY

In Chapter 8 I referred repeatedly to *entropy* and used it as some kind of synonym for *disorder*. I'd like now to introduce entropy more formally and show you how it works as a mathematically defined physical quantity. The concept arises first in the context of thermodynamics, which is the study of thermal properties of matter. This shouldn't surprise you, because I've talked so much about the second law of thermodynamics and its relation to disorder.

Historically there have been two rather separate approaches to thermal properties. The earlier, under the name of *thermodynamics*, originated long before the microscopic nature of heat was understood and developed as an independent branch of physics with its own laws and definitions. Then, as we probed the atomic structure of matter, it became clear that what we knew as heat was simply a manifestation of random motion at the microscopic level and was subject to the same laws of mechanics and electromagnetism that we were already familiar with. The study of thermal physics from this more fundamental basis is known as *statistical mechanics;* the former axioms—the laws of thermodynamics—have been demoted to theorems in statistical mechanics, results derived from the mechanical laws governing the atoms and molecules that matter is composed of. Thermodynamics has three of these basic laws.

The *Zeroth Law* (so numbered because, although it was an afterthought, it is logically more basic than the others) starts with the idea of *thermal equilibrium* and enables us to define temperature as an indicator of whether two systems can be in equilibrium with each other. Originally it wasn't realized that such an additional assumption was needed in order to justify logically the concept of temperature. The Zeroth Law simply states:

> *If each of two systems is in thermal equilibrium with a third system, then they are necessarily in thermal equilibrium with each other.*

If this were not so, you couldn't use a thermometer to tell whether two different objects are at the same temperature.

The *First Law* recognizes heat as a form of energy and states the conservation of energy in terms of the total energy, called the *internal energy,* of a given system. (All these systems are normally sitting still, so the potential energy and kinetic energy associated with the motion of large-scale objects don't enter in.) The statement is

$$\Delta U = Q - W \qquad (1)$$

where ΔU is the change in the internal energy U during some arbitrary process, Q is the heat energy transferred to the system, and W is the mechanical work done *by* the system during that process.

The *Second Law,* the one we are primarily concerned with, is basically the statement that all physical processes are irreversible: that there is no way to get back to the state of the universe that you started from. Of course you can get a given system back to its original state, but you pay an irreversible price somewhere else and there is always a net increase in the total amount of disorder. Historically there were two standard statements of the Second Law, independently arrived at and later proved to be equivalent:

> Kelvin-Planck statement: *It is impossible to devise a process whose only effect is to draw heat from a thermal reservoir at a single temperature and convert it entirely into mechanical work.*

That is, you can't solve the energy problem by converting the thermal energy in the ocean into electric power, even though the ocean is not at absolute zero and contains a huge amount of thermal energy.

> Clausius statement: *It is impossible to devise a process whose only effect is to transfer heat from a reservoir at a lower temperature to a reservoir at a higher temperature.*

In other words, heat can't flow uphill. That's why it takes power to run a refrigerator even though all it is doing is removing energy from the things inside and transferring it to the air outside.

(There is also a fourth law, know as the third law of thermodynamics, which states the impossibility of reaching absolute zero. It is deeply important but doesn't play quite such a fundamental role as the others.)

An amazing number of consequences can be derived from these three laws, including complicated mathematical relationships involving all sorts of unexpected thermodynamic quantities, starting with entropy and going on to things like "enthalpy" and two or three varicties of what is called "free energy." My aim is to try to probe the basic ideas and give at least a hint of the richness of the subject. Now the First Law, energy conservation, adds nothing to what we already know, and the Zeroth Law, depending on how you look at it, is either trivial or so sophisticated that I don't want to get involved with it. Thus I want to direct your attention primarily to the Second Law.

The Zeroth Law, as I said, makes it possible to define temperature, so that you can say that temperatures must be equal for there to be thermal equilibrium. The Second Law then enables you right away to define the concepts *hotter* and *colder,* because heat always tends to flow from the hotter to the colder object rather than the reverse. Up to this point the temperature *scale* is more or less at your disposal as long as you keep hot and cold straight. To define a scale you can use any physical property you like that varies with temperature (the density of mercury, for example, or the stiffness of butter); you just assign numbers to different values of that property and call that the temperature.

But then you look more closely at the detailed physical consequences of the Second Law and you find—well, Kelvin found—that the Law permits you to give precise and unique definitions to two physical quantities, the *absolute temperature T* and the *entropy S*, that have the following properties:

1. The absolute temperature is always positive, and the state of lowest possible internal energy corresponds to $T = 0$. This temperature, referred to as *absolute zero*, corresponds to $-273°C$, approximately, or $-460°F$.

2. The entropy of a system depends only on its actual state, not on how it came to be in that state. At the molecular level, the entropy tells how *disorderly* the state is.

3. When the system is at a definite temperature and a small amount of heat dQ (which may be positive or negative) is added to it, the change in its entropy is given by

$$dS = \frac{dQ}{T} \tag{2}$$

4. In any real physical process, the total entropy of the universe always increases. The increase of entropy captures the Maxwell's baby effect: The disorder always increases.

To see how this works, look first at the Clausius statement of the Second Law. If a small amount of heat dQ flows from an object at temperature T_1 to another object at a lower temperature T_2, then by eq. 2 the total change in the entropy is

$$dS = \frac{-dQ}{T_1} + \frac{dQ}{T_2} \tag{3}$$

$$= \left(\frac{1}{T_2} - \frac{1}{T_1} \right) dQ \tag{4}$$

Now, we chose T_1 to be greater than T_2, so the quantity in parentheses is positive: dS is therefore positive for the allowed process in which dQ is positive, and dS is negative for the forbidden process in which dQ is negative—that is, where heat flows "uphill" from the colder to the hotter object. Thus the net entropy always increases in every allowed heat flow process.

In like fashion, when work is turned into heat at a given temperature (by friction, for example), there is just a single positive dQ, and dS is positive regardless of the temperature of the object heated. If you try to turn heat into work, on the other hand, in violation of the Kelvin–Planck rule, you have a single negative dQ, and dS is negative. In every case, no matter how complicated the process, when you add up all the entropy changes, the sum must turn out positive. The proof consists in showing that if ΔS is negative for any process, then you can exploit it to construct a machine that violates the Second Law, either in the Kelvin–Planck form or in the Clausius form.

It's often said that entropy is a measure—a negative measure—of useful energy. Thermal energy stored at high temperature is useful energy: When it is stored there's only a small increase in the entropy because of the factor $1/T$, corresponding to the fact that such energy can easily be extracted and used to drive engines. Thermal energy stored at low temperature, on the other hand, as in the ocean or a block of ice, is not useful energy: Storing energy at low temperature corresponds to a large increase in entropy, so the energy is not easily extracted and used.

I have to add that in each of these examples, I've stated the case too simply. In the first case it's true that the heat stored at high temperature can be used to do work, but only at the price of dumping some of it into a low-temperature reservoir in order to satisfy the Second Law (Kelvin–Planck version): The efficiency can never be 100%. In the second case the heat in a block of ice *can* be used (in heating your home with a heat pump, for example), but only at the price of some electrical power to run the heat pump. The heat you get out, though, is greater than the amount of electrical power you put in, because you do make use of thermal energy from the ice (see prob. 2).

PROBLEMS

1. Try to describe a world in which the zero[th] law of thermodynamics doesn't hold. Look for logical consistency—or inconsistency.
2. In discussing thermodynamics, we often make use of an idealized *heat reservoir*, which is imagined as a very large container with its contents at a definite temperature T, holding so much thermal energy that you can add or withdraw finite amounts of heat without altering its temperature significantly. Typically, in the kind of simplified models that we use to think about thermodynamics, we take thermal energy from one reservoir at a high temperature in order to run some engine, and we dispose of leftover thermal energy by transferring it to a cooler reservoir, the atmosphere for example. The fact of energy conservation makes the bookkeeping very simple: If heat Q_1 is extracted from the hot reservoir, and heat Q_2 is dumped into the cooler reservoir, then the work that the engine can do is simply $W = Q_1 - Q_2$.
    (a) Figure out the maximum efficiency of an engine run from a hot reservoir (temperature = T_1, heat extracted = Q_1) if you

dump only enough heat Q_2 into a cooler reservoir at temperature T_2 to satisfy the Second Law—that is, to avoid a decrease in the net entropy of the two reservoirs. Use eq. 2 to determine the entropy change of each of the reservoirs. The efficiency is defined as W/Q_1, the mechanical work performed divided by the heat energy extracted from the high-temperature reservoir.

(b) Figure out the maximum *coefficient of performance* of a heat pump that extracts an amount of heat Q_2 from the ground at temperature T_2 in order to supply heat Q_1 at the house temperature T_1. The difference $Q_1 - Q_2$ must be supplied as mechanical work W to run the heat pump. The coefficient of performance is defined as Q_1/W; explain why this is a reasonable definition. Note that in each case—parts (a) and (b) of this problem—the measure of performance is essentially benefit/cost.

(c) Work out some reasonable numerical examples for parts (a) and (b). You get the absolute temperature T by adding 273 to the number of degrees centigrade (also known as degrees Celsius).

(d) Find the maximum possible coefficient of performance for a heat pump to be used as an oven for baking. A moderate oven is around 175°C, and you can take room temperature as 20°C. After the oven cools off again, will the room have warmed up more or less than if a conventional oven had been used? By what factor?

9.2   ENTROPY AND DISORDER

Section 9.1 shows how a common-sense statement of the Second Law in terms of the familiar irreversible character of thermal processes can be made to yield a powerful mathematical tool—the entropy—and can be exploited to give all sorts of precise information about the behavior of physical systems. What hasn't emerged yet is how this is related to the microscopic disorder discussed in §8.2 and to the description of the Second Law as a statement of the tendency toward disorder. The connection lies in statistical mechanics, the discipline that derives the thermal properties of matter from the mechanical laws governing their microscopic behavior.

The First Law is easy to understand in terms of the microscopic laws, because we know already that energy is conserved and we understand that a transfer of heat is nothing but a transfer of energy—kinetic, potential, electrical, or what have you—in a microscopically random form. It's the Second Law that poses a problem. How do the microscopically reversible laws that govern the atoms and molecules give rise to a law of irreversibility? The answer lies with Maxwell's baby ($\S 8.2$).

I've probably given the false impression that you need only the microscopic laws of motion to do statistical physics, whereas the fact is that you need another assumption. For a complete description, you also need information about initial conditions. A complete specification of initial conditions—positions and velocities of all the particles in the universe—is clearly out of the question, and besides, we're not looking for the complete detailed predictions that such information might lead to. So what do we know about initial conditions that would provide useful additional information about the present behavior of things? The answer is in the matter of order versus disorder. The universe began in a state of great orderliness (the state produced by the Big Bang was extremely orderly, as I explained in $\S 8.4$) and is progressing toward an eventual state of total disorder. This is the fact that we must incorporate in some way into our microscopic laws and that will lead to the statements of irreversibility that we see in the Clausius and Kelvin–Planck statements of the Second Law.

The form that this extra assumption takes has been the subject of much discussion and philosophizing on the part of physicists over the years. The advent of quantum theory added to the confusion but did not, I think, alter the basic question. We usually make an assumption of microscopic randomness, in a form that goes something like this:

> *All possible microscopic states consistent with the given macroscopic information are equally probable.*

What does such an assumption have to do with the universe's running down, and how does it work?

What this assumption does is focus, not on the initial tidiness, but on the eventual untidiness, in this sense: It says that the present microscopic state does not contain any hidden information about some future state of microscopic tidiness, *as it would in any of the time-reversed scenarios mentioned in $\S 8.1$.* The present state certainly contains hidden information about *past* states of tidiness, so that if you run the equations of motion backward those tidy states reemerge. As far as the future behavior of the system is concerned, however, this hidden information makes no difference at all. Almost any randomly chosen microscopic state will behave in the same way in the future, provided only that the macroscopic conditions—volume, total energy, pressure and temperature distribution, and so on—are the same. The exceptions to this statement have the kind of fantastically small probabilities, like 1 in $10^{10^{100}}$, that I was discussing earlier and are of no practical importance whatever.

This basic assumption of microscopic randomness, then, says that nothing weird is *going* to happen, so it is a statement not about initial conditions but about *final* conditions. What happened to the initial conditions? It turns out not to be necessary to say anything about them; the importance of the statistical assumption is in what it says about the future, not in what it says about the past. The assumption would be equally valid in a totally rundown universe that was was also untidy to start with and untidy forever after. Thermodynamics would still be true in such a stagnant universe, though there'd be nobody around to know it.

Now, how does this assumption work? I'll try to give you a sketch of the logic—more verbal than mathematical. The first step is to study the laws of thermodynamics and scc how they arise out of the microscopic laws. I'll describe this and omit any discussion of all the hard work that comes after that by which physicists derive the detailed thermal properties of specific systems.

The Zero*th* Law and thermal equilibrium

You start by looking at the Zero[th] Law. You put two systems in equilibrium with each other, which means that you specify the total energy of the two together, but not how it's distributed between them, and you let them settle down for a long enough time that no further changes take place (we call this the relaxation time). Even when the two systems have totally settled down and the situation seems completely static, they are still interacting with each other at the microscopic scale, and small transfers of energy are constantly taking place in a very random fashion— what we call energy fluctuations. Now you use the statistical assumption to figure out the likelihood of different possible distributions of the energy between the systems, using some approximations that are extremely good when the number of atoms is large.

What you find is that the probability of a system's being in a state of energy E is proportional to a certain universal function of E called the *Boltzmann factor.* This factor is an exponential function of E, $e^{-\beta E}$, where β is a positive parameter, so the probability that E is much larger than its average value is extremely small. As it turns out, the probability that E is much smaller than its average value is also very small, because the number of states in a given energy interval is much smaller at low energy than at high energy. With this "density-of-states" factor included, the probability density in E for n free particles, for example, is given by

$$P(E) = CE^{(\frac{3}{2}n-1)}e^{-\beta E} \tag{1}$$

where C is a proportionality constant. This is shown in fig. 1 for several values of n. The energy scale is adjusted around E_{av}, the average value of E, which is in fact proportional to n. What you can see from these graphs, and what continues to astonish me, is the way the density-of-states factor, of the form E^N, combines with the simple exponential factor $e^{-\beta E}$ to produce such a very narrow distribution as the exponent N simply gets very large. For 10^{24} particles (a typical number on the laboratory scale), the spread in the variable E is around 10^{-12} E_{av}, far tinier than you could observe experimentally by any normal means. Thus the random fluctuations in the energy are, in fact, extremely small.

The important point for this discussion is that the parameter β *has exactly the same value for any two systems in equilibrium with each other.* This means (look back at the statement of the Zeroth Law in §9.1) that β can play the role of a temperature and the Zeroth Law is satisfied. You can see that when β is very small, large values of E are only slightly less probable than small values, corresponding to a hot system. When β is very large, on the other hand, large values of the energy are very unlikely, corresponding to a cold system; absolute zero corresponds to the limit in which $\beta \to \infty$ and the energy is at its absolute minimum.

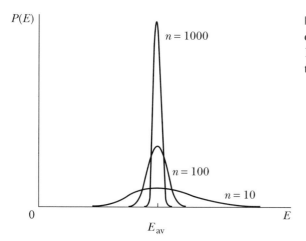

P(E)

n = 1000

n = 100

n = 10

0

E_{av}

E

Fig. 1 The probability distribution function (eq. 1), for different values of the particle number n.

   In other words, the temperature you're used to is low when β is large and high when β is small. In fact, β turns out to be exactly proportional to $1/T$, where T is measured on the absolute temperature scale due to Kelvin. There's a constant of proportionality that depends on your arbitrary choice of scale for defining temperature. The standard absolute temperature scale, called the *Kelvin scale,* uses the same units as the centigrade scale, and in these units the proportionality constant is $1/k_B$, where k_B, known as the *Boltzmann constant,* is given by

$$k_B = 1.38 \times 10^{-23} \, J/K \qquad (2)$$

Here K is the symbol for a kelvin, which is the same as a centigrade degree, as I said, even though the zero point is different for the two temperature scales.* In terms of the absolute temperature T, the Boltzmann factor $e^{-\beta E}$ can now be written in the form $e^{-E/k_B T}$, which is more familiar to most physicists.

The Second Law and the tendency to disorder

The First Law is obvious from the microscopic view (it's just the conservation of energy), so we go on to the Second Law, the interesting one. How does the Second Law follow from the statistical assumption? It doesn't exactly fall in your lap: You start constructing and studying the thermodynamic functions and their properties, you use the definition given in eq. 9.1.2 to define the entropy, and after a while all the mathematics falls into place. There's a little problem in that statistical mechanics deals for the most part with static properties of systems in equilibrium while the Second Law has to do with processes—with changes in time. It's not so obvious how you use relationships that don't involve time to figure out that the entropy must always increase with time.

   What you find, when you go back to looking at the microscopic states of a system, is that the entropy is very simply related to the number of

---

*A very theoretical theorist might choose units in which k_B is unity so that temperature would be seen as equivalent to energy and would be measured in joules. A temperature of 1 J would then be around 10^{23} °C and would be that temperature at which the mean thermal energy of an atom was around 1 J. A more useful energy unit for this purpose is the electron-volt (eV) introduced in §5.5, which corresponds to a temperature of about 10^4 K.

different possible microscopic arrangements, called configurations, that are consistent with the given macroscopic information. This is closely related to the untidiness that I discussed in connection with Maxwell's baby and our sense of time. There's some arbitrariness in how you count such configurations, but it turns out not to matter much, and you get something like

$$S = k_B \ell n \, P \qquad (3)$$

where P is proportional to the number of possible configurations. The expression $\ell n \, P$ in eq. 3 stands for the natural logarithm of P, or the logarithm to base e. (See *natural logarithm* in the Glossary.) An arbitrary factor on P just leads to an unimportant additive constant in S.

Now you can see the relationship to the Second Law: If all the different microscopic arrangements are equally likely and the changes in time involve some random process like Maxwell's baby (§8.2), then the system will progress with high likelihood toward a macroscopic state that corresponds to the largest number of possible configurations—that is, the state of largest entropy. In the case of Maxwell's baby, the macroscopic state of all the toys' being on the floor can be achieved in all sorts of different ways, corresponding to a large number of microscopic configurations, while having them all on the shelf counts as just *one* configuration. Where the number of particles is even modestly large, the likelihood of exceptions to this tendency to disorder becomes extremely small, and if the number of particles is itself really big, as in our universe, then exceptions to the Second Law become fantastically unlikely.

PROBLEMS

1. Find the exact value of the temperature corresponding to 1 eV.
2. **(a)** In the example of Maxwell's baby, the number of ways the room could be untidy, compared to its single tidy state, is around 10^{100}. Find the entropy difference in macroscopic units (J/K) between the untidy room and the tidy room.
   (b) Find the number of configurations P (eq. 3) that corresponds to an entropy of 1 J/K.
   (c) Find the entropy associated with Really Big Numbers (see §8.2) such as $10^{10^{100}}$.

9.3   CIVILIZATION AND EVOLUTION

Does evolution or the purposive activity of humans produce an increase in the orderliness of the universe? Does the parent picking up the toys violate the Second Law? I think not. Of course we as humans are capable of producing a kind of order—the kind that is important to us—out of the disorder around us, but the things we move around to produce order (toys, bricks, trees, and the like) are so large, and hence so few compared to the members of atoms and molecules involved in the increasing *disorder* of the world, that the effect is negligible.

Can we make some kind of quantitative estimate of the amount of order produced by the human race throughout its millenia of culture and civilization? What we need to calculate is the decrease in the total entropy associated with these organizing activities. Pursuing the model of Maxwell's baby, let's suppose that there have been around 10 billion human beings since the race began (a substantial fraction of whom are alive now because of the way exponential growth works), that each one of them has had the care of some 10 billion toys or other items (bricks and so forth), and that out of 10 billion possible places that each brick might occupy, the person responsible for it puts it in the one nice, tidy, elegant place that constitutes order. This picture can now be used to calculate the change in entropy by means of eq. 9.2.3. We follow the same logic as in the argument following table 8.2.1 in order to figure out the total number P of possible arrangements: The total number of toys is now 10^{20} (10 billion people with 10 billion toys each) and the number of possible places for each toy is 10^{10}, so the number of configurations in the *absence* of any ordering is

$$P_1 = (10^{10})^{(10^{20})} = 10^{10^{21}} \tag{1}$$

The corresponding entropy, then, is

$$S_1 = k_{\mathrm{B}} \ell \mathrm{n}\, P_1 = k_{\mathrm{B}} \left(\frac{\log P_1}{\log e} \right) \approx (10^{-23}\,\mathrm{J/K}) \times \frac{10^{21}}{0.43} = 0.02\,\mathrm{J/K} \tag{2}$$

a small value on the macroscopic scale. There's only one orderly configuration in this model, so S_2, the entropy of the ordered state, is zero (because $\ell \mathrm{n}\, 1 = 0$), and the change in entropy is therefore a decrease of 0.02 J/K. This is a decrease in entropy comparable in magnitude to the increase produced by, say, lighting a match. So much for civilization.

It's the smallness of the Boltzmann constant k_B that keeps ΔS small, and it's not just a coincidence that k_B is so small. It's the quantity $k_B T$ that gives the average energies on the atomic scale, and since 10^{23} is a typical number of atoms in a small macroscopic system, these energies are necessarily smaller by a factor of $\sim 10^{-23}$ than energies on the macroscopic scale, which are of the order of joules. It follows that k_B has to be of the order of 10^{-23} in typical macroscopic units. In these terms the number 10 billion, the crude number used in setting up the model, is a very small number. The important number turns out to be the total number of toys, 10^{20}, which is comparable to the number of atoms in the match head and therefore tiny compared to the number of atoms in our local environment. Thus the kind of orderliness that's important to us as human beings involves the rearranging of such a small number of objects that it simply doesn't begin to reverse the constant increase of disorder (involving huge numbers of atoms, photons, and so forth) in the universe around us.

One fact that emerges from the discussion of Chapters 8 and 9 is that civilization—and life of any sort—is possible only in a universe of constantly increasing disorder. The fluctuations associated with human activities are just a tiny ripple in the huge tide of increasing disorder in our physical environment. It is this environment that makes possible all the processes necessary to life, from the shining of the sun to the muscular and mental activities of our bodies and minds, and that gives us the framework of directed time in which alone we are able to think and act. The second law of thermodynamics is our friend, not our enemy.

PROBLEM

1. Make rough estimates, using the same kind of logic as in eqs. 1 and 2, of the change in entropy that occurs in each of the following situations.
   (a) Fifty grams of perfume evaporate and diffuse throughout a room.
   (b) Maxwell picks up all of his baby's toys off the floor and puts them back on the shelf.
   (c) Half the grains of sand on a mile-long beach are reddish in color and half are greenish. You sort the grains into two piles according to color.
   (d) The number of particles in the universe is something like 10^{100}. Suppose this is also the number of places that a given particle can occupy. The particles are all initially in the same location, and then they become dispersed randomly among all the locations.

QUANTUM PHYSICS

"Some smaller quantum of earthly enjoyment"

THOMAS CARLYLE (1795–1881)

At the beginning of this century classical physics—basically Newton's laws of mechanics and Maxwell's equations for the electromagnetic (EM) field—seemed to give a very adequate account of the physical world, except for a few small difficulties that no one expected to be major problems. One of these difficulties was the Michelson–Morley experiment performed in 1887, which seemed to show that the earth is standing still. I discussed in Part I how this experiment was the precursor of a major revolution in our understanding of space and time. There were also some other little warning signs that scientists were aware of, this time in the domain of the very small, and these too blew up finally into a raging storm of confusion and change—a revolution fully as significant as relativity in its shattering effect on how we view the physical world. The peculiarities that are characteristic of this new revolution show up especially at the atomic or subatomic level, but the implications extend all the way into philosophical discourse on the fundamental nature of reality.

This new understanding of the nature of physical reality is characterized by the word *quantum:* we say quantum theory, quantum mechanics, quantum physics, quantum electrodynamics, quantum just about anything. It's an old word in English meaning "a certain amount," having started as a Latin word meaning "how much." Its use in physics referred originally to the discreteness exhibited at the microscopic level, where energy and many other physical quantities were found to occur in definite discrete amounts—contrary to normal expectations. (Look up *discrete* in the Glossary, and note that it's not the same as *discreet.*)

The story of quantum physics has two sides, one descriptive and physical, the other analytical and mathematical. On the physical side we have to face the fact that the world behaves in an odd fashion that is deeply inconsistent with our classically based intuitions. The most obvious aspects of this odd behavior are these:

Waves behave like particles.
Particles behave like waves.

The matter goes much deeper than this, as you'll see, but this split personality of the fundamental entities of nature runs all through quantum physics. It's important to remember, since you hear so much about "wave–particle duality," that classically particles and waves bear no resemblance to each other at all, so this ambiguous behavior really shook up our habitual ways of thinking about physical reality.

The other side of the story is analytical: The mathematical machinery that has gradually been developed to deal with this odd behavior is itself rather odd—abstract and not at all intuitive. (Physicists have gotten quite used to the mathematics of quantum theory, of course, because it really does describe how things work; nevertheless, deep down, I think most of us still find it strange.) The most fundamental characteristic of the change in our mathematical description of nature could be summarized like this:

> *You can't predict the future.*
> *You can only state the odds.*

Predicting the future, in fact, was in essence the central task of physics: to find and understand the equations of motion that precisely determine how a system will behave in the future given its present state. If physics can't do that, what is it good for? And yet, this is just the situation we find ourselves in, and physics has by no means thrown in the sponge. The new equations of motion of quantum physics—the Schrödinger equation, basically—do not predict what the outcome of a given experimental observation will be; they predict only the *probabilities* for the different possible results. This makes physics at the small scale very fuzzy, but it has very little effect on physics on the macroscopic scale because the statistics are *so good,* with billions upon billions of microscopic events contributing to an ordinary observation at the scale of our human bodies, that the probabilities turn into virtual certainties. If you play roulette long enough, your winnings—or losings, rather—will be very accurately determined by the probabilities.

The mathematical structures that we've had to develop to deal with this nondeterministic state of affairs are quite abstract, as I said, and seem far removed from the physical realities of the laboratory—so much so that you need some very unobvious rules for interpreting the mathematics and relating it to experimental observations. We find that in practice the machinery—mathematics plus rules of interpretation—works extremely well, enabling us to describe any physical system and predict experimental results with great precision. Nevertheless, the physical reality behind this abstract structure that physicists have developed remains obscure. We can describe what we *see*, so to speak, but we can't describe what *is*. I'll discuss some of the ways people have tried to deal with this problem in Chapter 15.

CHAPTER 10

Odd Behavior of Particles and Waves

10.1 WAVES AND PARTICLES

I've said that one of the peculiar things about the quantum world is a confusion between waves and particles. Particles such as electrons act like waves, and waves such as light signals act like particles. This is to say that all of these things, whatever you call them, show *both* wave and particle properties. An early name for the quantum theory of particles, in fact, was wave mechanics. Both *wave* and *particle* are familiar words, of course, but their meanings are rather vague in our normal usage. We looked at both waves and particles in Part I, but I'd like to reiterate here something of what these words mean in classical (prequantum) physics.

Waves

In my description of waves in §1.3, which you should review now, I stressed the fact that a wave is a spatial pattern that varies with time. The spatial pattern typically fills all the available space (as sound waves do in a room), and the wave motion is extremely complex, compared to the motion of a particle, in that it involves some kind of oscillatory motion at every point in space—an infinite number of different motions taking place at the same time.

It's a very distinctive characteristic of wave motion that two or more waves of the same kind combine to give a resulting wave that is the algebraic sum of the separate waves. Two identical waves that are in step with each other combine to give a wave of twice the amplitude, while two identical waves that are out of step combine to give a wave of zero amplitude—no wave motion at all. The waves are said to interfere, and this interference can be constructive, as in the first example, or destructive, as in the second. The terminology we use to describe waves is introduced briefly in §1.3 and treated more fully in Chapter 11.

Particles

The concept of a particle is simpler than that of a wave and should be fairly familiar because we talked so much about particles in Part I. The word is used in several different senses, though, and we need to be as clear as we can about the idea because we talk so much about wavelike properties and particlelike properties in quantum theory. What I mean by a particle is simply a very small object or even, ideally, an infinitesimal object—a *point particle*—with a definite mass all concentrated at a single location in space and with definite values of velocity, momentum, and kinetic energy. These quantities—location, velocity, momentum, and energy—can all vary with time. You normally need only specify the location and velocity, and the momentum and energy are then determined. If you compare their descriptions, you'll see that a particle is a much simpler thing than a wave, because the description of the particle involves only six quantities (three components of position and three components of velocity), while the description of a wave involves an infinite number of quantities (the value of the field variable at every point in space). A particle may oscillate, to be sure, but while a wave consists of an infinite number of oscillations, a particle can show only one.

Despite the fact that waves and particles have so little in common that it seems almost pointless to compare them, I want to list here the essential differences between these two concepts, with an eye to the quantum confusions that I'll be talking about.

1. A wave fills all available space, while a particle has a single location at any given moment and is totally absent from all other points in space.

2. A wave can encompass many different types of motion at the same time, while a particle can do only one thing at a time.

3. A wave is characterized at any given moment by a wave pattern, which may show one or more different wavelengths, while a particle is characterized at any given moment by its position and velocity.

4. A wave is a pattern of variations in some physical variable, such as an electromagnetic field or gravitational field, while a particle exists as an object and involves no kind of field variable.

5. Waves interact by combining; two waves entering the same region of space can reinforce or cancel each other, as described in §1.3. Particles interact by what we call scattering: hitting each other and bouncing off in different directions.

I want you to see how unreasonable it is that there should be any confusion of identity between particles and waves, so that you'll appreciate the psychological stress that this confusion of identity caused when it showed up in the actual behavior of the waves and particles of nature.

PROBLEMS (refer also to §1.3)

1. Specify whether each of the following kinds of motion represents a wave motion or a simple oscillation, explain your answer, and give three more examples of each: **(a)** the moon going around the earth, **(b)** ocean tides, **(c)** a vibrating violin string, **(d)** a yoyo, **(e)** a cork bobbing on the ocean, **(f)** a bar on a xylophone, **(g)** a radio transmission.

2. (a) For each of the examples of wave motion in prob. 1, including the three extra examples, tell whether the wave is one-, two-, or three-dimensional, and specify the field variable.
   (b) Give three examples from your everyday experience of wave pulses, and again specify the field variable.

 **ODD BEHAVIOR**

Now, what's the difficulty? What are the oddities in the behavior of waves and particles that are supposed to point toward a major change in our basic understanding of physics? Historically, these oddities had mostly to do with light and other forms of electromagnetic radiation, and the

1880

1885 Balmer formula for hydrogen spectrum

1887 Michelson—Morley experiment

1890

1897 Discovery of the electron Thomson

1900 1900 Black—body radiation Planck

1902 Photoelectric effect Lenard

1905 Photoelectric effect Einstein (and special relativity)

1910 1911 Nuclear atom Rutherford

1913 Bohr model of hydrogen atom
1914 Franck—Hertz experiment

1914
       Photoelectric effect Milliken
1916

1920

1923 Compton effect
1924 de Broglie
1925 Quantum theory Schrödinger/Heisenberg/Pauli
1927 Uncertainty principle; electron diffraction Davisson—Gerber;
         quantum electrodynamics Dirac
1928 Dirac formulation of quantum theory; Dirac relativistic equation
1930   for the electron

1932 Neutron Chadwick
      Positron Anderson

Fig. 1 (opposite)   Rough chronology of the early history of quantum theory.

difficulty in each case was that the radiation was behaving in a particlelike way. The trouble wasn't at first identified with anything like particles—it was simply a puzzling violation of standard EM theory. The early stages in the development of quantum theory consisted in recognizing these violations as particlelike behavior, at which point the term *photon* was coined, and then learning how to make use of this photon picture to understand a lot of different experiments. The picture was full of contradictions, but it worked for many practical purposes. It took twenty years or so for this not very satisfactory picture to evolve into a proper and consistent theory.

Equally unexpected behavior of electrons was also observed from early on, and after a while it was recognized as wavelike properties of what should have been simple particles. This idea of matter waves played a very important part in the development of a proper quantum theory, as you'll see in due time. The first mathematically consistent quantum theory, in fact, was a theory of electrons advanced in 1925 (the quantum theory of EM radiation appeared within the following two years).

In the next section I'll describe two situations in which the behavior of EM radiation is at odds with classical physics. I'll take up a third example, the particularly important case of atoms and their spectra, which I want to go into a bit more deeply, in Chapter 12. A chronology of the major events in the development of quantum theory appears in fig. 1.

10.3 PARTICLES OF LIGHT

At the end of the nineteenth century we thought we understood light very well: It's an electromagnetic (EM) wave—wave patterns in the electromagnetic field—which means that there's an electric field and a magnetic field at every point in space, and the electric and magnetic field vectors vary in space and time so as to make a wave pattern. The equations that describe electric and magnetic fields, Maxwell's equations, describe light waves completely accurately, including getting their speed c exactly right.

What, then, were the physical effects that didn't fit this picture? In certain circumstances, as mentioned in §10.2, light insisted on behaving like particles, each with a definite energy and momentum, appearing and disappearing according to some unknown rules but not divisible into smaller bits. I want to give you a brief description of two of these effects, thermal radiation and the photoelectric effect, in order to give you a feeling for how this particlelike character shows up experimentally.

Thermal radiation

The first difficulty I want to tell you about is a frustrating discrepancy that physicists found between theory and experiment for the radiation emitted by hot objects, in particular at the high-frequency end of the spectrum. As you know, all hot objects emit radiation.* Hot stoves emit heat (infrared radiation), and very hot objects also emit light (becoming red-hot, white-hot, etc.). In fact anything that isn't at absolute zero emits radiation. Even a cake of ice emits infrared radiation. It may seem to exude coldness, but what is really happening is that it soaks up the radiation *you're* emitting and gives very little back (some, that is, but not much), while an object at room temperature—a wall, say—gives off just enough radiation that you don't feel either cold or hot standing near it. The colder the object the less radiation it gives back, but it always emits some radiation.

The reason these objects emit radiation is that all the molecules are in random thermal motion for any temperature above absolute zero. (Recall that the absolute, or Kelvin, temperature scale starts at absolute zero, the point of no thermal motion at all.) Thermal motion includes

---

*Physicists often talk about "black-body radiation" in connection with this problem; this is because there's a theorem in thermodynamics that says that a perfect *absorber* of radiation (black and not shiny, like soot) is also a perfect *emitter* of radiation. What this means is that when it's heated to a given temperature, a perfectly black object emits the maximum possible amount of radiation at every wavelength, whereas a hot object that's not perfectly black emits less radiation in an uneven spectrum peculiar to the particular object. This distinction makes very little difference for our discussion; the difficulty with the high-frequency part of the spectrum is there for any hot object. It's just that the problem is easier to discuss mathematically in terms of an ideal black body.

the motion of electrons, and because electrons are electrically charged particles, that motion in turn produces EM radiation. The classical theory of heat tells how much radiation should be emitted and how it should be distributed in wavelength: If $I(\lambda)\,d\lambda$ is defined as the rate at which energy is emitted in the small wavelength range $d\lambda$, then the classical theory says that

$$I(\lambda) \propto \frac{T}{\lambda^4} \tag{1}$$

where T is the absolute temperature. This implies two things that are in violent contradiction to our experience. First, it says that the distribution in wavelength is the same for all temperatures except for the overall factor T. The cool object would glow with exactly the same color as the hot object (blue, because the factor $1/\lambda^4$ strongly favors short wavelengths), just not so brightly. In fact, when you work out the intensity of radiation predicted in this classical picture, you find that an object at room temperature would be as bright as the sun—though not for long, because it would lose all its thermal energy almost instantaneously.

The second problem is that when you add the contributions from all different wavelengths [by doing the integral $\int I(\lambda)\,d\lambda$] to get the total intensity of radiation from a hot object, it comes out infinite. In practical terms this means that any object not already at a temperature of absolute zero would *immediately* radiate away all its energy and drop to absolute zero. The whole problem comes at the short-wavelength, or high-frequency, end of the spectrum, where the intensity becomes infinitely large as λ approaches zero, producing what's called the *ultraviolet catastrophe.*

What we see experimentally is quite different from this: There is a strong *suppression* of the short-wavelength end of the spectrum, so at low temperatures there is virtually no radiation in the visible region, and at any temperature there is a cut-off in intensity that keeps the total rate of emission finite. Fig. 1 shows what the actual intensity distribution looks like at different temperatures. [In the graph the intensity $I(\lambda)$ is shown rescaled by a factor $1/T$ in order to make it easier to compare different values of T; classically this makes the curve independent of T—see eq. 1 and the curve marked "classical distribution" in fig. 1.] Notice how the distribution across the visible region changes with temperature, peaking at the red end for lower temperatures and toward the blue end for higher temperatures.

I now want to explain how this suppression can be accounted for by supposing that light acts somewhat like particles. The photon hypothesis

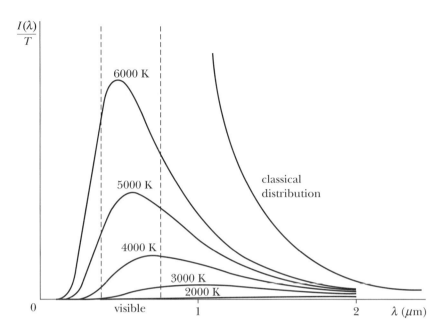

Fig. 1 The black-body spectrum at different temperatures. All the curves become the same at very long wavelengths. The curve labeled "classical distribution" is the same for all temperatures because of the scale used.

in a primitive form dates back to 1900, when Max Planck, who had been fooling around with this problem of the ultraviolet divergence (the infinite integral that I was talking about) hit on a formula that worked amazingly well. He then tried to *understand* what he'd done, and the best he could do was to say that for some reason the emitting atoms could gain or lose energy only in discrete amounts—in multiples of an amount ΔE that is related to the frequency of the emitted radiation by

$$\Delta E = hf \qquad (2)$$

I believe that Planck originally introduced the constant h with the idea of letting it go to zero at the end of the calculation (a standard mathematician's trick) to see if in that way he could get a finite answer. Well, the limiting procedure didn't help at all, but he found that if he picked an appropriate *nonzero* value for h, he got a beautiful fit to the radiation spectrum at every temperature. Planck was not thinking at all of making light behave like particles; he attributed the effect to the process by which the radiation was emitted by hot objects rather than to the nature of the radiation itself.

Let's try to understand why this trick would control the divergence. At absolute temperature T, the amount of energy available for any given microscopic purpose is loosely $k_B T$, where k_B, Boltzmann's constant (see §9.2), was already known and understood as the factor describing thermal energies on the atomic scale:

$$k_B = 1.38 \times 10^{-23} \, \text{J/K} \qquad (3)$$

$$= 8.62 \times 10^{-5} \, \text{ev/K} \qquad (4)$$

(The energy unit eV, the electron-volt, was defined and described in §5.5.) If radiation of frequency f can exist only in amounts hf, then when $hf \gg k_B T$ there isn't enough thermal energy to produce light of that frequency and so the higher frequencies are suppressed, as needed.

Planck spent a lot of effort trying to justify his assumption on more fundamental grounds, but it was still far too early in the game for the necessary kind of quantum thinking to be at all possible. Such an understanding came only with the passage of time. Planck wasn't thinking about these packets of energy as *particles* in any sense, but his result is just what you'd get if the radiation did consist of particles of energy hf. His constant h is now called Planck's constant, and its approximate value is

$$h = 6.626 \times 10^{-34} \, \text{J} \cdot \text{s} \ \text{(joule-seconds)} \qquad (5)$$

$$= 1.24 \text{eV} \cdot \mu\text{m}/c \qquad \text{(microscopic units)} \qquad (6)$$

(Note the use of the speed of light c in the units for h in eq. 6. This is a standard practice and means here that the product hc, which has the units of energy $\times$ length, has the value 1.24 eV $\cdot$ µm.)

The photoelectric effect

If x-rays or ultraviolet (UV) radiation strikes a metal surface (the target), electrons are knocked out, as shown crudely in fig. 2. This effect was first studied by Philipp Lenard in 1902. The *photoelectrons*, as they're called, can be collected by an electrode nearby and detected as a *photoelectric*

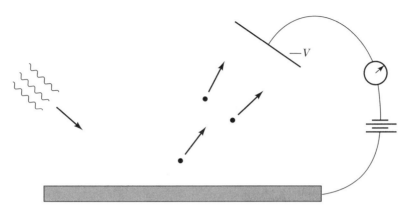

Fig. 2   The photoelectric effect.

current. The energy of the electrons can be studied by putting the collecting electrode at a negative potential $-V$ relative to the target. That is, you introduce an electric field that opposes the motion of the electrons toward the collecting electrode; only electrons with enough energy can overcome this *potential barrier.* It was found that if you increase V to a certain point, the photoelectric current goes to zero, which means that there is a definite maximum energy for the electrons emitted.

It was known that it takes a certain minimum amount of energy to shake an electron loose from the metal (the *work function* φ_0 of the metal), so it was believed that a sufficiently strong EM wave could shake the electron and give it enough energy to come loose. It was expected, for any given wavelength of radiation, that if the intensity I were great enough, photoelectrons would be produced, and that if the intensity were further increased, both the number and the kinetic energy K of the electrons would also increase. As the intensity was made small, K was also expected to become small.

What Lenard found was in fact very different. For long-wavelength radiation no photoelectrons are seen *no matter how great the intensity.* There is a sharp threshold λ_0, beyond which there is no effect. For $\lambda < \lambda_0$, on the other hand, photoelectrons are produced no matter how low the intensity. The rate of emission of photoelectrons is proportional to I, reasonably enough, but their energy doesn't depend on I at all, even down to extremely low intensities. There is instead a definite maximum kinetic energy that depends only on the wavelength of the radiation used and that increases as λ decreases. There was no way to understand any of these features in terms of the classical theory of EM radiation.

It was Einstein who in 1905 thought of applying Planck's idea to the photoelectric effect. He picked up on Planck's formula and began to think in terms of the radiation itself being quantized—divided into discrete, indivisible bits. In a paper that proposed this particlelike concept quite generally, one of his examples was the photoelectric effect. If the light can be thought of as particles with energy hf,

$$E = hf \tag{7}$$

hitting the surface [Planck's formula (eq. 2)], and if the electrons are bound in the metal with an energy φ_0, then hf has to exceed φ_0 before any electrons can be knocked loose. Thus we have the threshold frequency

$$f_0 = \frac{\varphi_0}{h} \tag{8}$$

below which no photoelectrons are produced. The wavelength λ of an electromagnetic wave is related to its frequency f by

$$f\lambda = c \tag{9}$$

c being the wave speed (see §11.2, especially eq. 11.2.12, for details), so the threshold wavelength associated with f_0 is

$$\lambda_0 = \frac{c}{f_0} = \frac{hc}{\varphi_0} \tag{10}$$

Note that wavelength is inversely proportional to frequency, so it's the shorter-wavelength photons that have higher energy and produce photoelectrons.

High intensity just means more photons, not more energetic ones, and if none of them is energetic enough to kick out electrons, then no matter how many of them there are, no photoelectrons can be produced. (Even at very high lab intensities, the likelihood of two photons hitting the same electron at the same time to knock it out is extremely tiny.) For $f > f_0$, the maximum kinetic energy the electrons can have is the energy left over:

$$K_{\max} = hf - \varphi_0 \tag{11}$$

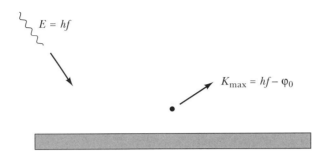

Fig. 3   The Einstein picture. Each photon has energy hf, and the photoelectron that is produced has maximum kinetic energy $hf - \varphi_0$.

as shown in fig. 3, exactly consistent in fact with the observation. The most precise confirmation of Einstein's description of the process was actually obtained later, by Robert A. Millikan, around 1915.

Note especially that while you might say that the photon hypothesis explains black-body radiation and the photoelectric effect, *we have not explained photons.* What we've found is that EM radiation sometimes behaves like particles, but we still have no idea why.

What is the momentum of a photon? You learned in §5.4 that ordinary particles can never reach the speed of light but that a particle with zero rest mass *always* travels at speed c. The photon, then, is a particle with zero rest mass, and eq. 5.4.14, rewritten as

$$E = pc \tag{12}$$

(because the photon has no rest energy), relates its momentum and energy. Note that this corresponds to the general energy–momentum relation 5.7.17 for the case $m = 0$. Even before photons appeared on the scene, it was known that classical EM radiation carried momentum as well as energy, and in fact the relation between energy and momentum for the classical field is exactly that given by eq. 12. There is thus an interesting—and essential—consistency between the classical theory of electromagnetic fields and the particle dynamics described in Chapter 4.

Eq. 12 now enables us to get the momentum of a photon in terms of the frequency or wavelength of the corresponding radiation. Inserting the Planck relation $E = hf$ into eq. 12 yields

$$p = \frac{E}{c} = \frac{hf}{c} \tag{13}$$

which can be made more concise by putting it in terms of the wavelength by the relation $c = f\lambda$ (eq. 9):

$$p = \frac{h}{\lambda} \qquad (14)$$

This, together with the Planck relation, gives the rules for relating wave properties, frequency and wavelength, with the two basic particle properties, energy and momentum. Just as the particle energy is associated most naturally with the wave frequency, by the Planck relation, you see here that the particle momentum is associated most naturally with the wavelength. This is deeply related to the fact, discussed in §4.1, that energy and momentum are the conserved quantities of nature associated with the basic displacement symmetries in time and space, respectively.

PROBLEMS

1. Find the temperature at which photons in the visible range ($\lambda = 500$ nm, say) would have energy $5k_B T$—the approximate point where $I(\lambda)$ has its maximum). Check your answer for reasonableness against fig. 1.

2. (a) Find the wavelength at which a photon would have an energy equal to (i) the kinetic energy of a dust speck floating in the air (rough estimate), (ii) the rest energy of an electron, (iii) 1 eV.

 (b) How many FM radio wave photons would it take to have a total energy of 1 eV? (See §11.5, particularly fig. 11.5.1, for information about different kinds of electromagnetic wave.)

3. (a) For a metal surface with work function φ_0, draw a careful graph of K_{max}, the maximum energy of photoelectrons emitted (eq. 11), versus f, the frequency of the incident radiation. Show clearly the intercepts and slope of the function.

 (b) Draw a careful graph of K_{max} versus λ, the wavelength of the radiation ($\lambda = c/f$). Show clearly the behavior of K_{max} as λ approaches zero.

 **10.4** WAVELIKE ELECTRONS

Interference, as I described it in §1.3, is a wave phenomenon; it arises when two or more waves can be combined at the same location and either cancel or reinforce each other. Interference is seen with sound waves, light waves, radio waves, and indeed every variety of wave. The new and puzzling thing is that the same effect is observed with *electrons,* and the interference patterns observed are just like those you see with electromagnetic waves.

Electron interference is the only example in this chapter where you actually see the wave properties of particles of matter; as I'll explain later, these wave properties of electrons are also directly responsible for the character of atomic states (Chapter 12) and are the main point of the Schrödinger equation—the wave equation that we use to describe quantum particles. Electron interference is a special case also in that the effect was predicted before it was observed, so it didn't produce the same kind of puzzlement that photons did. In 1923 Louis de Broglie,[*] considering the fact that light has both wave and particle properties, suggested that electrons, though known to be particles, might in similar fashion show wave properties as well. He proposed that the appropriate wavelength to attribute to an electron should be related to its momentum ($p = mv$) in the same way a photon's wavelength is related to *its* momentum (eq. 10.3.14), namely,

$$\lambda = \frac{h}{p} \tag{1}$$

Eq. 1, relating wavelength to momentum, does nicely for a non-relativistic particle because it doesn't involve the speed of light *c,* whose appearance always has to do with relativistic effects.

The earliest direct observations of electron interference were made in 1925 by G. P. Thomson and in 1927 by C. J. Davisson and L. H. Germer. The predicted wavelengths are extremely small—much smaller than the wavelength of visible light and, in fact, something like the size of a single atom. In this respect they are similar to x-rays, a form of EM radiation with wavelengths ranging roughly from 0.01 to 10 nm, where 0.1 nm is a typical atomic size. You'll recall that to produce interference you have to provide alternative paths that differ by just a few wavelengths, so in order to see interference with waves of such small wavelength, you'd need a soap film only a few atoms thick—and soap

---

[*] Prince Louis-Victor de Broglie, pronounced "duhBROY."

films that thin are too fragile to survive. Fortunately, nature provides the equivalent of this in crystals, whose atoms are arranged in a completely orderly array, in layers whose separation is typically just that small. Crystals had already been in use for some time for producing interference with x-rays; waves reflected from the different layers either reinforce or cancel each other, depending on whether they come off in step or out of step, just like the waves reflected from the surfaces of the soap film described in §1.3. (The effect in a crystal is more intricate than in a soap film, because in a crystal you can get a lot of different possible layer separations depending on the direction in which you slice it.)

When the same trick was tried with electrons, interference patterns were seen that were a lot like the ones produced by x-rays, with wavelengths agreeing nicely with de Broglie's formula—*a direct violation of the classical conception of the electron as a simple particle.* Similar effects have since been seen with neutrons and other kinds of particles, and even with bigger objects like atoms and molecules, indicating that *all* kinds of matter have some sort of wavelike character. The split personality appears to be universal.

You may ask how electrons bouncing in odd directions off a crystal can prove that there are interference effccts. What you see is lots of electrons bouncing in some directions and very few in other directions; the allowed directions are just the directions in which waves of wavelength $\lambda = h/p$ (eq. 1) would reinforce each other, while the forbidden directions are those in which waves of that wavelength would cancel. You can't say you've proved that this effect is caused by wave interference, but it's certainly just like an interference pattern, and there's no other reasonable explanation. It's important to realize that this conclusion—that electrons have wave properties—is only an *indirect* inference; you don't actually see the waves.

PROBLEM

1. Find the de Broglie wavelength and energy in electron-volts of each of the following:
   (a) An electron of kinetic energy 1 eV (determine its momentum from the given value of the energy)
   (b) An electron traveling at 1 m/s
   (c) A uranium atom traveling at 1 m/s
   (d) A baseball traveling at 1 mm/h

10.5   WAVE INTERFERENCE

We must now begin to come to grips with the paradox. You can't just throw up your hands and say there's a contradiction so it can't be; you have to have some kind of faith in the logical consistency of the universe and try to find out how it *can* be. The first thing to do is to examine your presuppositions and see if you've assumed more than is justified by the evidence. I think the thing to look at first is the assumption that if we see electron interference, then an electron must *be* some sort of wave. Is it possible that there is a wave of some kind that is causally related to the electron behavior but is not itself the electron? If so, how can we tell what kind of wave it is and how it's related to experiments in which we see electrons behaving like particles? In the remainder of this chapter, I'll try to give you an idea of how these ideas begin to be reconciled. In this section I want to tell you in a little more detail how wave interference works, so that in the following two sections we can study the interplay between particle properties and wave properties.

As I explained in §1.3, interference is possible when two or more waves enter the same region of space and combine to produce either cancellation or reinforcement. A typical way for this to happen is to have a wave that can travel to its destination by two or more different paths. If the different waves arrive in phase (in step with each other), then they reinforce each other and give a stronger wave, while if they arrive out of phase, they can cancel each other and produce a wave of reduced, or even zero, amplitude. The example discussed in §1.3 is the color pattern that you see on a soap bubble or an oil slick on water, where the light can be reflected from either the front or the back surface of the thin film, and the two reflected waves can then be either in step or out of step, depending on the wavelength of the light.

Another simple example, more convenient for this discussion, is the *two-slit interference* effect.[*] Let a light beam (or any kind of wave) of a single wavelength shine on a screen with two slits very close together and, passing through the slits, arrive at a white wall (see fig. 1). On the

---

[*]This is often called two-slit diffraction. *Diffraction* is a word that refers to the fact that waves spread out when they pass through a small opening, much as sound waves spread and fill all available space; I'm using the word *interference* instead to emphasize the way the two waves interfere with each other to produce cancellation or reinforcement.

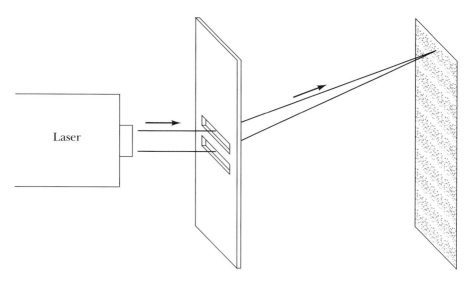

Fig. 1   Two-slit interference.

wall you'll see a series of equally spaced light and dark fringes, some-thing like those shown in fig. 1 but much closer together.

The angle of bending in fig. 1 is *much* exaggerated. It would typically be a fraction of a degree, with the light spreading out just a bit as it passes through the slits and illuminating a small region of the wall; fig. 1 just shows the two ways the light can get to one particular point on the wall. The light can arrive at that point by two different paths of slightly different length, and the path difference depends on the angle. If ΔL, the difference in the two paths, is zero or an integer multiple of the wavelength—that is,

$$\Delta L = n\lambda, \quad n = 0, \pm 1, \pm 2, \ldots \tag{1}$$

then as they get to the wall the two waves are in step with each other, just as in fig. 1.3.3, and the wall appears bright. This way of adding two (or more) waves to get a single combined wave is called *superposition,* a con-cept that plays a major role in quantum physics (and is discussed in §13.4). If the path difference is $\lambda/2, 3\lambda/2, \ldots$, on the other hand—that is,

$$\Delta L = (n + \frac{1}{2})\lambda, \quad n = 0, \pm 1, \pm 2, \ldots \tag{2}$$

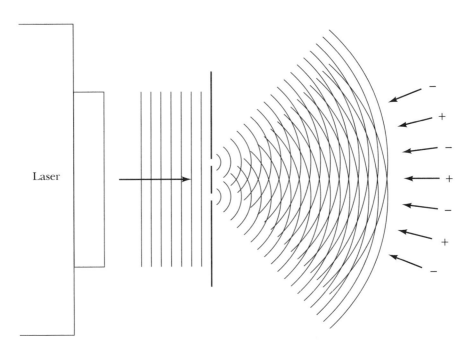

Fig. 2   Wave pattern for two-slit interference.

then the waves arrive out of step and cancel as shown in fig. 1.3.4, and the wall at that point is dark.

Another view of wave interference is shown in fig. 2, a close-up look at the waves emerging from the two slits. The vertical lines to the left of the screen and the circular lines to the right represent wave crests. The actual wave pattern to the right is the sum (superposition) of the two waves shown. The arrows marked + indicate the directions along which the two waves are in phase and therefore reinforce each other, and the arrows marked – show where the two waves are out of phase and therefore cancel.

By looking at the geometry of the light rays near the two slits, as in fig. 3, you can get a formula for ΔL in terms of the angle θ at which the light emerges from the slits. Applying simple trigonometry to the right triangle in the figure, you get

$$\Delta L = d \sin\theta \qquad (3)$$

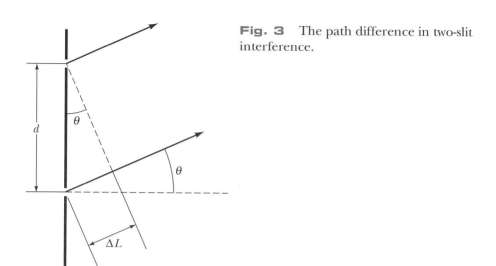

Fig. 3 The path difference in two-slit interference.

where d is the slit separation. Combining this with eqs. 1 and 2, you can figure out the angles θ that correspond to dark and light areas on the wall. For a bright fringe you need $\Delta L = n\lambda$, so the angles are given by

$$\sin\theta = \frac{n\lambda}{d}, \quad n = 0, \pm 1, \pm 2, \ldots \tag{4}$$

As you look at increasing values of θ, corresponding to points farther up the wall, ΔL increases smoothly through the different integer and half-integer multiples of λ, corresponding to reinforcement and cancellation of the waves. If you graph the intensity I of the wave against the angle of deflection θ, your graph looks like fig. 4. This graph exactly corresponds to the *picture* of the pattern shown on the wall in fig. 1.

   Although two-slit interference is just one special case of wave interference, it is nonetheless typical and is representative of the different ways in which waves can combine to reinforce or cancel. The confusion arises when you see *particles* behaving this way—the typical quantum identity crisis. We'll explore the meaning of this confusion in the next section.

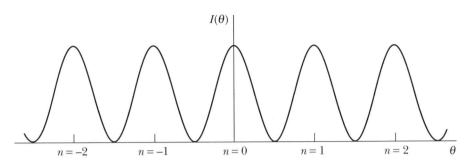

Fig. 4   The two-slit pattern: intensity as a function of angle of deflection.
The angle θ is assumed to be small, so $\sin \theta \approx \theta$ (Appendix A, eq. 6.3).

PROBLEMS

1. Eq. 3 gives the path difference for the two-slit experiment in terms
    of the angle of deflection of the beam.
    (a) Write down the formula for θ_n, the angle corresponding to the
         n^{th} bright fringe.
    (b) Find the angles θ corresponding to the first intensity mini-
         mum (away from the central maximum) and to the first three
         maxima, in the case where $\lambda = 500$ nm and $d = 0.5$ mm. Find
         the separation of the fringes on a wall 3 m away. What would
         the slit separation d have to be in order for the first maximum
         to be at $\theta = 45°$? Can you use the approximation $\sin\theta \approx \theta$ (Ap-
         pendix A, eq. 6.3) for either part of this problem? Explain.
2. A monoenergetic beam of electrons strikes a crystal surface at nor-
    mal incidence (at right angles to the surface).
    (a) The lowest electron energy at which strong reflection (con-
         structive interference) at normal incidence is seen is 18 eV.
         Find the momentum and de Broglie wavelength of the elec-
         trons and the separation of the first two crystal layers. (Only
         the first two layers need be considered here.)
    (b) At what other energies would you expect to see strong reflection?
    (c) At what energies would you expect to see minimum reflection?
         (This problem is somewhat artificial because most of the ef-
         fects of electron diffraction from a crystal are actually due to
         scattering from the array of atoms on the surface layer.)

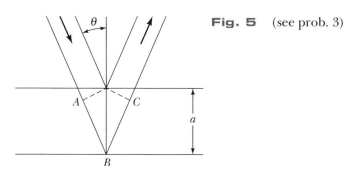

Fig. 5   (see prob. 3)

3. A wave is reflected from two parallel surfaces at an angle of incidence θ from the normal, as shown in fig. 5 and is reflected at the same angle.
   (a) Show that the path difference ABC between the two possible paths is given by $\Delta L = 2a\cos\theta$, where a is the layer separation.
   (b) Neutrons incident on a crystal with layer separation 0.15 nm are reflected strongly when the angle of incidence θ is exactly 89°. Find the de Broglie wavelength, momentum (in eV/c), and energy (in eV) of the neutrons.
   (c) Find the angle of incidence at which 25-eV electrons would be strongly reflected.

10.6   WHERE'S THE PARTICLE?

We've looked now at several quite different physical phenomena that violate classical notions of how things should behave: thermal radiation, the photoelectric effect, and electron interference. In each of these the peculiar behavior boils down to *duality:* Everything in nature seems to share the dual—and apparently contradictory—character of both waves and particles. The essence of the contradiction is this: A wave generally fills all the space available to it and can interfere with other waves, while a particle occupies only a small volume in space and cannot interfere (in the sense of wave interference) with other particles.

To probe this paradox, let's look at the two-slit interference experiment of §10.5 from a particle point of view. In the classical wave picture the waves fill up the space, as suggested in fig. 10.5.2, and the wave pattern to the right of the slits consists of a combination of the waves coming through the two slits. In some areas the two waves cancel and in other areas they reinforce each other, so that different areas on the wall are more bright or less bright, producing the pattern of fringes shown in fig. 10.5.1.

From the particle point of view, the beam must be thought of as a stream of particles—photons or electrons as the case may be. Each particle, landing on the wall, makes a tiny spot (we can make a permanent record of the spots by putting a photographic film on the wall), and the fringe pattern that you see represents the distribution of spots: lots of spots where the beam intensity is high and no spots where the beam intensity is zero, duplicating fig. 10.5.1.

Now imagine that the intensity of the beam is turned way down until there's just one particle per second passing through the slits. Each particle makes a spot on the film, and it takes a long time before you can tell whether the spots still make up a fringe pattern. This has actually been tried experimentally (different details, same principle), just to make sure, and no matter how slowly you send the particles through, the same pattern of dark and bright fringes finally emerges. The fact that the particles get deflected a bit is no problem—they can bounce off the edge of one of the slits—but there seems to be no reason at all why they should avoid the dark parts of the pattern. If a particle goes through the upper slit, for example, then its behavior can hardly depend on where the lower slit is, and yet the location of the fringes is determined by *both* slits, as you can see from eqs. 10.5.1–10.5.3, which show that the angle θ for a bright or dark fringe depends on the slit separation *d*.

You might suppose, in the case of light, that the individual photon is just a little wave pulse with part going through each slit. However, if you put the photographic film very close to the slits, at a distance much less than the slit separation (see fig. 10.5.2), then you still find a single sharp spot produced by each photon, either at one slit or the other—never at both. You could do a photoelectric-effect experiment (see §10.3) to see whether the photon has its full energy *hf*, and you'd find that it does. The message is that if you look to see which slit the photon went through, you get a definite answer: Half the photons go through the upper slit, and half go through the lower. If you back off to the wall, on the other hand, where the photon can get to a particular spot by either route, then you see the interference fringes and are forced to use a wave picture to describe them.

As I've said, this whole discussion applies equally well to any kind of supposed particle—photon, electron, neutron, atom, or whatever. The essence is this:

It travels like a wave and hits like a particle.

PROBLEM

1. Explain, for a reader unfamiliar with quantum ideas, why there is a paradox for the case of **(a)** electrons reflected from a crystal (§10.4 and §10.5) and **(b)** the Michelson interferometer (§1.3).

 PROBABILITIES

In talking about the two-slit experiment in §10.6, we looked at the question "How does the particle know where it's allowed to land?" There is a deeper question related to this: "What *determines* where it lands?" which is embarrassing because it has no clear answer, and physical laws are supposed to answer questions of that sort. In the situation where the film is placed very close to the slits, we find that half the particles appear at the upper slit and half at the lower one, but we find also that *there is no way to predict for a particular particle which slit it will choose.* It's the same when the film is moved back to the wall and fringes are seen; you cannot predict ahead of time which fringe a particular particle will land in. As far as physicists have been able to tell—and they have thought long and hard about this—the behavior is intrinsically unpredictable.

What quantum theory does, in fact, is enable us to predict *probabilities* and probability distributions with a high degree of precision. This means that when there are lots of particles you can make statistical predictions about their behavior that are amazingly precise. The fact that *half* the particles will be found at the upper slit is a statistical prediction, just like predictions about heads and tails when you're flipping a coin. The two-slit interference pattern shown in fig. 10.5.4 is to be thought of as a probability distribution. Thus $I(\theta)\,d\theta$ is proportional to the probability that a particle will be found within the range $d\theta$ when it lands on the wall, and it is therefore proportional also to the fraction of the particles, when there are lots of them, that fall within that range.

All this is very strange. It is the business of physics to predict the behavior of things—to predict them without uncertainties, given complete information about the state of the system at some earlier time. The only reason I don't know whether a coin will come up heads or tails is that I can't control the initial throw with enough precision, but there's no reason in principle why that shouldn't be possible. We had always believed, in other words, that physics is completely deterministic. As you can imagine, physicists did not accept easily the loss of determinacy, and many of them, including Einstein, went to their graves unconvinced. A lot of effort was given to what are called hidden-variable theories, in which submicroscopic variables inaccessible to experiment are supposed to determine the behavior in a completely classical kind of way, mimicking indeterminant behavior in the same way as a flipped coin or a roulette wheel. Such theories have been abandoned by most physicists, and our present understanding is that physics is indeed fundamentally probabilistic, rather than deterministic, in character. (You may have heard of Bell's theorem, which pointed the way to a direct experimental proof that no physically reasonable deterministic picture could be correct.)

It is something of a puzzle to me why a nondeterministic theory should be so hard to swallow. For many physicists, even some who were very intelligent and very thoughtful, the idea was not just difficult but *impossible* to accept, presumably because of deep intuitive feelings about how nature has to behave. The laws of nature simply *had* to be deterministic in character, because that's what laws of nature are: rules for predicting the behavior of things. Those of us who have grown up in the quantum age, though, have had to learn how unreliable intuition is as a guide to fundamental truth—unreliable simply because our intuition develops initially through observations of the nonrelativistic macroscopic world alone. As a consequence of this lesson, perhaps, I can't see any philosophical necessity for the future to be determined by the past, any more than the past is determined by the future.

I can imagine a universe in which events are simply sprinkled randomly over space and time (recall those spacetime diagrams from Part I), with no correlation between those at one end (the past) and those at the other (the future). In that case, of course, there would be no continuity in time, no laws of motion at all, and no people to worry about the lack of them. What we seem to have is something in between the two extremes: a world in which the events of past and future have a high degree of randomness, but also with strong statistical correlations that are described by the probabilistic laws of quantum theory.

Actually, I do have some trouble with the concept of probability itself. It corresponds to an intuitive idea and to our practical experience, but it's remarkably hard to pin down with a rigorous definition. When I say that the probability of a coin's coming up heads is 50%, I'm trying to say that if you flip the coin a large number of times—call this number N—it will come up heads half the time and tails half the time, and the bigger N is, the better this prediction will be. But there's no *necessity* for the 50% prediction to get more accurate as N gets bigger. There is always a possibility that the fraction of heads will deviate substantially from this prediction no matter how large N is. All you find you can say is that the probability of a big deviation from 50% gets smaller as N gets bigger, which brings you around in a circle, because probability is what you were trying to define in the first place.

What you'd like to use for a definition is

$$P = \lim_{N\to\infty} \frac{H}{N} \qquad (1)$$

where H is the number of heads. The fact is, though, that there is no assurance that the ratio will have a limit as N becomes large, because the laws of chance don't rule out *any* possible distribution of heads and tails. You can still throw all heads no matter how big N is—it just gets more and more improbable! I think the best way out of the dilemma is to *postulate* (as a student once suggested to me when I was discussing this difficulty) that in real life the limit will always exist, and then you're free to use the limit to define the probability.

What you can begin to see from our discussion of the two-slit interference problem is that the waves themselves are not observed directly but that they govern the probabilities for *particlelike* observations. Where the wave amplitude is small, the probability of finding the particle is small, and where the wave amplitude is big, the probability of finding the particle is big. This relationship between wave amplitude and probability, which we'll come back to repeatedly, is a very basic characteristic of quantum theory and is true for *all* kinds of particles, including photons.

All of the puzzles I've been discussing—the particle character of light, the wave character of electrons, and the probabilistic character of the theory—make their appearance in the example of the one-electron atom (the hydrogen atom) and its radiation spectrum, which are the subject of Chapter 12.

PROBLEMS

1. To calculate the probability that a coin will come up heads twice if it's thrown 4 times, say, you enumerate all the ways it might fall in 4 throws and take the fraction that include just two heads, like this:

HHHH	HTHH	THHH	TTHH
HHHT	HTHT	THHT	TTHT
HHTH	HTTH	THTH	TTTH
HHTT	HTTT	THTT	TTTT

There are $2^4 = 16$ possible sequences, and 6 of them (underlined) correspond to two heads and two tails, so the probability of throwing two heads is $6/16 = 0.375$. (This calculation rests on the assumption that for a single throw, heads and tails are equally likely.)

(a) Use this counting method to find the probability of throwing exactly 50% heads in 2, 4, 6, and 8 throws.

(b) The probability $P_n(m)$ of getting m heads out of n throws is given in terms of the *binomial coefficient:*

$$P_n(m) = 2^{-n}\binom{n}{m}$$

where

$$\binom{n}{m} = \frac{n!}{m!(n-m)!} = \frac{n(n-1)(n-2)\cdots(n-m+1)}{m(m-1)(m-2)\cdots 1}$$

Check that this formula gives the right answer in each case in part **(a)**, and go on to calculate $P_{2m}(m)$ for $2m = 20$ and $2m = 30$. (The central expression uses the *factorial* function $n! = 1 \cdot 2 \cdot 3 \cdots n$. If your calculator has the factorial function, you can use this expression; if not, use the equivalent expression on the right.)

(c) Find the probability of throwing all tails in 4, 10, 20, and 50 throws.

(d) Draw a careful graph of $P_n(m)$ for each of the cases $n = 4, 10,$ and 20, using m/n for the abscissa (x coordinate) and $P_n(m)/P_n(n/2)$ for the ordinate (y coordinate) so that the different distributions can be compared.

(e) Check the case $n = 10$ experimentally. It's easiest if you have 10 coins and can throw them all at once. Do this about 20 times, recording the number of heads in each throw, and make a bar graph of your averaged results. Use the same scale as in part **(d)** so that you can compare your experimental result with the theoretical curve.

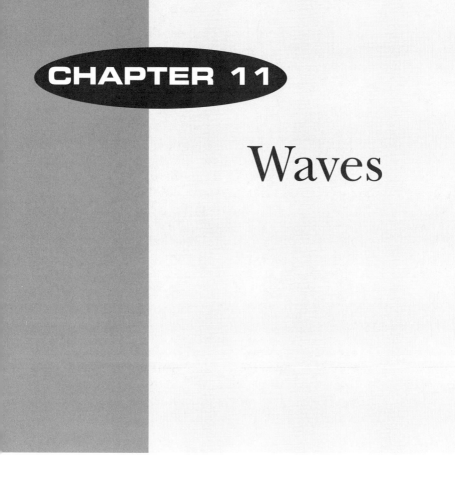

CHAPTER 11

Waves

11.1 SIMPLE WAVES

In Chapter 10 you learned that in the quantum world every kind of matter exhibits wave properties and that in fact the motion of what we've considered particles must now be thought of in terms of wave motion, with well-defined wavelengths and the possibility of wave interference. The waves, as I described them in Chapter 10, are still rather mysterious and ill-defined, though we've begun to see that they must have something to do with probabilities: The probability of finding a particle in a particular place is related to the amplitude of the wave motion at that place.

A proper quantum theory must necessarily be a theory of waves.

In the early days, as I've said, the theory that was developed was often referred to as wave mechanics.

In §1.3 I introduced you to the basic notions of waves. I want in this chapter to build up the mathematical description of waves so that you'll be familiar with such terms as *wavelength, frequency, wave speed, sine waves, superposition,* and *normal modes.*

A wave motion in general is a complicated, irregular thing, as I've mentioned, with all sorts of different motions going on in the different parts of the space that is filled by the wave. In typical physicist fashion, we reduce the complex problem to simple components, because an arbitrarily complex motion can be described as a combination of a large number of these components, and each simple component, taken by itself, can be described and understood quite easily. The basic simple wave, the building block of this operation, is called a *sine wave* and is described in terms of the sine and cosine functions: $\sin\theta$ and $\cos\theta$ (see Appendix A, §1, for details on trigonometric functions). The spatial pattern of one of these simple waves is a sinc function, and the oscillations in time are also sine functions. Let's proceed, then, with a brief description of sine waves.

Most of the ideas we need can be illustrated with one-dimensional waves, which you can visualize as waves in a horizontally stretched string constrained to vibrate only in the vertical plane. Start with a snapshot—a picture of the wave in space at a single instant of time—and introduce the time variation of the pattern as a later step. The simple sine wave is given by

$$y(x) = A \sin kx \tag{1}$$

where $y(x)$ represents the displacement y of the string at a point a distance x along it, and the constants A and k are used to specify the size and shape of the wave pattern. The graph of $y(x)$ is shown in fig. 1.

You may not be used to this kind of graph for a trigonometric function. The argument kx in eq. 1 (that is, the quantity whose sine is evaluated) is just an angle expressed in radians.* The range from 0 to 2π radians (0° to 360°) is what you're probably familiar with for trig functions; it corresponds to just the first complete cycle to the right of the

---

*A circular arc of radius r and arc length s subtends an angle, in radians, equal to s/r. When the arc length s is just equal to the radius r, then the angle is 1 radian, or about 57.3°.

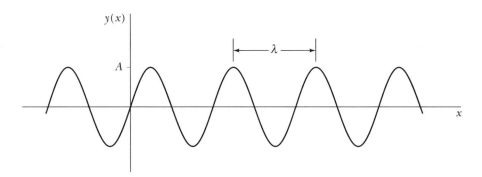

Fig. 1 A sine wave. A is the amplitude and λ is the wavelength.

origin. The sine increases to a maximum where the argument kx is $\pi/2$ radians, or 90°, drops again to zero at 180°, and then goes to negative values for angles between 180° and 360°. For the purpose of looking at waves and oscillations, it's useful to extend the trig functions over a range much larger than one cycle, making particular use of the fact that trig functions are periodic—that is, they simply repeat themselves every 360°, or 2π radians.

The constant A in eq. 1 is called the *amplitude*. It represents the maximum value that y can take, because the sine function itself varies between +1 and -1. The constant k, which I'll call the *angular wave number*, tells how rapidly the wave oscillates as x increases. (See §12.1 for a discussion of the term *wave number*.) Since kx is an angle in radians, the angular wave number k is measured in radians per meter (rad/m). It is commonly written simply in units m^{-1}; an angle in radians is just the ratio of two lengths, so radians properly have no units at all. When x increases by $2\pi/k$, the product kx increases by 2π, which means that the sine wave has gone through one complete cycle. Thus $2\pi/k$ is the *wavelength*, called λ:

$$\lambda = \frac{2\pi}{k} \qquad (2)$$

the length of one complete cycle of the wave. Large k corresponds to small wavelength.

A more general form, including an arbitrary *phase shift* in the x direction, is

$$y(x) = A \sin(kx - \varphi) \qquad (3)$$

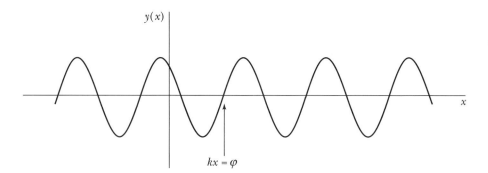

Fig. 2   Sine wave with arbitrary phase shift.

where the phase angle φ indicates a shift of the picture to the right or left, for φ positive or negative. This is shown in fig. 2, which is simply a shifted version of fig. 1. The phase shift corresponds to a shift of the pattern through a distance

$$\Delta x = \frac{\varphi}{k} \qquad (4)$$

because the primary zero of the sine function, where the argument $kx - \varphi$ is equal to zero, is at $x = \varphi/k$ instead of at $x = 0$ as in eq. 1. The shift can be described in different terms: the size Δx of the displacement, the number of cycles (or fraction of a cycle), or the phase angle in radians. The description in terms of a shift in phase, represented by φ, is very standard because it gives a qualitative picture of the displacement that doesn't depend on the wavelength.

PROBLEMS

1. **(a)** Find a value of the phase shift φ in eq. 3 that corresponds to $y = A \cos kx$.
    (b) Find a value of φ that corresponds to $y = -A \sin kx$.
    (c) In each case, are any other values of φ possible? Explain.
    (d) Show that the right side of eq. 3 can be written in the form $y = y_1 \sin kx + y_2 \cos kx$, and find y_1 and y_2 in terms of A and φ, using the trigonometric relationship

$$\sin(A + B) = \sin A \cos B + \cos A \sin B$$

2. On a single graph, sketch carefully the function $\sin(kx - \varphi)$ for $k = 4\pi$, and for $\varphi = -\pi/8, 0, \pi/8, \pi/4, 3\pi/8$, and $\pi/2$, with the range of x from 0 to 1. (The units for x are meters and the units for k are m^{-1}.) For each value of φ, find the value of x nearest to $x = 0$ at which the sine function is equal to zero.

11.2 WAVE MOTION

Now we're ready to look at the mathematical description of sine waves that vary in time. Here the whole spatial pattern (as in fig. 11.1.2) varies in time and repeats itself periodically with a definite frequency. It may do so through the pattern's moving up and down like a vibrating guitar string (a standing wave), or moving to the right or to the left (a traveling wave).

Standing waves

A *standing wave* is illustrated in fig. 1. This is the typical behavior for a vibrating system that is constrained at some boundary points, such as a vibrating string fastened at the ends. The dotted curves represent the

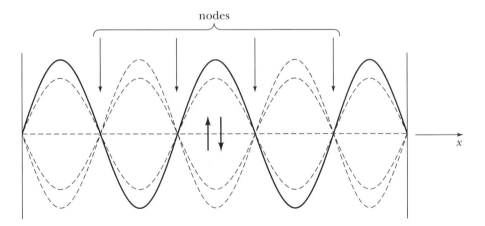

Fig. 1 A standing wave. The motion of the pattern is up and down only, and there are fixed nodes where the displacement y is always zero.

pattern at successive moments of time, as the pattern oscillates up and down. The formula for a typical standing wave is

$$y(x, t) = A \sin kx \sin \omega t \tag{1}$$

The displacement is zero at all times at those values of x for which $\sin kx$ is zero; these points are called the *nodes*. Note that the distance between nodes is $\lambda/2$: There are two nodes per wavelength. At any given time t, the *shape* of the pattern is given by the factor $\sin kx$, with amplitude $A \sin \omega t$. This amplitude varies in time according to another sine function; in this case the argument of the sine involves the time variable t. Writing $A(t)$ for this time-varying amplitude, you have

$$A(t) = A \sin \omega t \tag{2}$$

The constant ω, which is called the *angular frequency*, tells how rapidly the amplitude oscillates as t increases. Because ωt, like kx in eq. 11.1.1, is an angle in radians, the angular frequency ω is measured in radians per second (rad/s). When t increases by $2\pi/\omega$, the argument ωt of the sine function increases by 2π, which means you've gone through one complete cycle, so $2\pi/\omega$ is the *period* of the oscillation, called T:

$$T = \frac{2\pi}{\omega} \tag{3}$$

the time for one complete cycle of the oscillation. Large ω corresponds to small period.

The ordinary frequency f is the number of cycles per second:

$$f = \frac{1}{T} \tag{4}$$

and since there are 2π radians per cycle, the relation between ω and f is

$$\omega = (\text{radians/cycle}) \times (\text{cycles/second}) \tag{5}$$

$$= 2\pi f \tag{6}$$

The standard unit for frequency is the hertz (Hz), which is simply 1 cycle per second. See §1.3.

Traveling waves

The other standard kind of wave motion is the *traveling wave;* a typical pattern is shown in fig. 2. The dotted curves represent the pattern at successive moments of time, as the pattern moves to the right. The formula for a typical traveling wave is

$$y(x,t) = A \sin(kx - \omega t) \tag{7}$$

where ω is again referred to as the angular frequency. If you compare this with eq. 11.1.3, you'll see that ωt is just the phase shift φ of the spatial pattern at a given time t. As t increases the shift increases, so this truly represents a wave pattern moving in the $+x$ direction if k and ω are positive. (See prob. 11.1.2.) The actual shift of the pattern at time t, according to eq. 11.1.4, is proportional to the time t:

$$\Delta x = \frac{\omega t}{k} \tag{8}$$

so the velocity of the wave pattern, called the *wave speed* is given by

$$v_0 = \frac{\Delta x}{t} \tag{9}$$

$$= \frac{\omega}{k} \tag{10}$$

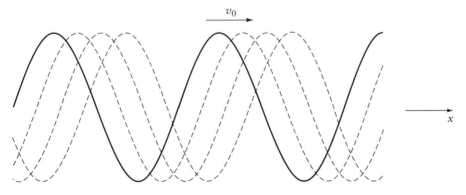

Fig. 2 A traveling wave. The wave pattern moves to the right, and there are no fixed nodes.

Eq. 10 is one of several standard expressions for the wave speed. Another expression for v_0 can be gotten by using eq. 11.1.2, which relates k to the wavelength, and eq. 6, which relates ω to the frequency:

$$v_0 = \omega \times \frac{1}{k} = 2\pi f \times \frac{\lambda}{2\pi} \tag{11}$$

$$= f\lambda \tag{12}$$

or

$$v_0 = \frac{\lambda}{T} \tag{13}$$

from eq. 4. This last form corresponds to the fact that in one period T the wave pattern has to travel a distance λ in order to look the same as it did to start with, and

$$\text{speed} = \frac{\text{distance}}{\text{time}} \tag{14}$$

Note that in a traveling wave in a stretched string or rope, the rope itself doesn't move to the right or left; if you make a mark on the rope, the mark just moves up and down, with a speed that has nothing to do with v_0, as the wave pattern goes by. You see this also in water waves, where, if you float a cork on the water, the cork just oscillates in place as the waves go under it. *Only the pattern travels.*

The kind of traveling sine wave given by eq. 7 is usually referred to as a *plane wave* for the not very logical reason that the corresponding wave form in *three* dimensions has wave crests along parallel planes traveling with the wave speed v_0 in a direction perpendicular to the planes. Such planes are usually referred to as *wave fronts.*

What I've given you so far is a description of wave motion, both verbal and mathematical, but I haven't told you anything about the dynamics—the physical laws that govern the motion. For waves in a stretched string and for sound waves in matter, the motion is governed by Newton's laws: $F = ma$, in fact. The medium is made up of a very large number of particles—the molecules of the string, for example—and there is an $F = ma$ equation for every particle. The dynamical description is more complicated than for a single particle, though not so bad as you might expect. You end up with a partial differential equation, called a *wave equation,* that amounts to a differential equation (analogous to $F = ma$) for every value of x, the position along the string. For waves of other types, such as EM waves, there are of course other dynamical laws that govern the motion, yielding in each case an appropriate wave equation. The waves that we need for quantum theory are of a new type, and new wave equations had to be invented to describe their motion.

PROBLEMS

1. **(a)** Find the frequency f and period T for an oscillation with angular frequency ω equal to **(i)** 1 rad/s, **(ii)** 200π rad/s, **(iii)** $2\pi \times 10^{-6}$ rad/s.
   (b) Find ω (in rad/s) and f (in Hz) for **(i)** the ocean tides ($T = 12$ h), **(ii)** the rotation of the second hand on a clock, **(iii)** the hour hand.
2. **(a)** Which of the following are possible wavelengths (in cm) for standing waves in a vibrating wire of length 24 cm?
   $$8, 9, 12, 16, 24, 36, 48, 72$$
   (*Hint:* Draw a picture, like fig. 1, and determine whether there can be a node at each end of the wire.)
   (b) Find the angular wave number k for each allowed wavelength.
3. In prob. 11.1.2 you graphed the sine wave $\sin(kx - \varphi)$ for increasing values of the phase angle φ and worked out the successive positions of the first zero of the function.
   (a) If φ increases from 0 to $\pi/2$ in a time 0.05 s, what is the speed of travel—distance/time—of the point where the function is zero?
   (b) Using the relation $\varphi = \omega t$, work out the angular velocity ω, and compare the result from part **(a)** with the formula (eq. 10) for the wave speed in terms of k and ω.

 **11.3**   SUPERPOSITION

In §10.5 I explained the property of superposition: how two waves can be added together to reinforce or cancel each other. The point I want to emphasize now is that any two or more simple waves can be added together to give a resulting wave with all sorts of different possible shapes. Here are a few examples. In fig. 1 two waves of the same wavelength but different amplitude and phase are combined. The resulting

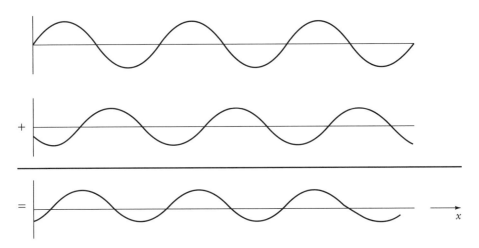

Fig. 1   Superposition of waves of the same wavelength with different amplitude and phase.

wave is a sine wave that isn't in phase with either of the original waves
and whose amplitude is less than the sum of the separate amplitudes.
Let's look at the math. First note that the sine wave (11.1.3) with arbitrary phase can be seen as the superposition of a simple sine and cosine,
using the formula [see prob. 11.1.1(d)] for the sum of two angles. Replacing $-\varphi$ by φ (just to simplify the steps), we get

$$y(x) = A \sin(kx + \varphi) \tag{1}$$

$$= A(\cos\varphi \, \sin kx + \sin\varphi \, \cos kx) \tag{2}$$

$$= A_1 \sin kx + A_2 \cos kx \tag{3}$$

where

$$A_1 = A \cos\varphi \tag{4}$$

$$A_2 = A \sin\varphi \tag{5}$$

Next note that if you are given the form shown in eq. 3, you can reexpress it in terms of the amplitude and phase as in eq. 1 by solving eqs. 4 and 5 for A and φ, which you may recognize as the basic equations for a right triangle. The result is simply

$$A = \sqrt{A_1{}^2 + A_2{}^2} \tag{6}$$

$$\tan\varphi = \frac{A_2}{A_1} \tag{7}$$

and the signs of A_1 and A_2 determine the correct quadrant for φ.

Now the two wave patterns shown in fig. 1 are (in some arbitrary units)

$$y_1 = 5 \sin x \tag{8}$$

$$y_2 = 4 \sin(x - \frac{3\pi}{4}) \tag{9}$$

Eqs. 1–5 enable you to rewrite eq. 9 as

$$y_2 = -2.828(\sin x + \cos x) \tag{10}$$

(see prob. 2), so the sum $y_1 + y_2$ can also be written in the form of eq. 3 with $A_1 = 5 + (-2.828) = 2.172$ and $A_2 = -2.828$. From eqs. 6 and 7, then, you get

$$A = 3.57 \tag{11}$$

$$\varphi = -0.92 \text{ rad} = -52.5° \tag{12}$$

Try to see the reasonableness of these figures as a description of the amplitude and phase of the bottom curve in fig. 1.

Fig. 2 shows some examples of superposition of waves of the same amplitude but different wavelengths. The character of the resultant wave pattern depends especially on the ratio of the wavelengths. The resultant wave repeats itself exactly (we say it's *periodic*) when the ratio is rational,

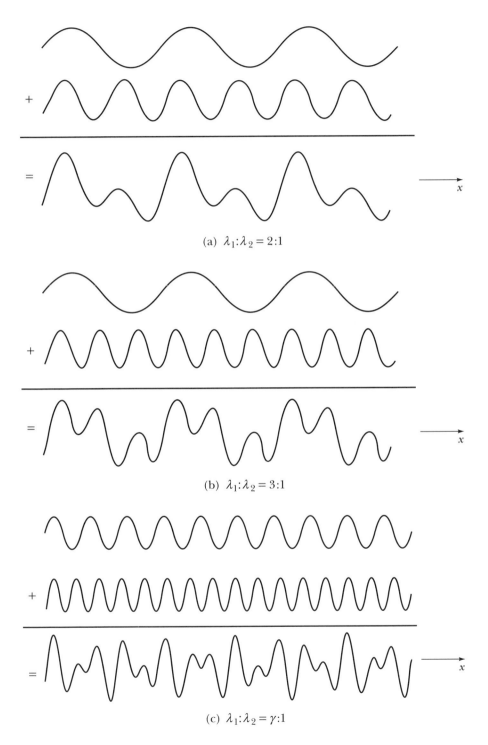

(a) $\lambda_1 : \lambda_2 = 2{:}1$

(b) $\lambda_1 : \lambda_2 = 3{:}1$

(c) $\lambda_1 : \lambda_2 = \gamma{:}1$

Fig. 2 (opposite)   Superposition of waves of different wavelength. In (c), $\gamma = (1 + \sqrt{5})/2 = 1.618\ldots$, the "golden ratio." Note that the pattern repeats itself in (a) and (b) but never quite repeats itself in (c).

and not otherwise. As a final example, fig. 3 shows the result of combining six sine waves according to the formula

$$y(x) = \sum_{m=1}^{6} \frac{1}{m} \sin mx \tag{13}$$

As you might guess, you can reconstruct a wave pattern of any reasonable shape by appropriately combining sine waves—that is, sine and cosine functions. This process is known as *Fourier analysis.*

Fig. 3   Resultant of six sine waves.

PROBLEMS

1. The formula (eq. 7) for the phase angle φ in terms of the coefficients A_1 and A_2 of eq. 3 doesn't determine φ completely, because the angles φ and $\varphi + \pi$ have the same tangent. An unambiguous formula for φ (apart from multiples of 2π) is

$$\tan\frac{\varphi}{2} = \frac{A_2}{A + A_1}$$

with A given by eq. 6. Prove this relation, and show that it does determine φ uniquely.

2. (a)   Check the accuracy of eq. 10, and draw graphs of y_2, of $-2.828\sin x$, and of $-2.828\cos x$, to check the consistency of eqs. 9 and 10.

    (b)   Check the resulting values of A and φ (eqs. 11 and 12), and compare your results with fig. 1 for reasonableness. Remember that positive ϕ now represents a shift to the left, because I changed the sign convention in eq. 1.

(c) Follow the same procedure to get a general formula for the amplitude and phase of the wave that results from superposition of two waves of the same wavelength but of arbitrary amplitude and phase:

$$y(x) = y_1(x) + y_2(x)$$

$$y_1 = A_1\sin(kx + \phi_1)$$

$$y_2 = A_2\sin(kx + \phi_2)$$

Taking $\phi_1 = 0$, make graphs of the amplitude and phase (optional) of the resultant wave as functions of ϕ_2, first for the case of equal amplitudes A_1 and A_2, and then for a case in which $A_1 \neq A_2$ (much the more common circumstance). Interpret the maxima and minima in the resultant amplitude.

3. In connection with fig. 2, I mentioned that the resultant of two sine waves is periodic (keeps repeating itself exactly) when the ratio of the two wavelengths is a rational number, and not otherwise. Can you explain why this is? Find a formula for the length Δx *of the repeating segment.*

 **NORMAL MODES**

Another characteristic feature of wave motion, one that is very important for quantum theory, is the fact that when the motion is confined to a finite region, only certain discrete frequencies can occur. These frequencies are associated with definite standing wave patterns, known as *normal modes of oscillation,* or simply *normal modes,* and it is these normal modes that I now want to direct your attention to.

The usual way for a wave to be confined is that the region where the wave motion takes place has boundaries beyond which the wave cannot travel. If you set up a localized disturbance in this case, the resulting waves can travel around within the confined region and in fact interfere with each other in a complicated way, but you can't have a regular sine wave traveling always in the same direction. The only kind of regular sine wave possible is a standing wave. You can see this with a guitar string; because it's fastened at the ends, the ends can't oscillate—that is, the wave has to have a node at each end, as in fig. 11.2.1. A traveling wave (fig. 11.2.2) cannot be used for waves with the endpoints fixed, because every point of the wave oscillates.

Look at the standing wave of fig. 11.2.1. How far apart are the nodes? A wavelength corresponds to a complete cycle of the wave (fig. 11.1.1), and the distance between neighboring nodes is only half a cycle, or half the wavelength, $\lambda/2$. Such a half-cycle in a standing wave is often referred to as a *loop*. We know there has to be a node at each end of the string, but there's no restriction on how many nodes there are in between, so it should be possible to have any integer number of loops in the standing wave, as illustrated in fig. 1. These different standing-wave patterns are called *normal modes* of the vibrating string, and the first one, with only one loop, is called the *fundamental mode*. For any system that can have standing wave patterns like this, the mode of lowest frequency—the fundamental mode—is a pattern with no nodes except at the edges.

It's easy to see now what the possible wavelengths are for a vibrating string of a given length L, because the length of each loop is just $\lambda/2$ and there are n loops:

$$n \times \frac{\lambda}{2} = L \tag{1}$$

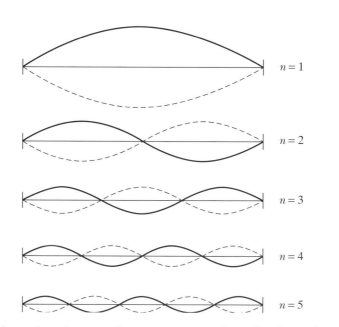

$n = 1$

$n = 2$

$n = 3$

$n = 4$

$n = 5$

Fig. 1   Normal modes: standing wave patterns in a vibrating string.

Hence

$$\lambda = \lambda_n = \frac{2L}{n}, \quad n = 1, 2, 3, \ldots \tag{2}$$

Note that the wavelength of the fundamental mode, with $n = 1$, is $2L$, corresponding to the fact that the single loop represents just half a wavelength.

The next step is to figure out the frequencies of the normal modes. For many types of wave, the wave speed v_0 is the same for all wavelengths, and if you know v_0 you can get the frequency f from eq. 11.2.12. In the case of the guitar string, v_0 depends on the tension and the density of the string, but not on λ or L. The frequencies, then, using eq. 11.2.12, are given by

$$f_n = \frac{v_0}{\lambda_n} = \frac{nv_0}{2L} \tag{3}$$

$$= nf_1 \tag{4}$$

Here f_1 is the frequency of the fundamental mode, given by

$$f_1 = \frac{v_0}{2L} \tag{5}$$

and the frequencies of the other modes are simply integer multiples of the fundamental frequency.

It's this fact that makes stretched strings useful for making musical instruments. Frequencies that are related as the ratios of small integers correspond to harmonious musical intervals, so the sounds produced by the different modes of a string blend harmoniously when the string is made to vibrate. [You can get a glimpse of how this works by comparing (a) and (b) with (c) in fig. 11.3.2.] Normally a string is vibrating in all modes at once when it is plucked, bowed, or struck; the musical *pitch* you hear corresponds to the fundamental frequency, and the other frequencies that are present affect only the quality, or timbre, of the tone. To hear the other modes as distinct tones, you can suppress the fundamental by touching the string lightly at the position of one of the possible nodes shown in fig. 11.2.1—for example, at a distance $L/2$ or $L/3$ from the end of the string. Then the modes that have a node at that position are free to vibrate, while the modes that don't have such a node, including the fundamental, are suppressed. This technique is commonly used in tuning a guitar (to compare the pitch of the different strings) and in playing harmonic tones on a violin.

The normal modes of two- and three-dimensional vibrating systems, like drumheads and bells, don't usually have simple frequency ratios. A simple case is that of a square vibrating membrane, like a square drumhead, for which the first five distinct normal modes are shown in fig. 2. The dotted lines represent nodal lines. Because of the symmetry of the square, there are additional modes with the same frequencies as (b), (d), and (e), but with x and y directions interchanged. (If the shape of the membrane is not symmetrical, the nodal lines, curiously enough, don't usually intersect.) If the displacement is represented as $z(x,y)$, where x and y both range from 0 to L, then the normal modes have the simple form

$$z = A \sin k_1 x \sin k_2 y \sin \omega t \tag{6}$$

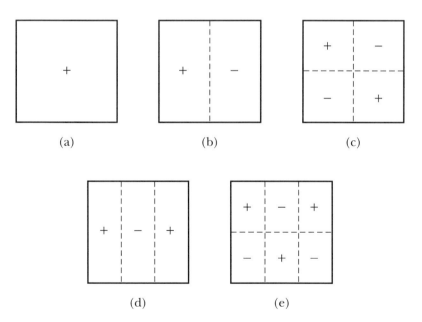

Fig. 2   Some of the normal modes of a square membrane. The regions marked + and − oscillate 180° out of phase with each other.

with

$$k_1 = \frac{n_1 L}{\pi}, \quad n_1 = 1, 2, 3, \ldots \tag{7}$$

$$k_2 = \frac{n_2 L}{\pi}, \quad n_2 = 1, 2, 3, \ldots \tag{8}$$

and

$$f = \frac{\omega}{2\pi} = f_0 \sqrt{\frac{1}{2}(n_1{}^2 + n_2{}^2)} \tag{9}$$

where f_0 is the frequency of the fundamental mode. Note that f/f_0 is irrational for most modes.

Let me conclude this lesson on waves with a brief summary.

> Definition: *A wave is a spatial pattern that varies with time.*
> Distinctive characteristics:
> *1.* Superposition —*Two or more waves in the same region combine to make a single wave.*
> *2.* Interference —*Superposed waves can reinforce or cancel each other or combine in a variety of other ways.*
> *3.* Normal modes —*Wave motion in a confined region can occur only in certain definite standing-wave patterns, with definite associated frequencies called normal frequencies.*

PROBLEMS

1. **(a)** Show that for standing waves on a stretched string for which the wave speed is v_0, the angular frequency ω and the angular wave number k (eqs. 11.2.1 and 11.2.6) are related by $\omega = v_0 k$.
   (b) Find the values of λ, f, k, and ω for the three modes of lowest frequency in a string of length 0.5 m, where the wave speed is 200 m/s.
   (c) The same rules apply for a laser, in which particular normal modes of electromagnetic waves in a long, narrow rod or tube are very strongly excited. For a normal mode of wavelength 500 nm (in the visible region) in a laser of length 20 cm, find the mode number n (eq. 2) and the values of f, k, and ω. Note that the wave speed is c, the speed of light.

2. **(a)** The relation $\omega = v_0 k$ derived in part (a) of prob. 1 can be applied also to the square membrane shown in fig. 2 by making use of the effective angular wave number $k = \sqrt{k_1{}^2 + k_2{}^2}$. Work out the normal frequencies for the 10 lowest modes of the square membrane as multiples of the lowest frequency (the fundamental). Remember that $f = \omega/2\pi$ (eq. 11.2.6). Draw a diagram for each of these modes, showing the nodal lines as in fig. 2 and labeling the modes with the correct mode numbers n_1 and n_2.

   (b) Find the normal frequencies for the 10 lowest modes for waves in a three-dimensional box, using $k = \sqrt{k_1{}^2 + k_2{}^2 + k_3{}^2}$ for the effective angular wave number.

11.5 ELECTROMAGNETIC WAVES

The electromagnetic spectrum

Light is well known to be an electromagnetic wave. In this case the physical quantities that vary over space and time to form the wave pattern are the electric and magnetic field strengths $\underline{E}(\underline{r},t)$ and $\underline{B}(\underline{r},t)$, the fields that produce forces on electrically charged particles such as electrons. The total range of wavelengths that have been observed and studied, shown in fig. 1, varies over something like twenty orders of magnitude (twenty powers of 10) and the wavelengths that correspond to visible light cover only a very small part of this range, as you can see in the figure. Radio waves, microwaves, x-rays, and gamma rays are all EM waves, differing only in their wavelength. The energies shown in the right-hand column of fig. 1 are the energies of the corresponding photons, given in electron-volts (eV). The electron-volt is the standard unit of energy for the microscopic realm; it is equal to the energy gained by an electron when it is accelerated through a potential difference of 1 volt. When the photon energy gets to the range of 10 or more eV, it has enough energy to knock apart an atom or molecule and to damage body cells, which is why you can get sunburned by ultraviolet light but not by visible light or infrared radiation, and why x-rays can cause genetic damage.

The scales in fig. 1 are what are called *logarithmic scales,* where the actual distance along the vertical axis is proportional to the logarithm of the variable. This means that equal distances along the vertical axis correspond to equal *factors* in the variable being plotted.

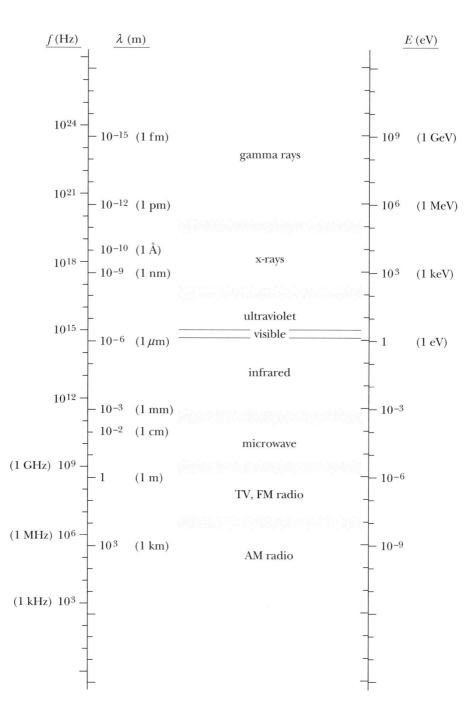

Fig. 1 (opposite)   The electromagnetic spectrum. The energies at the right are the corresponding photon energies. The boundaries are fuzzy because the definitions are fuzzy (except for the visible spectrum).

Electric and magnetic fields

Although I've talked about EM fields and EM waves a great deal, I haven't yet told you just what these fields *are*—how they're defined. The electromagnetic force is one of the basic forces that occur in nature, along with the gravitational force, the nuclear force, and the "weak force" (responsible for various kinds of radioactivity). We've learned that each of these forces is best described by a certain kind of field. We use the word *field* to describe the forces that particles experience as they move around through space. The force that a particle feels at a particular point in space is determined, according to some rule, by the field at that point. In order to specify the forces completely, the field has to be given at every point in space and for every possible time: Each field is therefore a *function* of the spacetime coordinates r and t. Different kinds of force are governed by different fields: the electromagnetic field, the gravitational field, and so on. The fields you commonly see are vector quantities, but in fact fields can have all sorts of different structures. Even if no particle happens to be present we still say that there's a field, and each field carries both energy and momentum, just as particles do.

The gravitational force is proportional to the mass of a particle and is given (if you're not thinking in terms of general relativity) in terms of a gravitational field g, as described in §6.2: $F = mg$. If you recall from relativity theory that mass is equivalent to energy (§5.6), you'll see that the gravitational force can be properly thought of as acting on energy, one of the fundamental conserved quantities of nature. The electromagnetic force on a particle, in a similar way, is proportional to a new type of conserved quantity, the electric charge, carried by the particle. It is the flow of electrically charged particles, the electrons, that produces our familiar electric currents, and it's the EM forces involving those charges that we exploit in using all kinds of electrical power. The electric charge is generally indicated by the letter Q and is usually measured in *coulombs* (C). The familiar unit of electric current, the *ampere*, consists of a flow rate of 1 coulomb per second: $1 A = 1 C/s$. And the familiar quantity *voltage*, or *electric potential*, defined as work or energy per unit charge, is measured in *volts:* $1 V = 1 J/C$.

There are two kinds of EM force on a charged particle: a force that's independent of the particle's velocity and another force that depends on that velocity. For this reason we need to define two fields, the *electric field* $\underline{E}$, giving the velocity-independent force (already introduced in §4.3, eq. 4.3.19), and the *magnetic field* $\underline{B}$, giving the velocity-dependent force. For a particle of electric charge Q moving with velocity $\underline{v}$, the force law looks like this:

$$\underline{F} = Q\underline{E} + Q\underline{v} \times \underline{B} \tag{1}$$

(The × will be explained in just a moment.) The two terms represent the electric force and the magnetic force, respectively. Various things about this formula may need explaining. First, you see that each of the forces is proportional to the charge Q on the particle, as promised. A particle with zero charge experiences no force, and a particle with negative charge feels a force in the opposite direction. Second, the force is the sum of two terms, each of which is a vector: an *"electric force"* $Q\underline{E}$ and a *"magnetic force"* $Q\underline{v} \times \underline{B}$. The electric force has the same form as the gravitational force $m\underline{g}$—a constant times a field vector—and is simply a force in the same or opposite direction as the vector $\underline{E}$. The units of electric field are N/C (newtons per coulomb), consistent with eq. 1; the units are often given also as volts per meter, or V/m, which turns out to be just the same as N/C (prob. 4).

The magnetic force—the second term in the force law of eq. 1—is more complicated and involves a new kind of product, $\underline{v} \times \underline{B}$, called the *vector product* or *cross product*, whose magnitude and direction follow a rule that's explained in detail in Appendix A, §2. The reason this complicated product is needed is that the direction of the magnetic force doesn't depend simply on the direction of the magnetic field but varies also according to the direction of the velocity vector $\underline{v}$. What the expression $Q\underline{v} \times \underline{B}$ says is that the magnetic force is *perpendicular* to both the velocity $\underline{v}$ and the field $\underline{B}$ and that the strength of the force is given by

$$F_{\text{magnetic}} = QvB \sin\theta \tag{2}$$

where θ is the angle between the velocity $\underline{v}$ and the field vector $\underline{B}$. If the velocity is parallel to the direction of $\underline{B}$, then $\sin\theta = 0$ and the force is zero.

The unit of magnetic field is the *tesla*, abbreviated T and chosen to make the numbers agree in eq. 1 when metric units are used for the other variables. Because the magnetic field is (force)/(charge × velocity), the tesla has to be given by

$$1 \text{ T} = 1 \text{ N}/(\text{C} \cdot \text{m}/\text{s}) \tag{3}$$

$$= 1 \text{ N}/(\text{A} \cdot \text{m}) \tag{4}$$

where 1 ampere (1 A) is a current of 1 C/s. (The tesla is actually a very large unit—it takes a really big electromagnet to produce a field of 1 T. A more modest unit in common use is the gauss, which is 10^{-4} T.)

Electromagnetic waves

Look back now at what an EM wave is like. The electric field distribution in space forms a wave pattern like the ones I talked about in §11.2, and the magnetic field also forms a similar pattern; the two patterns are very closely related. Both the electric and magnetic field vectors always point in directions perpendicular to the direction the wave is traveling, and they are also perpendicular to each other. Thus if the wave is traveling in the x direction, then the electric vector might point in the y direction and the magnetic vector in the z direction. Such a wave, called *plane-polarized radiation* because the electric field pattern lies in the x–y plane, would be written like this:

$$\underline{E}(\underline{r},\, t) = E_0 \,\hat{y} \sin(kx - \omega t) \tag{5}$$

$$\underline{B}(\underline{r},t) = B_0 \,\hat{z} \sin(kx - \omega t) \tag{6}$$

where $\hat{y}$ and $\hat{z}$ are unit vectors in the y and z directions, and the magnitudes E_0 and B_0 are related (on account of Maxwell's equations) by

$$B_0 = \frac{E_0}{c} \tag{7}$$

or, in sensible theorists' units,

$$B_0 = E_0 \tag{8}$$

It's also possible for the electric field vector to *rotate* as a function of x and t, so that it points sometimes in the y direction and sometimes in the z direction. Such a wave is called *circularly polarized* or, more generally, *elliptically polarized.* The magnetic field, naturally, is rotating also in this case, in a closely related way.

PROBLEMS

1. How much (by what factor) is a typical UV photon more energetic than a photon of visible light? How about a typical x-ray photon?
2. On graph paper, make a *linear* plot (distance proportional to the variable, unlike fig. 1) of the EM spectrum, first in terms of wavelengths and then in terms of frequencies, in each case going from zero through the visible part of the spectrum. Take the range of the visible spectrum as extending from 380 to 760 nm. On each plot, mark as well as you can the different parts of the EM spectrum, and for the parts that don't fit on the paper, tell how far away, in meters, they would be if the paper, were extended to include them.
3. Work out the vector force (magnitude and direction) in each of the following examples (you'll need the "right-hand rule" for the vector product, from Appendix A, §2; remember that a negative value for the charge Q reverses the direction of the force).
   (a) A particle of charge +5 µC moves due east at a speed of 10 m/s, at a location where the earth's magnetic field points due north and has the value 25 µT.
   (b) An electron ($Q = -e = -1.6 \times 10^{-19}$ C) moves at a speed of $0.6c$ in a direction 50° east of north in the same location as that given in part (a).
4. (a) Check the equivalence of the two units N/C and V/m used for electric field strength, given that the volt (V) is equivalent to J/C, energy per unit charge.
   (b) Show that the unit of magnetic field strength, the tesla (T), given in the text as equivalent to N/A · m, can also be written as V · s/m².

CHAPTER 12

The One-Electron Atom

In Chapter 10 I emphasized that most of the difficulties that were moving us in the direction of a quantum theory had to do with duality: the split personality of every physical entity that makes it behave sometimes as a wave and sometimes as a particle. As I tried to make clear, it isn't possible within any classical kind of picture to reconcile these two—that is, for something to have both wavelike and particlelike properties at the same time. Although the early work, especially that of Niels Bohr starting in 1913, leaned toward living with the contradiction of duality and learning to use it effectively to describe specific phenomena, the situation was still recognized as unacceptable; it was the effort to *reconcile* the wave and particle characteristics in a logically consistent way that gave rise to a proper quantum theory, developed by Schrödinger, Heisenberg, Born, Dirac, and a number of others, later in the 1920s.

A key testing ground for the new ideas and the new mathematical methods as they appeared was the hydrogen atom, the simplest bound system in nature, consisting of one proton and one electron. These were primitive pointlike objects so far as we knew, held together by the simple attractive electrical force between their two charges. In this chapter we'll look at that particular system in some detail.

 SPECTRA OF ATOMS

It had been a mystery for some time why atoms, when stimulated, emitted radiation of only certain definite wavelengths and no others. Such a spectrum was said to be *discrete,* and the observed wavelengths were referred to as *spectral lines.* Until early in the present century, too little was known about atoms for people even to begin to discuss the problem of the discrete spectrum, but around 1909 this situation changed.

> The apparently illogical term *spectral line* is related to the way the different wavelengths are observed. The light is emitted in all directions, with all the wavelengths mixed together. In order to separate them, the light is analyzed with a *spectrometer:* First a narrow slit is used to block all but a well-defined beam; this beam is then bent by a prism or diffraction grating so that the different wavelengths are traveling in slightly different directions. If the wavelengths are discrete, then you have a distinct beam for each wavelength, and when these beams are viewed through an appropriate eyepiece, they form images of the original slit that look like narrow lines of colored light, each in a different location in the field of view. These lines are what are called spectral lines, and they look like lines only because the original slit looks like a line. If you broaden the slit to let more light through, then all the lines you're seeing get broader too and may overlap. For weak lines you have to find the right compromise between intensity and sharpness. The term spectral line has stuck and is now freely used to refer to the different wavelengths emitted by the atom without any reference to how they're measured.

Atoms had been postulated and used as an aid in understanding the nature of matter for a long time, but it wasn't until 1909 that Ernest Rutherford and his associates at Manchester University in England began to discover how the atom itself is constructed. In that year, on a suggestion from Rutherford, Hans Geiger and Ernest Marsden did experiments that probed the structure of matter—gold foil—with massive charged projectiles and found, by the way the projectiles bounced, that each atom has to have a tiny massive core. The projectiles were α particles, now known to be helium nuclei (two protons and two neutrons tightly bound together) but not very well understood at that time. What they observed was that the α particles, tiny and hard as they were known to be, were occasionally seen to bounce backward, to the astonishment of the observers. In a talk many years later Rutherford said, "It was quite the most incredible event that has happened to me in my life. It was almost as incredible as if you fired a 15-inch shell at a piece of tissue paper and it came back and hit you." Rutherford reasoned that this could happen only if a large part of the mass of the atom were concentrated in a volume much tinier than that of the atom itself. He concluded that the atom has to consist of light, negatively charged electrons moving about a very small, massive, positively charged nucleus like planets in orbit around the sun.

The atomic "solar system"

The parallel to the solar system is very close, for several reasons. In the first place the nucleus, like the sun, is very much more massive than the particles traveling around it and therefore, being very little affected by them, remains nearly stationary at the center of the system. Furthermore, the attractive electrical force that the nucleus exerts on each electron, like the gravitational force of the sun on a planet, is proportional to $1/r^2$, the inverse square of the distance from the nucleus or the sun. There is, however, this difference: Whereas the force that the planets exert on each other is very weak—nearly negligible compared to the force due to the sun—the electrons exert a comparatively strong force on each other that can by no means be neglected. This is one of the reasons why the hydrogen atom is so much easier to deal with than any other; because it has only one electron, the interelectron force is not a problem. There is one further difference: The interplanetary forces are attractive, while the interelectron forces are repulsive. (The consequences of this difference aren't important at this stage of our discussion.)

Because of these similarities—especially the inverse square force law—the electron in a hydrogen atom should travel 'n a nice elliptical orbit around the nucleus like a planet around the sun. Classical mechanics allows the orbit to be of any size and hence of any frequency. Being charged, each electron ought to act like a little transmitting antenna with an oscillating current distribution and should emit EM radiation at the same frequency as the motion of the electron. And *this* means that all possible frequencies of EM radiation (or, equivalently, all possible wavelengths) should be observed, which is just what we *don't* see. On top of that difficulty, the electron ought to lose energy as it radiates, and as it loses energy it should keep on emitting radiation at an ever-increasing frequency until all its energy has been radiated away and the electron has collapsed into the nucleus. The resulting object would be uncharged and extremely tiny and would bear no resemblance to the atoms we in fact observe. It would be very much like a neutron (neutrons hadn't been discovered yet), which has about the same size and mass as a proton but has no electric charge. Such objects could in no way form matter of the conventional sort. They would instead pack easily together to make a form of matter perhaps 10^{15} times more dense than ordinary matter, something like the nuclear matter of which the nucleus itself is made, or like a neutron star (see prob. 7.1.1). This classical prediction of the instantaneous collapse of matter, accompanied by large amounts of ultraviolet radiation, is consistent with the description in §10.3 of the ultraviolet catastrophe, the classically predicted conversion of all thermal energy into UV radiation.

What we see, however, is atoms that *don't* collapse, that emit only certain allowed discrete wavelengths, and that spend most of their time emitting *no radiation at all* regardless of the motion of the electron and in total violation of what we thought we knew about the mechanics of such systems.

The Rydberg formula

Let's look now at the wavelengths of the hydrogen spectral lines. These had been observed long before to obey a particularly simple formula, evidently related to the simplicity of the atom itself. They had been found by Balmer (1885) and others to fit the curious formula

$$\frac{1}{\lambda} = R\left(\frac{1}{m^2} - \frac{1}{n^2}\right) \tag{1}$$

where m and n are positive integers ($m < n$) and R is a single experimentally determined constant, $R = 1.0968 \times 10^7$ m^{-1}, the *Rydberg constant* (now known to eleven decimal places).

Johann Balmer was a Swiss schoolteacher interested in numerical puzzles. He was presented by a friend with the wavelengths of four of the visible spectral lines of hydrogen and came up with the equation you get by setting $m = 2$ in the formula (eq. 1), with the four lines corresponding to $n = 3$, 4, 5, and 6. The sequence of lines with $m = 2$ is therefore called the *Balmer series*. Balmer knew nothing of the physics but was clever enough to ask about the higher values of n, and in fact the corresponding wavelengths were already known. (See prob. 2.) Balmer was also led to speculate on extending the formula to other values of m. He had the formula in a somewhat different form than eq. 1, though, and his extension to other values of m was incorrect. Johannes Rydberg got it right in 1890, and the general formula goes by his name.

Other atoms have much more complicated spectra, with no simple algebraic formula like this, but in every case it still turns out that

$$\frac{1}{\lambda} = T_n - T_m \qquad (2)$$

with a single set of numbers T_n, called *term values,* giving the entire set of spectral lines. It is because of this relation (and the quantum physics that underlies it) that spectroscopists often work with the quantity λ^{-1}, known as the wave number,* rather than with the wavelength itself. Note that its units are m^{-1} (inverse meters) or cm^{-1} (inverse centimeters).

PROBLEMS

1. (a) Calculate to three decimal places, and display on graph paper, the wavelengths of the Balmer series (eq. 1, with $m = 2$) up to $n = 10$.
    (b) In what region of the EM spectrum (fig. 11.5.1) are the lines for $n = 7$ through $n = 10$? What happens to λ as n becomes extremely large?
    (c) The first four wavelengths of the Balmer series, in nanometers, are 656.5, 486.3, 434.2, and 410.3. If *you* had been given these wavelengths, could you have guessed the formula? What strategies would you have tried?

---

*The wave number $1/\lambda$ is equal to the number of wave cycles per unit length; it's the spatial analog of the frequency f, the number of cycles per unit time of an oscillatory motion. The wave number is related to the angular wave number k, introduced in §11.1, by the relation $k = 2\pi/\lambda$ (eq. 11.1.2). The angular wave number is the number of *radians* of phase per unit length in the wave pattern and is thus analogous to the angular frequency ω, which is measured in radians per unit time. I coined the term *angular wave number* for k, by analogy with ω, because I am not aware of any other unambiguous term for it.

2. In what part or parts of the EM spectrum do the series for $m = 1$ (the Lyman series) and $m = 3$ (the Paschen series) lie? Make a plot of the wave numbers $1/\lambda$ (see the discussion following eq. 2) for these first three series, showing the range of each series and where they fall in the EM spectrum. What's the advantage of using $1/\lambda$ instead of λ?

 ENERGY LEVELS

Now, what about the idea that light comes in photons, which was already in the wind by the time Rutherford discovered the nuclear structure of the atom in 1911? What Niels Bohr did first of all was to follow through consistently on the photon idea and work out its implications when it is applied to the discrete spectra of atoms.

If atoms are emitting radiation, then the radiation must consist of photons, and you can use eq. 12.1.1 or eq. 12.1.2 to figure out the energy of the photon: Using the expression hf for the photon energy (eq. 10.3.7), you get

$$E_{\text{phot}} = hf = \frac{hc}{\lambda} = hcT_n - hcT_m \tag{1}$$

Bohr saw that this has to be some kind of energy conservation relation:

$$E_{\text{phot}} = E_i - E_f \tag{2}$$

where E_i and E_f, the initial energy and final energy, are seen as the energies of the atom before and after the photon is emitted. This means—and Bohr introduced it as a postulate—that every atom has only a discrete set of possible states of energy E_n:

$$E_n = -hcT_n \tag{3}$$

The energy values are negative, corresponding to the fact that it takes a positive amount of work to separate the particles that make up the atom. (See the discussion of binding energy in §5.6.) If you take $E_i = E_m$ and $E_f = E_n$, then eq. 2 gives the correct photon energy (eq. 1). Note that $T_m < T_n$, which means that E_f is more negative—lower—than E_i. The energy of the atom cannot change continuously in this picture; it can only jump from one value to another, with the emission of a photon of just the right energy to satisfy the conservation law. For the hydrogen atom, the energies would have to obey the simple rule

$$E_n = -\frac{hcR}{n^2} \tag{4}$$

in order to satisfy the Rydberg formula (eq. 12.1.1). The principle is the same for more complicated atoms with two or more electrons, but you can't expect a simple formula for E_n. It's wonderful that such a simple assumption makes sense of the discrete spectrum and of the unusual mathematical form (eq. 12.1.2) for the allowed wavelengths. You have to recognize, though, that big problems remain: In addition to the still unexplained particlelike character of the photon, you now have also the totally unexplained discrete energy levels of the atom and the lack of any description of the *process* by which an atom emits a photon.

The validity of the Bohr picture got a further boost in 1914 when Franck and Hertz observed a corresponding absorption effect, in which electrons bombarding an atom lose in the collision an amount of energy that exactly corresponds to the atom's jumping from a lower to a higher energy level. (The Franck–Hertz experiment is still used as a standard undergraduate lab experiment.) The atomic energy levels that this effect reveals are just the ones needed to describe the wavelengths of emitted radiation according to eq. 1. A similar effect is seen when EM radiation is absorbed by a gas of atoms; only those wavelengths are absorbed that correspond to photon frequencies given by eq. 1. A further, quite detailed confirmation of Bohr's picture was provided by this result: For a cold gas, only those absorption lines are seen that correspond to transitions from the ground state, while if the gas is hot enough for some of the atoms to be excited, absorption lines corresponding to transitions from these excited states can also be seen. All of this makes no sense in the classical picture, where light of any wavelength would necessarily lose energy to such a collection of charged particles simply by causing them to move more energetically.

PROBLEMS

1. **(a)** Calculate the energies of the four lowest levels of the hydrogen atom (eq. 4), both in joules and in electron-volts. (Use the

value of the Rydberg constant from §12.1; other physical constants are collected in Appendix C, §4).

(b)   It turns out (see §12.3) that the energy calculated in part (a) is equal, apart from the sign, to the average kinetic energy of the electron in its orbit. Use this fact to find the average velocity of the electron when the atom is in its ground state. Is it all right to use the nonrelativistic formula for kinetic energy? Explain.

2. (a)   Use the approximation formula $(1 + \varepsilon)^{\alpha} \approx 1 + \alpha\varepsilon$ (eq. 1.4.12) to show that for large values of n, $1/(n + 1)^2 \approx 1/n^2 - 2/n^3$.

(b)   Use this result to show that for large n, the transition $n + 1 \rightarrow n$ gives a photon with frequency proportional to $1/n^3$, and find the proportionality constant (give both a formula and a numerical value).

 **THE BOHR ATOM**

The next problem Bohr faced was to find a way to predict atomic energy levels. Since it wasn't until 1925 that a proper theory was developed, radically different in character from any ideas available to Bohr, it's not surprising that he did a lot of groping in the dark. The *Bohr atom,* as it's called, is nonetheless an impressive achievement. He played with a number of ideas, building on the expected classical orbits but devising ad hoc rules for restricting the allowed orbits so as to get only the energies given by eq. 12.2.4. The simplest such rule involves a physical quantity called *angular momentum,* defined later in this section (eq. 24). Angular momentum is analogous to the kind of momentum discussed in Chapter 4 but is associated with rotational motion. If you postulate that the angular momentum can have only values that are integer multiples of $h/2\pi$ (a rule that has stood the test of time and is found to be a *consequence* of a proper quantum theory), then you find that the energies you calculate for the orbits have exactly the values given by eq. 12.2.4. Bohr experimented with many "quantization rules," including this angular momentum rule, but it wasn't appreciated until later that the angular momentum rule is the way of completing the Bohr picture that makes the best sense physically. For Bohr the rules were arbitrary, but he hoped, rightly, that his experimentation with ad hoc rules would provide a basis for the later development of a proper theory.

What I want to show you now is what you might call a derivation of the Bohr formula (eq. 12.2.4) for the possible energy values for the hydrogen atom:

$$E_n = -\frac{hcR}{n^2} \tag{1}$$

The physics is mostly classical, with the one added assumption about the angular momentum that I mentioned in the last paragraph. I want to stress again that this kind of reasoning is a precursor of a true theory; the intuitions are very close to the mark, and the answer is as accurate as it can be without bringing relativity in, but the actual physics is wrong because most of it is classical. The only correct part of the derivation turns out to be the angular momentum quantization rule—the part that Bohr had the most trouble with!

The first version proposed by Bohr involved circular orbits only (it was later generalized by Bohr and Sommerfeld to include elliptical orbits). The proton is 2000 times more massive than the electron, so it's a good approximation to neglect its motion altogether (like neglecting the motion of the sun when you study planetary orbits). The way the logic works is this: The energy consists of kinetic energy plus potential energy (see §5.3), which depend on the electron's velocity v and on the radius r of its orbit. Newton's second law, $F = ma$, when you use Coulomb's law (eq. 4.3.20) for the force between two charged particles, gives you one relation between v and r, but this by itself isn't enough to determine the energy. When you add the condition on angular momentum that I've mentioned—the quantization rule—then you have another condition on v and r, and you can solve for both of them and determine the energy of the atom. The quantization rule involves an arbitrary integer, so the energy involves that integer also, giving the discrete allowed values needed for the Bohr picture (eq. 1).

The basic equation of motion, when you treat the problem nonrelativistically, is $F = ma$ (eq. 4.3.10); we need to know the electrical force between the two charges, and we also need to know the acceleration of a particle moving in a circle. The force between two charges is given by Coulomb's law (eq. 4.3.20):

$$F = \frac{kQ_1Q_2}{r^2} \tag{2}$$

where Q_1 and Q_2 are the charges, r is the distance between them, and k is a known constant whose value depends simply on the units used. The force is attractive if (like the electron and proton in our hydrogen atom) the charges have opposite signs. In the hydrogen atom, the proton and the electron have equal and opposite charge, $\pm e$, where $e = 1.609 \times 10^{-19}$ C (C represents coulombs, the units of charge—see §11.5). Thus the force is

$$F = -\frac{ke^2}{r^2} \tag{3}$$

The minus sign indicates that the force is attractive.

Let's look next at the acceleration of the electron. Acceleration is the rate of change of velocity (§4.2 and §4.3), and in the case of motion at constant speed, the velocity changes only in direction. For circular motion, the acceleration turns out to be a vector pointing toward the center of the circle, with magnitude given by

$$a = \frac{v^2}{r} \tag{4}$$

You can relate eq. 4 to the force that acts on you when you're in a car going around a tight curve. Say you're in the passenger's seat on the right side, and the car is turning to the left. The extra force you feel is a sidewise force from the door on your right (as you press against it), and it's that force *to the left,* due to the door, that is just what's needed to change the direction of your velocity as the car turns toward the left. The force increases with the speed of the car (the factor v^2) and with the tightness of the curve (the factor $1/r$).

To derive eq. 4, let x and y be the coordinates of the electron as it moves in a circle, and look at the motion right near the point $x = 0$, $y = r$, as

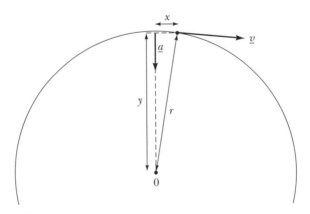

Fig. 1   Acceleration in circular motion at constant speed.

shown in fig. 1. The Pythagorean theorem says that

$$x^2 + y^2 = r^2 \tag{5}$$

and we can use our handy approximation formula (eq. 1.4.12) to get an approximate value for y when x is very small:

$$y = \sqrt{r^2 - x^2} = r\sqrt{1 - \frac{x^2}{r^2}} \tag{6}$$

$$\approx r\left(1 - \frac{x^2}{2r^2}\right) \tag{7}$$

$$= r - \frac{x^2}{2r} \tag{8}$$

Where x is very small it's approximately the distance traveled, vt, so we get

$$y \approx r - \frac{v^2 t^2}{2r} \tag{9}$$

If you compare this with the corresponding term $-g_0 t^2/2$ in eq. 4.4.9 (where the acceleration g_0 is due to gravity), you see that in this case there's a vertical component of acceleration of magnitude

$$a = \frac{v^2}{r} \tag{10}$$

in agreement with eq. 4. At the moment $t = 0$, x is neither speeding up nor slowing down, so the horizontal component of acceleration at this moment is zero and the acceleration is purely vertical—at right angles to the motion and pointing toward the center of the circle. In circular motion at constant speed, all points are equivalent, so this formula gives the correct acceleration at any time.

To apply $F = ma$ to the electron going around the proton, you take the force from eq. 3 and the acceleration from eq. 4, and you get

$$\frac{ke^2}{r^2} = \frac{mv^2}{r} \tag{11}$$

(Both the force and the acceleration point toward the center of the circular orbit, so we need worry only about the magnitudes in this equation.) Eq. 11 is our first relation between v and r. It enables us to express the speed v, and hence the kinetic energy (eq. 5.2.8), in terms of the yet-unknown radius r:

$$K = \frac{1}{2}mv^2 = \frac{ke^2}{2r} \tag{12}$$

The potential energy is already known in terms of r (eq. 5.3.12). It is

$$U = -\frac{ke^2}{r} \tag{13}$$

Thus if we knew r, we could work out the total energy of the atom.

The potential energy (see §5.3) is the energy associated with the electrical force (eq. 3) between the particles; it's defined as the work that the force is capable of doing as the electron moves infinitely far away and is therefore obtained as the integral of the force. From eq. 5.3.6 and eq. 3, it is

$$U = -\int F\,dr + U_0 \tag{14}$$

$$= -\int \left(-\frac{ke^2}{r^2}\right) dr + U_0 \tag{15}$$

But $r^{-2}dr = -d(r^{-1})$ by eq. 4.2.23, so

$$U = -\int d\left(\frac{ke^2}{r}\right) + U_0 \tag{16}$$

$$= -\frac{ke^2}{r} + U_0 \tag{17}$$

We conventionally take the potential energy to be equal to zero when the particles are infinitely far apart (that is, when $r = \infty$, so you just choose $U_0 = 0$, and you get the standard result, eq. 13.

This shows the logical connection between the expression 4.3.20 for the force between two charges and the expression 5.3.12 for the potential energy. The potential energy is negative in this case because you would have to *supply* energy to overcome the force and pull the particles apart. You'd have to say in that case, that the electrical force is doing a *negative* amount of work.

Bohr's picture requires that only certain values of r are possible and that all other values are somehow forbidden. He knew what values of the energy were to be allowed (eq. 12.2.3 and eq. 12.2.4), so he could figure out what values of r were wanted as follows:

$$E_n = K + U \tag{18}$$

$$= \frac{ke^2}{2r} - \frac{ke^2}{r} \qquad \text{(eqs. 12, 13)} \tag{19}$$

$$= -\frac{ke^2}{2r} \tag{20}$$

$$= -\frac{hcR}{n^2} \qquad \text{(eq. 1)} \tag{21}$$

from which, by comparing eq. 20 and eq. 21,

$$r = n^2 a_0, \quad n = 1, 2, 3, \ldots \tag{22}$$

where

$$a_0 = \frac{ke^2}{2hcR} = 0.53 \text{ Å} = 0.53 \times 10^{-10} \text{ m} \tag{23}$$

The radii of the orbits, then, have to be proportional to n^2 with proportionality constant a_0. This characteristic distance a_0 is known as the *Bohr radius* because it's the radius of the first Bohr orbit. Note that a_0 is here obtained in terms of the Rydberg constant R, which was determined from the observed wavelengths of the hydrogen spectrum.

Having done all this, we still have no physical reason for picking these radii. All we've done is find the values of r that correspond to the *experimental* term values that go with the Rydberg formula (eq. 12.1.1). Bohr tried a variety of different rules that would give him these values of r, but none had any obvious basis in physics—the necessary physical ideas were still some way off. The one nearest to the true reason was the rule I mentioned for quantizing—assigning discrete values to—L, the angular momentum. Angular momentum is a very basic conserved quantity, representing the amount of rotational motion a system has in much the same way the linear momentum, discussed in Chapter 4, represents the amount of straight-line motion a system has. For a particle in steady circular motion, the angular momentum is

$$L = mvr \tag{24}$$

which is r times the linear momentum mv (see §4.1). This says that a particle with a given speed contributes more to the angular momentum the farther it is from the center.

The quantization rule for L that gives the answer we want is

$$L = n\hbar \tag{25}$$

where the symbol $\hbar$, given by

$$\hbar = \frac{h}{2\pi} \tag{26}$$

is a very commonly used form of Planck's constant. Introduced by Dirac, it's called the *Dirac h-bar*, or simply *h-bar*. Its usefulness is related to the usefulness of radians, as compared with revolutions, for measuring angles. It's a funny thing that Planck's constant has just the same units as angular momentum; otherwise we wouldn't have such a tidy rule (see prob. 1). Eq. 25 gives us our second relation (the first was eq. 11) between v and r:

$$mvr = n\hbar \tag{27}$$

which enables us to solve for the actual values of r and hence (from eq. 20) the values of the energy.

We need now to combine eq. 27 with the first relation between v and r, eq. 11, and solve for r. It helps to start with L^2, in order to be able to substitute for v^2 using eq. 11:

$$L^2 = m^2 v^2 r^2 \equiv mr^3 \left(\frac{mv^2}{r} \right) \tag{28}$$

$$= mr^3 \left(\frac{ke^2}{r^2} \right) \tag{29}$$

$$= kme^2 \, r \tag{30}$$

Putting in the allowed values of L, $L = n\hbar$, we get the allowed values of r,

$$r_n = \frac{L^2}{kme^2} \tag{31}$$

$$= \frac{(n\hbar)^2}{kme^2} \tag{32}$$

$$= n^2 a_0 \tag{33}$$

in exactly the desired form, along with an explicit expression for the Bohr radius that is almost exactly right:

$$a_0 = \frac{\hbar^2}{kme^2} \tag{34}$$

You can't measure the radius itself, but it gives the right value for the Rydberg constant R (eq. 23). The biggest approximation I've made has been to neglect the effect of the motion of the nucleus. With that included (an 0.05% correction), the agreement is extremely good. I should remind you that we assumed only circular orbits for the electron. The generalization to elliptical orbits makes the calculation much more complex but doesn't change the energy values.

A standard way of displaying the energy levels visually is shown in fig. 2 for the case of hydrogen.

After correction for the motion of the nucleus, this result and the corresponding result for the Rydberg constant (eq. 12.1.1) are accurate to the order of 1 part in 20,000. The biggest corrections after this involve what is called the *fine structure:* a splitting of each energy level into several levels very close together. These tiny energy shifts are due to a force called the *spin–orbit interaction* that the Coulomb field of the proton exerts on the electron's magnetic moment (the electron is a little magnet, and the magnetic moment is the strength of the magnet) as it moves in orbit around the proton. The corrections are smaller than the energy values themselves by a factor of the order $(v/c)^2$. In fact, the spin–orbit interaction is properly regarded as part of the relativistic correction, which is given with superb accuracy by the relativistic quantum theory of the electron devised by Dirac. With the value of v determined by eqs. 11, 33, and 34, the dimensionless factor $(v/c)^2$ turns out for $n = 1$ to be equal to α^2, where α, known as the *fine-structure constant,* has the value

$$\alpha = \frac{ke^2}{\hbar c} \approx \frac{1}{137} \tag{35}$$

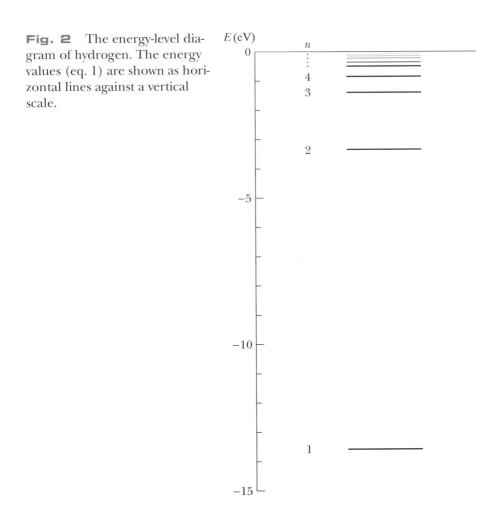

Fig. 2 The energy-level diagram of hydrogen. The energy values (eq. 1) are shown as horizontal lines against a vertical scale.

Further corrections arise from the magnetic moment of the proton, from its finite radius (around 0.5 fm—the proton is *not* a point particle), and, most significantly, from interactions with quantum fluctuations in the space around the electron. This last comes under the heading of QED—*quantum electrodynamics.*

All we have to do now is explain where eq. 25, the rule for quantizing angular momentum, came from. It's again wonderful that such a tidy rule should contain all the complexities of the Rydberg formula. Even if you didn't know what would come next, this rule would have the ring of truth.

PROBLEMS

1. (a) Check that the dimensions of $\hbar$ are the same as those of angular momentum.
    (b) Show that the value of v/c for the lowest Bohr orbit ($n = 1$) of hydrogen is equal to the expression for the fine-structure constant α (eq. 35), and check the numerical value given there.
2. (a) Find the force on your body, as a multiple of your weight mg, that must act to keep you in your seat as your car takes a curve of radius 500 m at 25 m/s. (You may approximate g, the acceleration of gravity, as 10 m/s².)
    (b) Find the speed, in meters per second and in miles per hour, at which the extra force you feel is 1/10 your body weight on a curve of radius 100 m.
    (c) Find the speed of a satellite in low earth orbit (radius of orbit ≈ radius of the earth; force = mg, the gravitational force on the satellite). Find the period of the orbit.
3. (a) Find the radius of the first Bohr orbit of a muonic atom, in which a μ^- replaces the electron in orbit about a proton. The mass of the muon is $m_\mu = 207\ m_e$.
    (b) Find the quantum number n for the Bohr orbit of the muonic atom that has the same radius as the lowest orbit of ordinary hydrogen.
4. The logic of this section applies in principle to the gravitational force (eq. 6.2.1), which has just the same form as the Coulomb force (eq. 2), but with ke^2 replaced by Gm_1m_2. Again, the problem is simplified if one of the masses is much less than the other, so that the motion of the larger mass can be neglected. Find the energy and radius of the lowest gravitational Bohr orbit for each of the following:
    (a) An electron in orbit about a neutron
    (b) A neutron in orbit about a 1-kg point mass
    (c) A 1-kg mass in orbit about the earth (with the earth's mass concentrated to a point for convenience)

 **TRAPPED WAVES**

In fact, the feature that Bohr was missing was the wave nature of the electron, an idea that wasn't conceived until 1923 (§10.4). Because the electron is confined by the electrical attraction of the nucleus, it follows that any wave motion associated with the electron must also be confined within a small volume about the size of the atom. It is the nature of waves, as I explained in §11.4, that when a wave motion is confined it can have only certain allowed frequencies of oscillation, like those of the normal modes on a stretched string. We haven't yet faced the question of how to determine those frequencies for electron waves, or of how to translate frequency into energy [it turns out to be the Planck relation (eq. 10.3.7), just as for photons], but in any case these discrete allowed frequencies will correspond to the kind of allowed energies we're looking for.

Let me start with a simple model to illustrate how the idea works: An electron is trapped in a long, narrow box (narrow so that we can treat the problem as one-dimensional, like the string in §11.4). The possible wavelengths are the same as for the string (eq. 11.4.2):

$$\lambda = \frac{2L}{n} \tag{1}$$

The de Broglie relationship (eq. 10.4.1) then gives us the corresponding value for the momentum of the electron:

$$p = \frac{h}{\lambda} \tag{2}$$

$$= \frac{nh}{2L} \tag{3}$$

We don't know how to get the frequency from the wavelength, because the wave velocity v_w may not be constant, so I'll use a different approach that avoids working with the frequency altogether. Because we're supposing that the electron moves freely inside the box, its energy is just the kinetic energy associated with this momentum (see eq. 5.7.20), and we're there. Working this out yields

$$E_n = \frac{p^2}{2m} \qquad \text{(eq. 5.7.20)} \tag{4}$$

$$= \frac{(nh/2L)^2}{2m} = E_1 n^2 \quad \text{(eq. 3)} \tag{5}$$

where

$$E_1 = \frac{(h/2L)^2}{2m} \tag{6}$$

Thus you see that when the electron is confined to a box, it has discrete energy levels. The formula is unimportant; the point is the discreteness of the levels. To get the hydrogen spectrum we'll need to do the same thing for the hydrogen atom, where the basic principle is the same: Confinement of the matter wave produces discrete energy levels. The problem is more difficult because when a particle is moving in a force field, its momentum keeps changing and so its wavelength keeps changing too. Furthermore, there isn't a sharp confining boundary—it's the force field that does the confining, and how far the particle can move depends on its energy. To do it properly you need to describe much more accurately the dynamics of the waves, and so far I haven't discussed this. What's needed is a wave equation, analogous to Maxwell's equations for the EM field, that gives a complete description of the physics of the electron waves. This wave equation turns out to be the Schrödinger equation, which will be introduced in Chapter 14.

You can, however, get a simplified version of the wave picture for the hydrogen atom by performing a standard trick using the simple circular orbits of our discussion in §12.3. (This is one of the first things de Broglie proposed when he conceived the idea of a wavelength for electrons.) What you do is try to stretch a wave of wavelength λ around a circular path, or orbit, of radius r, as shown in fig. 1. Because the wave must come back to where it started, the circumference of the orbit, $2\pi r$, has to be an integer multiple of λ:

$$2\pi r = n\lambda, \quad n = 1, 2, 3, \ldots \tag{7}$$

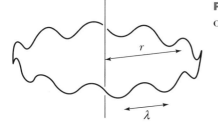

Fig. 1   De Broglie waves on a circular orbit.

Translating this by the de Broglie relation (eq. 10.4.1) into a statement about the momentum yields

$$2\pi r = \frac{nh}{p} \tag{8}$$

or

$$rp = \frac{nh}{2\pi} = n\hbar \tag{9}$$

The quantity rp is just the angular momentum (eq. 12.3.24), and you can see then that eq. 9 is just the quantization rule (eq. 12.3.25), which, as we found, gives the correct levels for the hydrogen atom.

You must recognize that this is only a primitive version of a derivation of the hydrogen spectrum, because the wave in this model is only one-dimensional, stretched around a circle of definite radius r, while any real wave has to fill the whole three-dimensional space around the nucleus. Nevertheless, even this primitive derivation represents a major break-through: The de Broglie hypothesis gets us on exactly the right track and shows us finally just why the atomic levels are discrete. The puzzling nature of atomic spectra is a direct consequence of *both* the particle character of light and the wave character of electrons!

PROBLEMS

1. **(a)** A one-dimensional electron of mass m is enclosed in a one-dimensional box of length ℓ. Obtain a formula for the possible wavelengths of photons emitted as the electron drops from one level to another.

    (b) Find the energies (in electron-volts) of the first four levels for a one-dimensional electron in a one-dimensional box of length 1 Å, find the six possible wavelengths of emitted photons, and locate them on the EM spectrum of fig. 11.5.1.

2. Repeat prob. 1 for an electron in a three-dimensional cubical box of side ℓ [see prob. 11.4.2(b)]. For part (b), "the first four levels" refers now to the four lowest distinct energies; indicate how many different states belong to each level. Each state has a different assignment of the quantum numbers n_1, n_2, and n_3, and several states may have the same energy.

CHAPTER 13

The Historic Principles

It's important to remember that in the early years of quantum theory, there was a lot of confusion and argument among physicists. The universe at the microscopic level was behaving in an impossible way, and there seemed to many to be no hope of reconciling the contradictory features that were emerging. Light could *not* consist of particles and still show the familiar interference effects that signal the presence of waves; neither could electrons be waves and still have the sharply defined momentum and energy that characterize particles of definite mass; and surely the laws of physics could not have given up their allotted function of predicting the future with arbitrary precision once initial conditions are known. Yet all these things appeared to be happening.

During this period, people were trying very hard to make some kind of sense of this new quantum world, and they achieved many deep and valid insights even before they managed to develop a proper mathematical quantum theory. Some of these insights took the form of "principles," which served as a guide to living with the contradictions; they did not in themselves constitute anything like a true theory, but they enabled people to do physics in many of the situations ordinarily encountered. In a sense these principles represent a distillation of the most distinctive features of the new physics: characteristics that would have to be—and indeed became—part of the proper theory when it finally appeared.

In this short chapter I want to introduce the most notable of these historic principles, reminding you though of their *preliminary* character. From our present perspective, these are no longer fundamental principles or axioms but are rather consequences or features of a fully developed theory.

13.1   CORRESPONDENCE

The *correspondence principle* is the requirement that, when a physical theory is overthrown, the new theory must correspond to the old theory—that is, the new and the old must agree—*in those realms of physics in which the old theory has been tested*. Bohr formulated this idea as a guide in the development of quantum theory, but it is pretty obvious that it's much more general in scope. You saw this logic in Part I when we talked about relativity, which clearly has to agree with nonrelativistic physics when velocities are small compared to c. The point now is that the new quantum theory—whatever theory that turns out to be—must agree with classical physics for large-scale events (baseballs, planets, and so on).

How can this be? Is a baseball supposed to have a wavelength and to exhibit interference fringes when you throw it through two holes in a wall? Are sound waves supposed to be quantized and hit the ear in a little bang as each quantum of energy arrives? When you think about it, you find that there's no real contradiction of our experience, basically because Planck's constant, the fundamental constant that describes all quantum effects, is so very small. In fact, $\hbar$ is about 10^{-34} J $\cdot$ s (eqs. 10.3.5 and 12.3.26), and its smallness relative to everyday units is much more extreme, for example, than the bigness of c, 3×10^8 m/s. That is to say, we expect quantum effects to be *much* less noticeable in our everyday experience even than relativistic effects.

To see how this works in practice, consider the baseball. Its wavelength is given by the de Broglie relation (eq. 10.4.1), $\lambda = h/p$, which means that if $p = 1$ in standard units (kg $\cdot$ m/s) then the wavelength is 10^{-34} m, which is far smaller than an atom (10^{-10} m) or even a proton

(10^{-15} m). [A thrown baseball might have $p = mv \approx$ (0.1 kg) × (30 m/s) = 3 kg · m/s, which doesn't affect the discussion—a factor of 10 more or less hardly matters.] What's the meaning of a wavelength this small? It can produce no observable effect. For one thing, it sets the scale of any possible interference pattern, which is therefore far too tiny to see even with the fanciest instruments. For another, it means that the fuzziness of a wave pulse, which is the fuzziness in the location of the ball at any given time, is also unobservably small. The trajectory of the ball has to be correctly predicted also—this issue is addressed in more detail in §14.3.

Now, what about classical waves? If you're dealing with a wave at the everyday level, with a wavelength of the order of 1 m (or 1 mm or 1 km—it doesn't matter), then it may consist of quanta, but the momentum and energy of the individual quanta are at the scale of 10^{-34} in everyday units and far too tiny to be observable. Any ordinary wave would have to comprise ~ 10^{34} quanta per second, and only the collective effect of all of them, purely wavelike, would be observed.

This discussion shows that the quantum nature of matter doesn't cause any problem in principle with our everyday experience, but it doesn't show in detail how quantum theory can reproduce the detailed *laws* of macroscopic physics, Newton's laws of motion and Maxwell's laws for EM fields in particular. This is one of the things you have to look out for when you start considering an actual mathematical quantum theory. I think it's amazing that a theory can be so radically different in its basic character from our familiar classical physics and still reproduce that classical behavior in all its detail.

PROBLEMS

1. (a) Explain why, in looking for photons in EM waves at the every-day level, it doesn't make much difference whether you use wavelengths of millimeters or kilometers.
    (b) Find the number of photons per second in a 1-mW beam of (i) 600-nm visible light waves, (ii) 500-m radio waves, and (iii) 1-nm x-rays.
2. Estimate the mass of a dust particle (say the density is comparable with water, 1 g/cm^3), and find the velocity at which its de Broglie wavelength would be around the diameter of a proton. Does the wavelength get bigger or smaller as the velocity increases?

13.2   UNCERTAINTY

Werner Heisenberg, one of the pioneers in the development of quantum theory, devised the *uncertainty principle* as an explicit example of the loss of classical certainty. The "uncertainty" referred to is not just human uncertainty but intrinsic uncertainty built into the laws of physics: There are physical variables that simply cannot be known simultaneously with complete precision. In its original form, the uncertainty principle refers to a particular example of this kind of incompatibility, that of the position and momentum of a particle. The reason for this incompatibility is simple: The momentum of a particle is related to an associated wavelength by the de Broglie relation, eq. 10.4.1, and your knowledge of that wavelength gives a measure of the fuzziness in your knowledge of the particle's position.

The principle is stated (in one dimension for now) in terms of Δx and Δp, which are defined as the indefiniteness of the particle's position and of its momentum, respectively. The Heisenberg relation is an inequality that says that Δx and Δp can't both be made arbitrarily small:

$$\Delta x \Delta p \gtrsim h \tag{1}$$

That is, if x is known—or measured—to a precision Δx, then p cannot be known—or predicted—to a precision better than about $h/\Delta x$. Equivalently, if the momentum is known to a precision Δp, then the uncertainty in position can be no smaller than about $h/\Delta p$.

When Heisenberg presented this result in 1927, he made no use of the wave character of matter (which he opposed at that time) but looked at the problem entirely in terms of the process of observation itself. His approach exploits the dual nature of light, using the Planck relationship, but avoids admitting the dual nature of electrons. A rough version of the argument is this: Suppose you use light to observe the position of an electron. The picture is fuzzy because the wavelength of the light sets a limit on how precisely you can locate the electron, so the uncertainty in the position of the electron, Δx, will be of order λ. The photons have momentum h/λ, or $h/\Delta x$, and the direction of the photon after it bounces off the electron is uncertain (because of the aperture of the observing lens). The photon therefore gives the electron an uncontrollable impulse (transfer of momentum) of that order, making the electron's momentum uncertain by that amount. Soft, gentle photons (photons of very low energy) have long wavelengths and produce very fuzzy pictures, while short-wavelength photons, the kind you need for sharp pictures, are very energetic and shake the electron in an uncontrollable way.

It turns out that eq. 1 can be made precise by defining the uncertainties Δx and Δp more accurately and by making use of basic properties of matter waves, along the lines of the argument mentioned in the first paragraph of this section. The momentum of a particle is associated with the wavelength of the associated wave, so a particle of well-defined momentum has to have a wave of definite wavelength. On the other hand, the wave associated with a well-localized particle, whose position is sharply determined, must be a localized wave—a wave packet—spread out only over a region of length Δx, as shown in the examples in fig. 1. Note that how well the wavelength is determined depends on how large Δx is compared to the wavelength λ. In fig. 1(a), Δx is much larger than λ, and you can see that λ is quite definite, though not completely so. The indefiniteness in λ is associated with the overall shape of the pulse: To get a localized packet, you have to superpose waves of slightly different wavelength, in the spirit of §11.3, in order to get them to cancel outside the region of the packet. In fig. 1(b), Δx is small, and it's much harder to make a good estimate of λ. The relation between Δx and the uncertainty in λ is a theorem in Fourier transforms (the theory of the analysis of arbitrary wave shapes in terms of pure sine waves), which is expressed

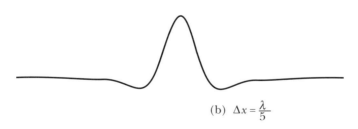

(a) $\Delta x = 5\lambda$

(b) $\Delta x = \dfrac{\lambda}{5}$

Fig. 1 Wave pulses.

most conveniently in terms of the angular wave number $k = 2\pi/\lambda$ (eq. 11.1.2). If Δx and Δk are defined as the standard (or root mean square) deviations* of x and k, respectively, from their mean values, then they must satisfy the exact inequality

$$\Delta x \, \Delta k \geq \frac{1}{2} \tag{3}$$

It is then a simple consequence of the de Broglie relation (10.4.1),

$$p = \frac{h}{\lambda} = \hbar k \tag{4}$$

that

$$\Delta x \, \Delta p \geq \frac{\hbar}{2} \tag{5}$$

the exact statement of the uncertainty principle. Note that this is an *inequality.* You can be *more* uncertain than this about both x and p but not less. In fact, in most normal experimental situations there is much greater uncertainty than this; remember that $\hbar$ is a very small quantity.

The variables x and p of the preceding discussion are referred to as *complementary variables.* In fact, every variable in physics is found to be one of a pair of complementary variables, related to each other by an uncertainty relation similar to eq. 5. (One example is angular momentum and angle of orientation in the case of rotational motion.) In a three-dimensional world, the position and momentum of a particle are vectors. The uncertainty principle applies to only one component at a time, because it is only p_x that is complementary to x, and so forth. There is no such restriction relating Δx and Δp_y for instance.

---

*The standard deviation of a variable is the square root of the mean of the squares of the deviations. That is, if y_i, $i = 1, 2, 3, \ldots$, are a set of measured values of y, and the true average value is $\bar{y}$, then the standard deviation is given by

$$\Delta y = [\langle (y_i - \bar{y})^2 \rangle_{av}]^{1/2} \tag{2}$$

You should examine carefully how the formula reflects the verbal definition.

PROBLEM

1. **(a)** Find the minimum momentum uncertainty Δp for an electron
    confined to a region of length 1Å (about the size of the hydro-
    gen atom), and compare it with the momentum of an electron
    in the first Bohr orbit (§12.3).
    (b) Make crude estimates of experimentally reasonable limits on
    Δx and Δp (work from Δv, the uncertainty in speed) for a
    thrown baseball, and compare it with the limit imposed by the
    uncertainty principle. Work out a few additional examples.

13.3   COMPLEMENTARITY

The principle of *complementarity,* which, like the correspondence principle, we owe to Bohr, is much more qualitative in character than the
others discussed here and more difficult to use in a systematic way.
A generalization of the ideas leading to the uncertainty principle, the
complementarity principle attempts to confront the contradictions of
quantum theory head-on and, in a sense, to formulate the fact of contradictions as a principle. The essence of the principle is that two different
kinds of experiment, or two different kinds of physical variable, can be
complementary* in that they are mutually exclusive (incompatible), but
both are needed for a complete picture.
    The prime example of this is wave–particle duality itself. The wave
properties and the particle properties of light are complementary, then,
because you need both to give a complete description but you can't
observe both at the same time. When one photon leaves its mark on a
photographic film or produces a photoelectron, there is no trace of the
wave properties—no extension in space, no observation of wavelength
or frequency. When you tune your radio to a particular station, on the
other hand, you are sorting out from all the waves in the atmosphere just
those of a particular frequency, by allowing them to produce a resonance
in an electrical circuit that will oscillate at just that frequency and no
other—a process that seems entirely meaningless in terms of localized
particles of radiation.

---

*Not *complimentary.* Two things *complement* each other when each supplies something that the other lacks.

The kind of observation you make determines which aspect of the radiation is relevant, and, in fact, complementary observations invariably obstruct each other. An example of this can be seen in the two-slit interference experiment of §10.5. You saw there that you could observe the photons at the wall where their distribution shows the effect of interference between the two slits, *or* you could observe them close to the slits where what you're determining is which slit the photon goes through. These again are complementary observations, and each is incompatible with the other. No matter how gently you do it, if you succeed in determining which slit the photon went through, you destroy the possibility of interference; conversely, if you do see interference, you have to have no knowledge at all of which slit the photon went through.

The early practitioners of the new quantum theory became masters of exploiting complementarity to get at physical results during the period when there seemed to be no satisfactory interpretation of the theory. The scattering of x-rays from electrons (the Compton effect) is an example of this: You employ wave interference to select x-rays of a definite wavelength and frequency; you convert this information into momentum and energy for a single quantum; you use relativistic particle dynamics to analyze the scattering of the two particles, the photon and the electron, and deduce the momentum of the scattered photon; and finally you shift gears again and convert this back to a wavelength, which is measured by means of wave diffraction from a crystal lattice as described in §10.4. The exact wavelength shift of the scattered x-rays is impossible to explain in terms of classical waves, but it comes out naturally as a momentum transfer in the photon picture.

Again, we must watch and see how the proper theory, as it is now interpreted, deals with this—how it can preserve the *effects* of complementarity while removing the contradictions.

 **SUPERPOSITION**

My final example of an historic principle, the *superposition principle,* extends the phenomenon of interference to a broad generalization about the quantum states of any system. You've seen (§10.5 and §10.6) how interference is a consequence of the way waves can be combined by simple addition (§11.3), so that where two waves have the same sign they reinforce each other and where they have opposite signs they cancel. The key here is the word *addition,* and it runs very deep—deeper even than the wave concept itself.

It works like this: Whenever a system can be in one or the other of two states, call them A and B, then it is possible for it to be in a state that is a combination of A and B, called a *superposition,* with properties analogous to those of a combination of waves. In such a combined state, if you do an experiment that tells whether the system is in state A or in state B, you will definitely get one answer or the other, with a certain probability for each, just like measuring which slit the photon went through in the two-slit experiment. On the other hand, if you do some other experiment complementary to the A-versus-B experiment (such as looking at the fringes in the two-slit experiment), then A and B can interfere—constructively, destructively, or in a variety of other ways—just as if A and B could somehow be added together to get the resulting state.

What this is giving us is an idea of the *state* of a physical system that is radically different from anything we've seen in classical physics. To describe the state of a classical particle at a certain moment of time, you give its position, r and its momentum p, and a different state would have a different set, r' and p'. How could you combine these states to get a third? If you tried, for example, to add the variables $(r + r', p + p')$, you'd have a state that has nothing in particular to do with the two states you started with, a state in which there's no possibility at all of the particle's being in one of the original states—and certainly no possibility of interference.

In the form of quantum theory developed by Schrödinger in 1925–1926, the superposition principle is satisfied by using waves to represent all the possible states of any physical system, in contrast to the classical way of giving positions and momenta of particles. The wave is called the Schrödinger wave function, and its behavior is governed by a wave equation, the *Schrödinger equation.* You'll learn more about the wave function and about the problem of what it means in Chapter 14.

Actually, waves are only one example of things that show superposition—that can be added and subtracted in a meaningful way. The most general sort of entity that shows superposition is a vector, which can be one of our familiar three-dimensional vectors or an abstract vector in a vector space of arbitrary—or even infinite—dimension. In the development of quantum theory, it was Dirac who took this step and made such abstract *state vectors* the building blocks of a beautifully general formal structure (known as a *linear algebra*) for quantum mechanics. For a system consisting only of nonrelativistic particles, this turns out to be equivalent to using a Schrödinger wave function to represent the state, but for more general systems, especially relativistic theories involving fields such as the EM field, the wave function picture is no longer adequate, and Dirac's way of thinking about the physics seems to represent its essential nature much better.

Each of the four principles I've introduced in this chapter is somewhat vague at this stage, but they become much more solidly based when a proper theory is available. Indeed, they become logical consequences of such a theory. I've given some indications here of how that theory will develop and include these principles as features, but it's important to remember that the principles were set down before quantum theory was properly formulated.

CHAPTER 14

The Schrödinger Wave Equation

You can't hope to understand the  *until you've understood the* $\left(\begin{array}{c}math\\physics\end{array}\right)$.

With the understanding gained from the de Broglie relation (eq. 10.4.1) that electrons and other particles have wave properties came Schrödinger's immediate realization (1925–1926) that you have to have a *wave equation*—an equation of motion describing the dynamics of the wave—to tell how it will behave in the future given its present state. Maxwell's equations are the (fairly complicated) wave equations that describe electromagnetic waves, and there are wave equations—all different—to describe waves of every sort: standing waves in a guitar string, seismic waves in the earth, sound waves in air, and so on. This way of approaching quantum theory, making it into a *wave* theory, was resisted vigorously by Bohr and Heisenberg, so it was a time of many arguments and confrontations that I won't describe here. The confusion was compounded by the fact that Schrödinger's own understanding of the relation between the wave and the particle that the wave is supposed to describe was by present standards seriously deficient: The process of getting it right, or even just logically coherent, wasn't easy.

14.1 THE WAVE FUNCTION

How do you find a wave equation when you don't know what's waving? These new *matter waves* are not like anything physicists had ever seen before. They aren't waves in a *material*, such as water, rope, or air. They aren't even wave patterns of a force field, like the EM field or the gravitational field, that can produce observable forces on particles of matter. Even if we don't know what matter waves *are*, we can still ask whether we know anything about how they ought to *behave*—and the answer is *yes*, we know quite a bit.

We know first of all that there must be *some* kind of wave associated with a particle, because we can see the interference patterns; and from the Planck and de Broglie relations, we also know something about the wavelength and frequency of the wave. Schrödinger himself thought that the electron *is* a wave, but as I've tried to show you, this view doesn't adequately deal with its truly particlelike properties or with the indeterminacy—the random behavior—that we see with electrons as well as with photons. The conclusion we've come to, already described in §10.7, is that a matter wave is not directly observable but gives information about the *probabilities* for where the electron might be found. You can also deduce from the wave function the probabilities for other types of measurements besides position: momentum, energy, and angular momentum, for example. I'll tell you more about what the wave function means in §14.4.

Following tradition, I'll call the wave function $\psi(x,t)$, where the arguments x and t indicate that the value of ψ (that is, the amplitude of the wave) depends on both position and time, in the way I talked about in §11.2. We don't know to start with what kind of variable ψ must be—it might be a simple number or it might be some kind of vector. One of the mysteries of quantum theory is that ψ turns out to be *complex,* which means it's a number that involves the imaginary quantity i:

$$i = \sqrt{-1} \tag{1}$$

One of my tasks will be to give you a rough idea of why we seem to need such numbers to do physics and why it doesn't violate common sense to use them.

One of the things a wave equation tells you is the relation between frequency and wavelength—how the time dependence is related to the spatial pattern. In fact, we already know the relation between frequency and wavelength from the Planck and de Broglie relations, so *the wave equation is just about determined.* The way we know how frequency and wavelength are related is simply this: The energy and momentum of a particle are related by $E = p^2/2m$ (eq. 5.7.20), and the Planck and de Broglie relations give E in terms of the frequency f ($E = hf$, eq. 10.3.7) and give p in terms of the wavelength λ ($p = h/\lambda$, eq. 10.4.1). Thus you get a relationship between f and λ:

$$hf = \frac{(h/\lambda)^2}{2m} \tag{2}$$

and you require that the wave equation give just that relationship for waves of all possible wavelengths. Recall that frequency has to do with rate of change, and wavelength has to do with the spatial pattern at a given time, so the frequency–wavelength relation tells you how the wave must change given the spatial pattern at a given time. That's in fact just what a wave equation does, because it's the equation of motion for a wave.

Finding a wave equation

Well, you go through the process I've described, trying to make the wave equation work for a simple sine wave of the kind we looked at in Chapter 11: $\psi(x, t) = A \sin(kx - \omega t)$, which has a definite frequency and wavelength. Because the wave function $\psi(x,t)$ is a function of more than one variable (x and t), the wave equation has to involve partial derivatives. A *partial derivative* is like an ordinary derivative applied to just one or the other of the two variables; it is written as $\partial\psi/\partial x$ or $\partial\psi/\partial t$. The time derivative $\partial\psi/\partial t$ produces a factor of ω on account of the chain rule of differentiation, and the spatial derivative $\partial\psi/\partial x$ produces a factor of k for the same reason, and you try to concoct a combination of these that duplicates the relation (eq. 2) between frequency and wavelength. [Remember that ω is proportional to the frequency f (eq. 11.2.6) and that k is proportional to $1/\lambda$, the inverse wavelength (eq. 11.1.2).] Then you run into frustrations because the sine and cosine functions keep getting in each other's way, and you find that the only way out of the difficulty is to combine them, using the imaginary number i in order to get the various factors right:

$$\psi(x,t) = \cos(kx - \omega t) + i\sin(kx - \omega t) \tag{3}$$

Such combinations of ordinary numbers and the imaginary number i are called *complex numbers*. (Because I think complex numbers are so interesting, I've given a more thorough description of them in Appendix A, §7. We'll use them very little, but they play an important role in the structure of quantum theory and I need to be able to talk about them from time to time.) With this kind of complex wave function, it becomes possible to construct a wave equation that works, and what you get is the Schrödinger equation for a one-dimensional free particle, shown in fig. 1(a). The reason I show these equations as a figure is that I want you to see what they look like because they're so famous, but partial differential equations are way outside the range of this book, so I don't want to present them as equations that need to be understood and used. The symbol ∇^2 that appears in the three-dimensional version [fig. 1(b)] is called the *Laplacian operator* and represents a combination of partial derivatives with respect to the three spatial coordinates x, y, and z.

Now why should physics be required to make use of a fictitious quantity like $\sqrt{-1}$? In the world of mathematicians, complex numbers have an astonishingly rich and fruitful existence, with a complete arithmetic that springs from this single additional assumption that

$$i^2 = -1 \tag{4}$$

The algebra of complex numbers is more complete and elegant than the algebra of real numbers, and the calculus also is richer and tighter than the calculus of real numbers. (We now use the word *real* to distinguish

$$i\hbar \frac{\partial \psi}{\partial t} = -\left(\frac{\hbar^2}{2m}\right)\left(\frac{\partial^2 \psi}{\partial x^2}\right) \qquad\qquad i\hbar \frac{\partial \psi}{\partial t} = -\frac{\hbar^2}{2m}\nabla^2\psi$$

(a)   (b)

Fig. 1   The free-particle Schrödinger equation. (a) One dimension. (b) Three dimensions.

ordinary numbers from imaginary or complex numbers.) Nevertheless, physics deals only in real numbers. There is no $\sqrt{-1}$ in the laboratory. Meters and counters and the wealth of other types of measuring device invariably yield real numbers—ordinary numbers—as their output.

Schrödinger himself resisted the introduction of i vigorously and allowed it to appear only after repeated efforts to avoid it had failed. What does it mean? I think myself that the rich significance of complex numbers in mathematics has perhaps obscured the more modest nature of their role in quantum theory. I'm inclined to view ψ as a vector in a two-dimensional vector space, as in fig. 2, with two *real* components—the cosine term and the sine term in eq. 3. In terms of its components, ψ is written as

$$\psi = \psi_1 + i\psi_2 \tag{5}$$

and the magnitude of ψ, written as $|\psi|$, is given by

$$|\psi| = \sqrt{\psi_1{}^2 + \psi_2{}^2} \tag{6}$$

The magnitude $|\psi|$ is also known as the *absolute value*, or *modulus*, of the complex number ψ.

The Schrödinger equation can be separated into a pair of real equations relating the two real functions ψ_1 and ψ_2, and quantum theory can

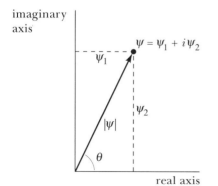

Fig. 2   Vector diagram for the complex wave function ψ.

proceed without hindrance. When a quantity like ψ is viewed as a two-dimensional vector, it is often referred to as a *phasor* because its orientation is given by a single phase angle.* In phasor language, i becomes a rotation operator, simply producing a 90° rotation of a phasor. If complex numbers with their elegant properties didn't exist, this phasor approach would still be perfectly possible and would be equivalent to the quantum theory we know. The role of complex numbers, as I see it, is to provide a surprisingly convenient *representation* of what can be viewed as an intrinsically real theory. As I'll show you in §14.4, the complex wave function ψ can be used to obtain real information about real probabilities for measurements of real physical variables.

Stationary states

Let me close this section with a word about the practicalities of solving the Schrödinger equation. Partial differential equations are difficult to solve in their full generality, and if we can, we always look for special solutions that contain the information needed. We come first to a class of special solutions, sufficient for a great many of the applications in physics, called *stationary-state solutions.* A *stationary state,* in quantum theory, is a state for which the wave function is constant in time except for a trivial complex *phase factor* (explained below). The probability distributions for the system remain constant, and the system seems almost static. It isn't static, though, because the particle momenta are not zero; it's more like a state of steady flow, similar to the flow of water through some complicated system of pipes. Each water molecule goes through a complicated motion, but the distribution of water and the flow pattern are exactly constant in time. In our case it's the probability distribution and an associated probability current distribution that are constant in time, while if you measure a particle's velocity, for example, you generally get nonzero values.

The time dependence of the phase factor in a stationary state has a definite frequency—it's a complex combination, in fact, of cosωt and sinωt just like that in eq. 3:

---

*Phasors appear elsewhere in physics when you need to deal with both amplitude and phase, a very common example being electrical circuit theory in the case where you have alternating currents and voltages of fixed frequency. The most compact representation of AC circuits is by means of complex amplitudes, just as with the Schrödinger wave function. Electrical engineers, though, use j to represent $\sqrt{-1}$, reserving the letter i for electric current.

$$\psi(x,\ t) = \varphi(x)[\cos\omega t - i\sin\omega t] \qquad (7)$$

This corresponds, in the picture given in fig. 2, to a simple clockwise rotation of the vector through an angle ωt, which means that the vector rotates at constant angular velocity ω. The phase angle θ in the picture takes the value $-\omega t$ at any given time t.

The fact that the frequency has a definite constant value also means that the energy has a definite constant value, by the Planck relation (eq. 10.3.7),

$$E = hf = \hbar\omega \qquad (8)$$

Thus a stationary state is a state of definite energy—what we call an *energy eigenstate*. Where the system is confined, as is the electron in a hydrogen atom, the stationary-state solutions are the normal modes of the wave motion (§11.4), and the frequencies are the normal frequencies. Thus the energy, like the normal frequencies, can take on only certain definite values and no other, just as Bohr wanted to see. When we solve for the stationary states of the nonrelativistic hydrogen atom, we get exactly the energies that Bohr needed for his model, but now they are derived from a proper theory.

PROBLEM

1. (a) Using fig. 2, show that the real part ψ_1 of the complex number ψ is given in terms of the phase angle θ by $\psi_1 = |\psi|\cos\theta$, and find the corresponding expression for the imaginary part ψ_2.
   (b) From part (a) show that you can write the complex number ψ as

$$\psi = |\psi|\ R(\theta)$$

   with $R(\theta) = \cos\theta + i\sin\theta$, as in eq. 3 and eq. 7.
   (c) Show that the derivative of $R(\theta)$ is given by

$$\frac{dR(\theta)}{d\theta} = iR(\theta)$$

   and, hence, that the time-dependent factor in the stationary-state wave function of eq. 7, $f(t) = R(-\omega t)$, satisfies

$$i\hbar \, \frac{df(t)}{dt} = Ef(t)$$

where $E = \hbar\omega$. This equation for $f(t)$ is the most primitive possible version of the Schrödinger equation and provides a mathematical basis for the Planck relation (eq. 8) between frequency and energy in quantum theory. [The complex function $R(\theta)$ plays an important role in the theory of complex numbers and is normally written (astonishingly) as a *complex* exponential, $R(\theta) \equiv e^{i\theta}$, where e, the base for natural logarithms, is the number $2.71828\ldots$, and the angle θ must be in radians. See Appendix A, §7.]

14.2   THE HYDROGEN ATOM

The first thing you want to do with a new theory is test it, and the most immediate problem of the time was the hydrogen atom. What are the solutions of the Schrödinger equation for the hydrogen atom? We're looking for the normal modes of the matter waves, which must correspond to solutions of the Schrödinger equation that have a fixed spatial pattern oscillating with a definite frequency, the stationary states described in §14.1. The wave isn't confined in a well-defined box, as it was in the normal modes of §11.4, but rather is constrained, somehow, by the potential $V(r)$, which represents the electrical force that keeps the electron from flying away (see the discussion of potential energy in §5.3). The frequency of the standing-wave solution fixes the energy of the electron by the Planck relation, and these standing-wave solutions thus correspond to physical states of definite energy $E = hf$, as discussed in §14.1. Such a state of definite energy is an energy eigenstate, as I said, and the allowed values of the energy are the *energy eigenvalues*.[*]

What do these standing-wave patterns look like? In a typical standing wave there are nodes—nodal points on a one-dimensional wave, nodal

---

[*] In fact, anything connected with an eigenstate is referred to with the prefix *eigen-*: eigenfunction, eigenvector, eigenwhatever. Any physical variable can have its eigenstates: A momentum eigenstate is a state of definite momentum, and you can have angular momentum eigenstates, position eigenstates, and so forth. The prefix *eigen-* was adopted from the German and means "proper," more or less.

lines in two dimensions, and nodal surfaces in three dimensions, as discussed in §11.4. In a spherically symmetrical system, some of the nodal surfaces are spherical, spaced at different distances from the center and showing wave behavior along the radial direction. Some of the nodal surfaces, on the other hand, are conelike or planar, going through the origin and showing how the wave function varies with angle. An example is a vibrating jelly: Imagine the possible standing waves in a hollow sphere filled with gelatin. (The sphere should be lined with glue to keep the gelatin from slipping.) If you start the oscillations by giving the sphere a quick squeeze, you'll get radial oscillation patterns with spherical nodal surfaces, while if you start the oscillations with a brief rotational motion, you'll get rotational patterns instead.

Quantized angular momentum

The rotational patterns are just the same for all spherically symmetrical systems and are referred to as *spherical harmonics,* functions that depend only on the angular coordinates and are known once and for all. For the Schrödinger equation, these angular patterns correspond to different possible values of the angular momentum of the particle (see §12.3) in much the same way that the wavelength in straight-line motion corresponds to the particle's momentum. The spherical harmonics are labeled by two indices, ℓ and m, which determine the nodal pattern on the one hand and the particle's angular momentum on the other. They're called the *angular momentum quantum numbers.* The index ℓ fixes the total number of nodal surfaces in the angular pattern and corresponds to the magnitude of the angular momentum according to the rule

$$|\underline{L}|^2 = \ell(\ell + 1)\hbar^2, \qquad \ell = 0, 1, 2, \ldots \tag{1}$$

The other quantum number, m, fixes the z component of angular momentum:

$$L_z = m\hbar \tag{2}$$

that is, the value of the angular momentum about the z axis. L_z is a maximum for an electron whose orbit lies in the x–y plane and is zero for an orbit in the x–z plane or the y–z plane. The value of m is thus related physically to the orientation of the angular momentum vector $\underline{L}$ (the orientation of the axis of rotation in a simple classical situation), and mathematically m is related also to the orientation of the nodal surfaces of the wave function. The allowed values of m are the integers between ℓ and $-\ell$:

$$m = 0, \pm 1, \pm 2, \ldots, \pm\ell \tag{3}$$

The way the integers ℓ and m fix the allowed values of the angular momentum realizes rigorously the idea of a quantized angular momentum that Bohr conceived (eq. 12.3.25) and that de Broglie derived crudely (eq. 12.4.9) using waves laid out along a classical circular orbit.

Energy levels

The radial distribution of the wave function, unlike the angular part, does depend on the details of the force that binds the electron to the nucleus. It depends also on the value of ℓ, the quantum number that fixes the magnitude of the angular momentum, but not on m, the quantum number related to orientation. For each given value of ℓ there is a whole series of solutions of the radial motion, labeled by a new index n, called the *principal quantum number*. The index n is defined so as to correspond to the Bohr labeling of the energy levels, with allowed values given by

$$n = \ell + 1, \ell + 2, \ldots \tag{4}$$

all the integers for which $n - \ell > 0$. The energy values that come out of this calculation depend in general on both n and ℓ, but for the Coulomb force law of the hydrogen atom, they turn out to depend only on the quantum number n and are just the familiar energy levels deduced by Bohr (eq. 12.3.21):

$$E_n = -\frac{ke^2}{2a_0 n^2} \tag{5}$$

where a_0 is the radius of the first Bohr orbit (eq. 12.3.23). The energy levels of hydrogen are displayed in fig. 1, which is the same as fig. 12.3.2, but with the different possible values of ℓ shown separately. For each value of n and ℓ there are $2\ell + 1$ possible values of m, as well as 2 possible

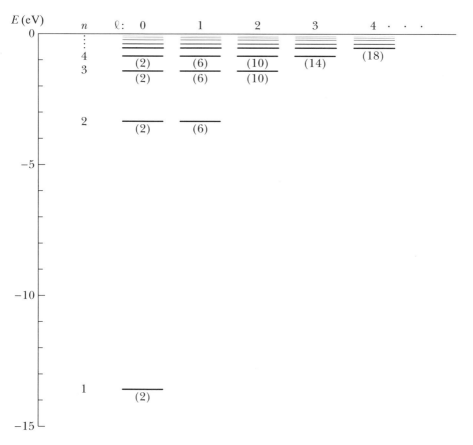

Fig. 1 The energy levels of the hydrogen atom. The numbers in parentheses are the degeneracies of the sublevels.

orientations of the electron spin,* so the *degeneracy* of each sublevel—that is, the number of different states for a given value of n and ℓ—is $2(2\ell + 1)$.

What I think is particularly striking about the solutions I've been talking about is how they reduce the states of the hydrogen atom—the electron going around the nucleus—to *standing waves,* normal modes of

---

*The spin angular momentum of the electron corresponds crudely to a rotation about its own axis.

oscillation very analogous to the modes of a stretched string or a vibrating membrane, as described in §11.4.

PROBLEM

1. (a) Using eq. 3, check the statement made in connection with fig. 1 that, for given values of the quantum numbers n and ℓ, the number of possible values of the quantum number m is $2\ell + 1$.
    (b) Derive a formula for the degeneracy of each energy level of the hydrogen atom—that is, the number of different quantum states for a given value of the quantum number n, which determines the energy. You'll need the formula for the sum of an arithmetic series:

$$a + (a + b) + (a + 2b) + (a + 3b) + \cdots + (a + nb) = (n + 1)\left(a + \frac{nb}{2}\right)$$

(There are $n + 1$ terms, with an average value of $a + nb/2$.)

14.3   THE CORRESPONDENCE PRINCIPLE

What has happened to the correspondence principle in all this? We were supposed to come up with a theory that would look just like ordinary Newtonian physics when applied on the large scale—baseballs as opposed to electrons and atoms. We've got a wave equation now, but how do these waves describe baseballs?

We don't have an interpretation for ψ yet, though in Chapter 10 you saw preliminary indications that it must have something to do with probabilities. In any case it seems pretty clear (and seemed clear from the early days in the development of this wave theory) that an object with a well-defined location has to correspond to a localized wave pulse: a wave pattern that is extremely small everywhere except in the vicinity of the one spot where the particle—the baseball—is known to be. Thus if the baseball follows a particular path, the wave pulse that is describing it had better follow that path also—namely, the path prescribed by Newton's laws.

Does the Schrödinger equation tell what path a wave pulse will follow? The answer is yes, of course, because as I tried to explain, it tells what *any* initial wave pattern will do as time goes on. We haven't yet shown how to include forces in the Schrödinger equation, so we'll have to be content for now with seeing whether the wave pulse has the right velocity to correspond to a free particle of given energy and momentum.

This looks a little tricky, as you may have noticed, because a wave pulse no longer has a well-defined wavelength, and hence the localized particle doesn't have a well-defined momentum. This lack of precision corresponds exactly to the Heisenberg uncertainty principle discussed in §13.2, which says just that: The position and momentum can't both be known to arbitrarily high precision. However, as I argued there, this isn't a practical problem for large-scale physics, because the required fuzziness in position and momentum is far tinier than the actual instrumental uncertainties in any measurement we can possibly make.

What we must do is look at a wave pulse that travels through space like a pulse traveling down a long rubber rope, as illustrated in fig. 1. Just as the tensions in the rope act on the pattern at any given time to determine how the pulse will move, the Schrödinger equation determines the motion of a wave pulse in ψ. The width of the pulse, Δx, is very small now compared to the macroscopic scale (feet or meters) that we're

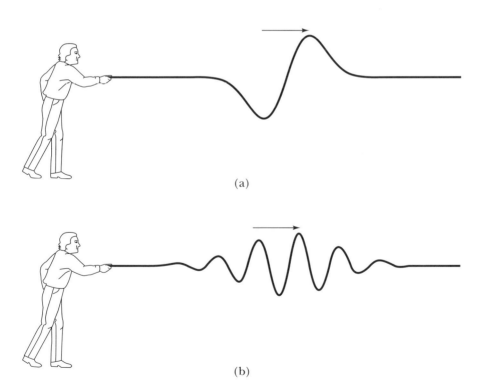

(a)

(b)

Fig. 1 Examples of traveling wave pulses.

talking about, and it represents the quantum uncertainty in the object's position. This width Δx is large enough, though, to allow the pulse to have a quite well-defined wavelength λ and hence a well-defined momentum $p = h/\lambda$ (see the discussion of fig. 13.2.1). Δx and Δp can both be extremely small by macroscopic standards and still satisfy the uncertainty relation (eq. 13.2.5).

The question simply is this: How fast does a wave pulse move? Does the pulse move at the same speed a classical particle would move? You might think we already know the answer to that: the wave speed vw—but that's usually wrong. The wave speed is the speed at which the wave crests move, and the fact is that the crests may move slower or faster than the pulse. This is often apparent in water waves, where a small traveling patch of waves is seen and wavelets travel across the patch. They form at the rear of the patch, traveling in the same direction but faster, so that they eventually reach the front and disappear. The speed of the pulse (the patch, for the water waves) is called the *group velocity*, *vg*, and the wave speed *vw* is often called the *phase velocity* because it is the speed of a point of the pattern with a definite phase, such as a crest or a trough.

Both of these velocities depend on how the frequency behaves as a function of the wavelength. We can represent that behavior by a function $\omega(k)$ (see §11.2), giving the angular frequency as a function of the angular wave number k. The wave speed, as you saw in §11.2, is given by

$$v_w = \frac{\omega}{k} \tag{1}$$

and this is *not* the same as the group velocity. The group velocity depends very much on how the angular frequency ω varies as a function of the wavelength, or angular wave number, and it is given, as it turns out, by the derivative

$$v_g = \frac{d\omega}{dk} \tag{2}$$

Is the group velocity always different from the wave speed? Only if the wave speed is a constant independent of the wavelength, as with light waves in empty space or sound waves in air (to a good approximation), are the two velocities the same. The derivation of eq. 2 involves expressing the pulse as a combination of plane waves that cancel everywhere except in the region of the pulse. You then look at how the plane waves change in a small time dt, and at how they then recombine to give a pulse shifted in location by an amount dx, which turns out to be equal to $(d\omega/dk) \times dt$.

To get the velocity of the wave pulse for the Schrödinger equation, all you have to do now is use eq. 14.1.2 (with eq. 11.1.2 and eq. 11.2.6) to get ω in terms of k:

$$\omega = \frac{\hbar k^2}{2m} \tag{3}$$

and differentiate:

$$v_g = \frac{d(\hbar k^2/2m)}{dk} \tag{4}$$

$$= \frac{\hbar k}{m} \tag{5}$$

$$= \frac{p}{m} \tag{6}$$

That is, the group velocity is just the same as the velocity of the classical particle, and we're all right. The correspondence principle is satisfied.

You can probe this idea a little deeper and see in it the rationale for the Planck and de Broglie relations themselves. It is a very general fact for a classical particle that where E is given as a function of p, the corresponding velocity has to be

$$v = \frac{dE}{dp} \tag{7}$$

because if dE is the increase in energy due to an applied force (eq. 5.4.1), then

$$dE = F\,dx = F(v\,dt) = v(F\,dt) = v\,dp \tag{8}$$

where I've used $dx = vdt$ (eq. 4.2.4), together with eq. 4.3.5:

$$dp = F\,dt \tag{9}$$

the basic defining relation for force. The requirement that the group velocity of the quantum wave pulse be equal to the classical velocity then gives you the condition

$$\frac{d\omega}{dk} = \frac{dE}{dp} \tag{10}$$

and the only reasonable way to achieve this (Is it the *only* way? I'm not sure.) is to require that E and p be proportional to ω and k, respectively. The constant of proportionality must be the same for both and is our old friend $\hbar$; thus the de Broglie and Planck relations (eq. 10.4.1 and eq. 14.1.8) are both linked directly to the correspondence principle, which is the requirement (see §13.1) that quantum physics look the same as classical physics on the large scale of our everyday experience.

PROBLEMS

1. For each of the following types of wave, the relationship between frequency and wavelength is given. Find in each case (i) the wave velocity vw in terms of λ, (ii) an expression for ω in terms of k, and (iii) the group velocity v_g in terms of λ. For each case, (iv) draw a graph showing both v_w and v_g as functions of the wavelength λ.
   (a) Electromagnetic waves: $f\lambda = c$
   (b) Deep-water surface waves: $f^2\lambda = g/2\pi$ ($\lambda \ll d$, d = depth)
   (c) Shallow-water surface waves: $f\lambda = \sqrt{gd}$ ($d \ll \lambda$)
   (d) Klein–Gordon wave: $f^2 = c^2/\lambda^2 + f_0^2$, with f_0 a constant
2. Prove the statement made following eq. 2 that if the wave velocity is independent of wavelength, then the group velocity is the same as the wave velocity.
3. Find a frequency–wavelength relationship for which the group velocity is always just twice the wave velocity. (Hint: Try $\omega(k) = Ak^n$, and look for n.)

14.4 THE MEANING OF THE WAVE FUNCTION

The Schrödinger equation provided at last[*] a firm mathematical footing for quantum theory and transformed it from an art to a legitimate science. You can set up the Schrödinger equation to describe even very complicated systems, like large atoms or molecules. The equation may

---

[*]Shortly before the Schrödinger equation appeared, Heisenberg introduced a mathematical formulation of quantum theory that involved matrices and also appeared to be a complete theory. At that time the matrix formulation had yet to evolve into a complete and systematic form; it looked totally different from Schrödinger's formulation but turned out in the end to be mathematically equivalent.

be very difficult to solve, but you know that it gives a complete and accurate description of the problem.

The result of solving the Schrödinger equation for a given system like an atom is a set of wave functions for the different possible stationary states (see §14.1) of the system, with an energy value for each state. The energy values are easy to compare with experiment through the measured spectrum of the atom (eqs. 12.2.1–12.2.3), but it was by no means clear at that time how the wave function itself ought to be interpreted. Its meaning was hotly argued among Bohr, Einstein, Schrödinger, Heisenberg, and others, including especially Max Born, who provided the interpretation that is now almost universally accepted. Born's probabilistic interpretation was in fact never accepted by either Einstein or Schrödinger, but they are no longer around to keep up the argument.

It had become clear that there is some kind of wave associated with the electron, but unlike all previously known types of wave, ψ does not seem to be directly observable. The places where ψ is large are where you are most likely to find the electron, but when you find it it still looks just like a particle. Since you can't tell for sure *where* you'll find the electron (like the photon in the two-slit experiment), and since ψ tells you only where it is most likely to be found, the natural interpretation of ψ is that it's related to *probabilities*. This is just what Born proposed, though many physicists found the implied loss of complete predictability very difficult to accept. We say that the probability density for finding the electron at a point r, at any given time t, is proportional to the magnitude of the wave function squared, and we write

$$P(r, t) = |\psi(r, t)|^2 \tag{1}$$

where the constant of proportionality has been adjusted (by means of a trivial scale factor on ψ) to equal 1. Recall that ψ is complex, which means it can be written

$$\psi = \psi_1 + i\psi_2 \tag{2}$$

with ψ_1 and ψ_2 real numbers. Then $|\psi|^2$ is defined (eq. 14.1.6) by

$$|\psi|^2 = \psi_1^2 + \psi_2^2 \tag{3}$$

a real positive number suitable for a probability density. We can't speak of the probability that the particle will be found *at* a point $\underline{r}$ because that probability is zero (like the probability of hitting the mathematical center of the bull's eye in darts), so we talk about the *probability density* $P(\underline{r},t)$. This means that if dV is a small volume that includes the point $\underline{r}$, then the probability that you'll find the particle in the volume dV is $P(\underline{r},t)\,dV$. If you do the two-slit experiment (as described in §10.5) with electrons, solving the Schrödinger equation with appropriate boundary conditions to represent the different paths the electron might take, then the behavior of $|\psi|^2$ at the wall will look like fig. 10.5.4, reflecting exactly the probability distribution for the different places where the electron might be found. See the discussion in §10.6 and §10.7, which is related directly to this interpretation.

Many-electron atoms

For a system of more than one particle (such as any atom, except hydrogen, with two or more electrons), the problem instantly becomes much more complicated, though the logic of stationary states and energy eigenvalues is still completely valid. The helium atom, for example, has two electrons, and the stationary-state wave function, which has to describe the probability distribution for both of the electrons, takes the form $\psi(\underline{r}_1,\underline{r}_2)$ (we are still neglecting the motion of the nucleus). The energy eigenvalue equation is now a partial differential equation in six variables, the components of $\underline{r}_1$ and $\underline{r}_2$, and can no longer be broken down into solvable parts as it can for a single electron. The problem can be solved numerically with considerable accuracy, but there is no solution in closed analytical form. As the number of electrons increases, the equations get prodigiously harder, and much cruder approximations are needed in order to make sense of the situation. Such approximate solutions are what chemists work with for the most part.

Probability amplitudes

The wave functions $\psi(x)$, $\psi(\underline{r})$, $\psi(\underline{r}_1,\underline{r}_2, \ldots)$, and so on that I've been talking about are special examples of a kind of quantity that runs all through quantum theory, probability amplitudes. A *probability amplitude* is any complex number that provides information about the probability of some measurement, given the state that the system is in. For a one-dimensional particle the complex number $\psi(x)$ determines the probability, given that the system is in the state ψ, of finding the particle in the neighborhood of the point x. For the helium atom, the complex number $\psi(\underline{r}_1,\underline{r}_2)$ determines the probability for finding one electron near $\underline{r}_1$ and the other near r_2. In general, a probability amplitude is used to determine the probability of finding the system in some state B when it's known to be in state A. The related probability amplitude is written $\langle B | A \rangle$, and the probability (or probability density) is the squared magnitude, $|\langle B | A \rangle|^2$. Thus the wave function $\psi(x)$, which is thought of here as a probability amplitude, is written $\langle x | \psi \rangle$, and the wave function for the helium atom (still neglecting the motion of the nucleus) is written $\langle r_1, r_2 | \psi \rangle$. There are general rules, of course, for calculating probability amplitudes, which I won't go into here.

Notice the rather odd way the probability is worded: "the probability of finding the system in state B when it's known to be in state A." If the system's in state A, how can it be in some different state B? This apparent paradox tells you something about the idea of a *state* in quantum theory: A quantum state is not an objective state of affairs that is either true or false. It seems a little closer to the truth to say that there is no objective state of affairs that you can put your finger on; a number of different possibilities are latent at any given moment, and any one of them can be realized as the outcome of some appropriate measurement. If you don't make a measurement, on the other hand, the different possibilities can interfere with each other and affect the probabilities for other measurements at some later time. Any measurement that you make can be cast in the form of a question: Is the system, or is it not, in some specified state B? The probability amplitude provides the answer in the form of a probability for that particular measurement.

Eigenstates

In §14.2 you saw several references to a situation in which some physical variable, such as energy or angular momentum, has a definite, precisely known value, and the probabilities for the measurement of that variable become certainties. The system is said to be in an *eigenstate* of the variable in question. Let me emphasize right now the *oddity* of the idea. Classically, variables *always* have definite values, and the way these definite values behave in time constitutes the substance of classical physics. In the quantum world, on the other hand, the situation is very different: The normal situation is that variables *don't* have definite values, but only some probability distribution of the kind I discussed earlier in this section. The wave function $\psi(r)$ of a particle tells you the probability for finding the particle at different possible locations r. The shape of the function ψ doesn't usually have a definite wavelength, corresponding to a definite momentum value, but is rather some combination of waves of different wavelengths that determines a probability distribution in momentum. There's typically a probability distribution in energy, too: The wave function is usually a combination of the energy *eigenstates,* or *stationary states,* described in §14.1, with an associated probability distribution to describe the likelihood of finding one or the other value.

The point of this discussion is to stress the fact that it's unusual in the quantum world to be certain about the value of any variable. If the wave function is a sine wave, with a single precise wavelength, then you do know what the momentum is and you're in a momentum eigenstate. When you're in a stationary state, a state of definite frequency (§14.1), then the energy is precisely determined and you're in an energy eigenstate. For every physical variable there is a special class of eigenstates for which the value is definite, but the common state of affairs is for the system to be in a mixture—a superposition—of such eigenstates.

The other distinctive feature of eigenstates that I want to mention is that most physical variables are restricted in the *values* that they may have. These possible values are called the *eigenvalues* of that variable, and there's an eigenvalue associated with each eigenstate. You've seen a restriction on possible energy values in the case of the hydrogen atom

Table 1   Eigenvalues of various physical variables

Variable	Spectrum of eigenvalues
Position x	All real numbers
Momentum p	All real numbers
Energy of free particle	All positive real numbers
Energy of hydrogen atom	The negative numbers $-hcR/n^2$, $n = 1, 2, 3, \ldots$, and all positive numbers (corresponding to unbound states, electron and proton far apart)
Energy of bound system	Discrete values, bounded below (minimum value, no maximum)
Angular momentum L	$|\underline{L}|^2 = \hbar^2 \ell(\ell + 1)$, $\ell = 0, 1, 2, \ldots$ $\underline{L}_z = \hbar m_\ell$, $m_\ell = 0, \pm 1, \ldots, \pm \ell$
Spin angular momentum of electron	$|\underline{S}|^2 = \hbar^2 s(s + 1)$, $s = 1/2$ $\underline{S}_z = \hbar m_s$, $m_s = \pm 1/2$

(Chapter 12), where the bound-state energies are given by the Bohr formula (eq. 12.2.4), and you've seen the effect also with angular momentum (eq. 12.3.25). Some variables, on the other hand, such as position and momentum, do not have restricted values. What the spectrum of possible eigenvalues looks like depends on the nature of the physical variable in question. Some examples are listed in table 1.

PROBLEM

1. Write a brief essay on each of the following questions.
    (a) Can a particle be real if its position is intrinsically indefinite?
    (b) What determines the place where a particular electron strikes the screen in a two-slit interference experiment?
    (c) Is the electron real at the moment that you observe it?
    In each case, you may need to analyze the question itself to see whether the terms are well defined and whether the question has meaning. If you find that the question is not meaningful, try to formulate a question that accurately addresses the problem—and then try to answer it.

CHAPTER 15

The Observer

Video ergo sum.
(I see, therefore I am.)

I come now to the very risky business of trying to describe what quantum theory is telling us about the ultimate nature of things. This is a game anyone can play, and you must understand that when I express my views, I'm not describing a generally accepted picture. There are at least as many notions of what quantum theory means as there are people writing about it—and there are plenty of those.

I'll start by reviewing what I believe are the accepted facts about quantum theory at a completely practical level: how it works in practice. This includes its probabilistic character, the standard understanding of how the quantum state—the wave function—is fixed in the first place, and the meaning of the concept of *quantum state* in practical terms.

After that I'll discuss the role of the observer, starting with the importance of observations in giving meaning to the physics and the inevitably subjective nature of what we know about observations. I'll argue that a physical system can't observe itself, and that without observations there's nothing left of physics. My personal conclusions are (1) that a conscious mind is essential to any real observation and (2) that it's impossible to understand the activity of the conscious mind in purely physical terms. You don't have to agree with my conclusions to pass the course—or to teach it. They're not intended to provide answers, but rather to stimulate thought and discussion on some very deep and murky questions.

15.1   THE RAW DATA

It's important at this stage to distinguish between the bare data of physics and the models that humans have constructed to give meaning to those data. The raw data are the totality of observations that people and other observers have made of the world about us, mediated through the senses. Consider the sense of sight, to be specific. Your immediate sensation as you look around is of a direct perception of things: There they *are!* When you think about it, of course, you realize your information is indirect; the photons from the sun bounce off some object, enter your eye, and hit the retina. You are not directly aware of the image on the retina, even, but only of the bits of information that finally arrive at the brain after being processed further by the optic nerve. Somewhere in the cerebral cortex, at the locus of conscious awareness, the signals arrive in vast abundance, millions per second, and present a pattern of information that is a reasonable replica of the image formed on your retina. These bits of information, which can be thought of as a very large number of binary bits (simple yes-or-no facts about the state of each of the relevant neurons), are the raw data. Everything else is interpretation. The world we think we see is our own reconstruction from these data. We've learned in the course of our development as individuals and as a species to fit these data—the data from all five senses, of course—to a three-dimensional model of an external world, filled with material objects and other things, and we interpret the data in terms of that model. The model works so well, so *extremely* well, that we simply take it for granted that it represents objective truth—the *real* world. (See prob. 1.)

Now all this soul-searching about how we perceive the world would be pretty academic if there were no reason to doubt that the real world is there and that it corresponds more or less to the mental image of it that most of us have. What makes these distinctions necessary is that there are indeed reasons to doubt those assumptions. Recall that quantum theory is full of the "fuzziness" of matter: intrinsic uncertainties in the location and velocity of particles, the coexistence and mutual interference of different possible scenarios for physical events, and other properties that are at odds with the classical model of the world. Quantum theory simply *refutes* the objective picture of things and events we've constructed. It does so primarily at the atomic and subatomic level, but this is only because that's where the fuzziness shows up most strongly. Fuzziness exists at every level, according to quantum theory, and the fact that the fuzziness is imperceptibly small on the scale of everyday life doesn't change in the least its significance for any basic understanding of the physical world. The question I want you to face is this: If the classical model of objectively existing things moving according to deterministic laws has to be abandoned, what is there to take its place? What has happened to our notions of reality?

To my mind, the common-sense assumption to make is that the only things we really know are the raw data of our sense perceptions, the elementary bits of information as they are directly perceived in the brain. You can argue about whether we know even this and about whether the conscious perceptions are truly raw data, but I feel this is a reasonable place to start, and perhaps the only reasonable place to start. I believe that you should question only what you are forced to question—and what we've been forced to question is the exterior world. The least we must do, then, is to separate the subjective world of perception from the objective world of things.

PROBLEMS

1. Write a careful analysis of how you deduce from your sensory data that the world around you is three-dimensional. Be careful to state any assumptions you need to make. Remember that the retinal image is two-dimensional. If you're thinking of using binocular vision as an argument, ask yourself whether one-eyed people would reach the same conclusion.

2. Give a critique of the argument in the last paragraph of this section. That is, examine the validity of assuming that the raw sensory data of our conscious perceptions provide an appropriate starting point for discussing physical reality.

15.2   THE WAVE FUNCTION

Now I want to review the basics of quantum interpretation, some of which you've already seen. The essential ingredient of this discussion is the wave function ψ of a given system, a complex-valued function with two important properties: (1) it obeys a deterministic equation of motion, the Schrödinger equation, and (2) it determines the probabilities for the results of any possible measurement. (In a more sophisticated treatment of quantum theory, ψ is replaced by a more abstract entity, the *state vector,* which nevertheless has exactly the same two basic properties.) In the language of quantum theory, ψ is said to represent the *state* of the system. (The *system* is the set of physical things, such as particles and fields, whose behavior you're studying; you normally assume that external influences on the system can be neglected. See *system* in the Glossary.)

> I won't say much in this chapter about stationary states, which were mentioned in §14.1. They are very useful in dealing with practical matters like the energy levels of an atom, but our fundamental interpretation has to do with how the system develops in time. A stationary state is just one very special case, and it isn't particularly relevant to the basic questions I'm talking about here.

There are two questions now that we need to look at:

(1)   What determines the initial form of ψ?

(2)   What happens to ψ when you make an observation on the system?

The answers to these two questions are closely related. For the first question, the answer is as follows: The wave function represents all the information you have about the system in the first place, and you get that information by making observations. The initial wave function is thus determined by the results of earlier observations. When you make another observation at a later time (and here we arrive at the second question), you gain new information, which *supplants,* in whole or in part, the information you had previously. You therefore replace the wave function at that time with a new one chosen to represent the new information. The sequence of events might go like this:

1.   At time $t = t_1$ you prepare the system in state A, which means that you make an observation and see whether the state of the system is A.

If not, you try again until the answer is yes. You therefore represent the state at that time by the correponding wave function $\psi_A(x)$:

$$\psi(x, t_1) = \psi_A(x) \tag{1}$$

(Here x represents all the coordinates needed to describe a configuration of the system.)

2. You then solve the Schrödinger equation (fig. 14.1.1),

$$i\hbar \frac{\partial \psi(x, t)}{\partial t} = \hat{H}\psi(x, t) \tag{2}$$

using eq. 1 for the initial condition, to obtain the wave function at any later time t.

The notation $\hat{H}\psi$ is a shorthand version of the right side of either of the equations shown in fig. 14.1.1, or some more complicated operation for a more complicated system. This operation may include multiplying ψ by various functions of x, differentiating ψ with respect to the variables x, or the like. The whole operation represented by $\hat{H}$ is called the *Hamiltonian operator* and is closely related to the energy of the system.

3. At time t_2 you decide to make another measurement to determine whether or not the system is in some state B corresponding to the wave function $\psi_B(x)$. According to a standard rule of quantum theory (see §14.4), the probability of finding that the system is in state B, given that the wave function is $\psi(x, t_2)$, is equal to

$$P_\psi(B) = |\langle B | \psi \rangle|^2 \tag{3}$$

where the *probability amplitude* $\langle B | \psi \rangle$ is a complex number determined by the two complex wave functions ψ and ψ_B, as discussed in §14.4.*

4. Now suppose you find that the system *is* in state B at time t_2. For predicting the behavior of the system after time t_2, you don't use the

---

*The probability amplitude $\langle B | \psi \rangle$ is given by the integral $\int \psi_B^*(x)\psi(x)\,dx$, which is called a *scalar product* in the language of quantum theory.

old function $\psi(x, t_2)$ for the wave function at time t_2; you use $\psi_B(x)$, because you now know that the system is in state B at time t_2:

$$\psi(x, t_2) \to \psi'(x, t_2) = \psi_B(x) \tag{4}$$

Other possible states that the system might have been in are now discarded and can no longer interfere with the state as it develops after time t_2. Thus, in the two-slit case, if you find the particle at slit 1, then the contributions to the wave from slit 2 are no longer to be included, and you don't see the two-slit interference pattern at the wall.

Perhaps the toughest thing to swallow in this picture is the abrupt change of the wave function when you make an observation. If you think of the state of the system as an objective state of affairs, then this discontinuous change is very indigestible. The wave function prior to the observation disappears completely and is replaced by another wave function that has nothing to do with the previous wave function (except the almost trivial fact that the probability amplitude $\langle B | \psi \rangle$ can't have been zero). The development of $\psi(x,t)$ with time in the absence of any observation is completely causal, governed by the Schrödinger equation, but this jump in ψ at the moment of observation has no causal description at all. The most straightforward way out of this dilemma is to suppose, as many physicists do, that ψ is *not* to be regarded as an objectively real field. In the first place, you can't measure it—and measurability is a standard criterion for the objective meaning of a quantity. You measure observables—quantities like position and momentum (see §15.3)—and the wave function enables you to determine the odds for different outcomes, but you can't measure the wave function itself. In the second place, different observers would use different wave functions to describe the system, depending on their perspectives. Thus, in the example we just worked out, another observer might first be informed that the measurement at time t_2 gave the result B, and then want to work out the probability of the system's being found in state A at the earlier time t_1. What she'd do is use $\psi_B(x)$ as the wave function at time t_2 and solve the Schrödinger equation for times prior to t_2, using its value at t_2 as the determining condition. The solution she gets is $\psi'(x,t)$, a solution that is governed by the Schrödinger equation (eq. 2) and the condition given in eq. 4 (second equality) and has nothing at all to do with the wave function ψ_A. The wave function at time t_1 according to the second observer is then $\psi'(x,t_1)$, so the probability of finding the system in state A at the time t_1, given that the state is B at t_2, is given by eq. 3 in the form

$$P_{\psi'}(A) = |\langle A | \psi' \rangle|^2 \qquad (5)$$

using $\psi'(x,t_1)$. When you work it all out, you find that the two observers predict exactly the same statistical correlation between the two measurements, even though they've used completely unrelated wave functions to describe the state during the time interval between t_1 and t_2. This strongly suggests that neither of these choices for $\psi(x,t)$ can be viewed as objectively correct to the exclusion of the other, and it supports the view that the wave function is a description of the information available to the observer, rather than an objective fact about the system itself.

I find I've violated my intention of restricting myself in this section to generally accepted procedures for using quantum theory in that I've argued for the nonobjective character of the wave function ψ—a view held widely but by no means universally. The digression into my own understanding of the meaning of ψ is intended to point up the fact that the rule works regardless of how you want to interpret ψ. Everyone agrees that you'll get all the right predictions if you simply throw out the old ψ whenever you make a measurement and introduce at that point a new ψ that incorporates what you now know about its state. (If your measurement doesn't completely fix the state of the system, there are rules for keeping as much of the old ψ as you need and throwing out only the part that is supplanted by your new information.)

15.3 OBSERVATIONS

The thing we now have to try to understand is the idea of an observation. In classical physics you don't pay any attention at all to the matter of observing a system. The system just exists, independent of any observer, and its behavior in time is well determined by its equations of motion. Observations serve only to get information about a state of affairs that is already well defined, and you can properly assume that the observations don't disturb the system enough to make any difference.

In relativity theory the observer begins to intrude in the sense that the state of a system is defined only with respect to a chosen observer, who serves to determine a frame of reference. There are no absolute frames of reference, and the same laws of physics work equally well for all observers—that is, in all possible frames of reference. Nevertheless, given a frame of reference, the existence and behavior of the system are still objective facts in the relativistic world, and they are still not disturbed by, or contingent on, the process of observation itself.

In the quantum world, however, the observer intrudes even more, as you saw in §15.2 and in a lot of the discussion of earlier chapters. The wave function is determined by the first set of observations, and the only predictions you can then make are in the form of probabilities for further observations. At a time when you're not observing the system, it's incorrect to talk about the positions of the particles, about their momenta, or even about how many particles there are; this is because particles are quanta (see §16.4), and states corresponding to different numbers of quanta of various sorts can be superposed. It's not just that you don't *know* what the system is doing, it's that it can't be said to be *doing anything*. All the different things the system *might* be doing can interfere with each other and are therefore all part of the reality of the situation: All the possible histories play a role in determining the final probability distribution. The only thing that you can rightly say describes the state of the system is the wave function, and what the wave function does is keep track of all the different interfering histories and provide a way of predicting statistical correlations among different possible observations at different times.

Without observations there's nothing left of physics. Even if you suppose that the wave function is known and that it develops nicely according to the Schrödinger equation, there is no objective state of affairs that it can be said to describe. All it gives is a set of probabilities about possible observations, and if no observations are ever made, then nothing can ever be said to really happen. To make matters worse, it doesn't make sense even to talk about a wave function if there's no prior observation to tell you what it should be. It's my view that for quantum physics to be meaningful *you have to regard observations as real events*. In fact, in view of the indefiniteness of events of the more ordinary kind, I would argue that observations may be the *only* real events.

That brings us to a very important question: When in fact do observations take place? The question is trivial at the practical level of labortory experiments, but I'd like to get to the bottom of this—the fundamental level where we're trying to probe the essential nature of things. The obvious answer for an experimental physicist is that an observation takes place when you take a measurement: when you use a measuring device to take data about a system—a voltage reading, a count of a particle counter, a spot on a photographic film, and so on. But at the conceptual level there's a problem with this, as I'll try to explain in the next few paragraphs.

To see or not to see

Imagine a sunbeam shining on a window pane. Say 10% of the light is reflected and 90% goes through (assuming this is the prediction of

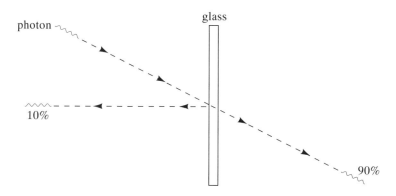

Fig. 1 Photon reflected from a window pane.

classical EM theory), as suggested in fig. 1. Now the sunbeam consists of photons—billions and billions of them—and each one may either bounce off the glass or go through. The different photons don't get in each other's way—the interaction between photons is *extremely* weak—but if you want to keep the situation clean you can turn down the intensity until there's just one photon per second, say, so that this isn't a problem. Anyway, you'll find that 10% of the photons bounce and the rest don't. For any given photon, physics is incapable of predicting ahead of time whether the photon will bounce (I hope you're used to that by now). The quantum calculation gives you only the probabilities, which agree exactly, however, with the relative intensities predicted by the classical wave theory. (That's the correspondence principle at work again.)

The question I want to address now is what really constitutes an observation of the photon. A standard way of measuring light is with photographic film, so we'll use a camera, as shown in fig. 2, to catch the photons. For our purposes it's a valid simplification to assume that each photon that enters the lens simply makes a spot on the film. (Of course, the process is actually a lot more complicated than that.) Does this constitute an observation? In practical terms it certainly does, but if you're trying to keep the logic pure there's a problem. Up to now we've taken the system as simply the photon, and the camera as the observing device. What happens, though, if we enlarge the definition of "the system" to include the camera, especially the molecules that make up the photographic film? When a photon hits the film, it produces a chemical change that can be thought of as a change in the quantum state of one

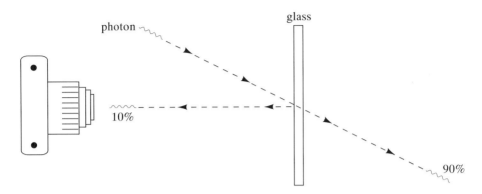

Fig. 2   The photon and the camera.

of the molecules. Now think about what happens to the quantum state of the whole system after the photon has had time to reach the camera. There are two possible states:

A: *No photon present, and a spot on the film*
B: *Photon going off to the right, and no spot on the film*

The state of the system at this time is

$$\psi = \alpha\psi_A + \beta\psi_B \tag{1}$$

The coefficients α and β are probability amplitudes, as in §15.2, and the probabilities are

$$|\alpha|^2 = 0.1 \tag{2}$$

$$|\beta|^2 = 0.9 \tag{3}$$

"Observation" versus "Superposition"

So here you have two ways of looking at the situation. In one, which I'll call Observation, the system is just the photon, an observation takes place, and after the observation the state is fixed one way or the other. In the other view, which I'll call Superposition, the photon and the camera together make up the system, no observation takes place, and the state of the system is still a superposition of the two possibilities. Now, which of these views is more accurate? Or are they completely equivalent? The critical difference between the two views is that *in a superposition of two states there is the possibility of interference*. In the state of affairs described by eq. 1, is it possible to construct an experiment to see interference between the two states A and B?

Now a camera is a macroscopic device, consisting of billions of billions of molecules, and the wave function for a camera must involve all of those degrees of freedom. For this reason, as I'll explain in §15.5, the experiment is a practical impossibility; and that, for most of us, is the end of it. In terms of experiments that will ever really be performed, the two views are therefore completely equivalent. Nevertheless, I don't believe you can stop there if you want to get a true understanding of the nature of things. At the conceptual level the interference experiment isn't impossible, it's just difficult. The number of molecules in the camera isn't infinite, it's just large. Quantum theory offers no reason in principle why you can't see the interference between states A and B. The two views are *not* equivalent, therefore, and the second view is the only justifiable one. No observation has in fact taken place.

This puts you in a dilemma. No matter what you do, the situation's just the same. Maybe you get a device like a photometer to scan the photographic film and count the spots, but that's not going to help because whatever device you use is itself made of molecules, so the same argument applies. Ultimately, it's the state of the whole universe that is in question, and it doesn't seem possible for the universe to observe itself. The wave function of the universe just goes on evolving with time according to some gigantic Schrödinger equation, and there's no entity external to the universe to observe it.

15.4 THE WATCHER

You may have noticed that I've avoided asking a very obvious question: What happens when you *look* at the photographic film or, more directly, at the photon itself, as in fig. 1? (Under optimal conditions it takes no more than two or three photons striking a retinal cell, or conceivably only one, to be perceptible. I'll idealize to one.) What happens to the argument? The difference is a subjective one, but I believe it's important. You *know* whether you've seen the photon—or the spot on the photographic film. The camera has no kind of self-awareness to enable

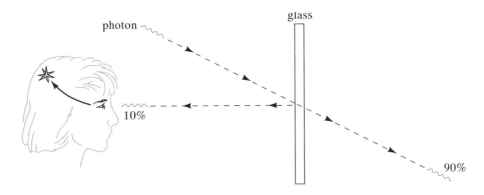

Fig. 1   You and the photon.

us to say that it has "seen" anything; it's useful as an observing device only in the sense that someone can look at the developed film and see what's on it. Otherwise it's no different from any other physical object that might be affected in some degree by the light that lands on it. The crucial question now is this: Does the fact of cognition—of conscious awareness—make a difference in the argument?

The two views of the situation, the Observation and the Superposition, now look like this:

Observation. The system consists just of the photon, and you have fixed the state by making an observation.

Superposition. You include yourself in your definition of the system and consider the state of the combined system, you and the photon, to be given by the superposition $\alpha\psi_A + \beta\psi_B$ (15.3.1), with the states A and B now defined as follows:

A: No photon present, and you know you've seen a photon
B: Photon going off to the right, and you know you haven't seen a photon

If the Superposition view is valid, then it must be possible in principle to produce interference between the two states.

There are real problems with both points of view, and I think you have to weigh these problems carefully against each other and form your own opinion on the matter. Remember that the possibility of interference shows that the two views are not equivalent, and one or the other must be right, but not both.

Observation versus Superposition again

If you accept the idea that an observation has taken place, then it has to be impossible to extend the definition of the system to include the observer. The observer therefore has to be external to the system, and the system, ultimately, is the entire universe. The brain is clearly made of atoms and molecules, so the conscious mind, in this view, would have to include an essential entity, the "watcher" you might say,* external to the physical brain and in fact outside the domain of physical description altogether. This view seems to me to lead to a dualistic picture: that there is some kind of *soul* distinct from the body and that some kind of interface between the nonphysical soul and the physical world occurs at the locus of consciousness in the brain.

A further complication of this view has to do with the state of the universe prior to the first observers. Unless you're willing to ascribe consciousness to inorganic matter (it's been suggested), then there has to have come a moment as the universe cooled off when the first conscious creature opened its little eyes and looked at things (fig. 2). Are we to suppose that up until then all conceivable universes coexisted (including huge numbers of them with no life at all, and no little creatures to observe them) and that at that exact moment the universe we know and love was picked out from among all the other possibilities as the one to continue with? Or should we suppose that the Big Bang itself, along with all its other mysteries, was set in motion by a cosmic, divine observation that fixed the quantum state to be just that and no other—and thereby set the stage for future observations by creatures like us when we came along to look at the world? You may take your choice, but if you buy the Observation picture I think it has to be one or the other.†

---

*My attention has been called recently to a similar picture in the *Yoga Sutras* of Patanjali, an ancient collection of aphorisms in the Hindu tradition. The essential self, referred to in these sutras as the "seer" (see-er) or "watcher," is in this picture external to the mind and beyond the level of immediate consciousness.

†I mentioned that some people have suggested the possibility of conscious observation on the part of inanimate matter as well. I believe that this is very unlikely, because it would mean that observations are continually being made at every stage of an experiment, by all the pieces of apparatus, and that in consequence we'd never see any interference at all. If you reject this thoroughly democratic view, though, the line between animate and inanimate things remains a serious puzzle.

Fig. 2   The first observation.

Now let's look at the alternative, the Superposition picture, in which the final state of the system is still a superposition of the two possible scenarios. In this view *no observation ever takes place*, with the embarrassing consequence that physics is entirely without content. The only things physics deals with, according to the quantum picture, are observations and the statistical correlations among them, and if there are no observations there's no physics. The wave function describes only possibilities—all the different things that might happen—but nothing can be said to *really* happen unless there's an observation to pin it down. What is perhaps more serious still is that there doesn't seem to be any wave function in the beginning, because it would take some prior observation to determine it. Out of all the conceivable states of the universe, none has ever been selected as the actual state at any time whatsoever, so you can't even talk about the *probabilities* for different possible future measurements; these probabilities exist only in relation to previous observations.

The argument I've just given doesn't have anything to do with consciousness. It just describes the difficulties you have if there aren't some kind of real observations.

Added to all this is the subjective argument that I *feel* myself to be a real observer. I know what I've seen, and I am therefore unwilling to accept the idea of being in a superposition of states in which I may or may not have seen it. Just the ordinary business of life involves the tacit assumption that things are as we see them to be, to the exclusion of all the things they might be otherwise. This argument carries quite a bit of weight with me personally, but I don't believe anyone knows enough about the nature of cognition to construct a reliable argument on this basis. I should make it clear, though, that it's only this intuitive sense that we do really observe the world, together with the lack of any other ways in which observations might occur, that persuades me that conscious observations are the real observations.

Other perspectives

You may of course find both of these alternatives grossly unacceptable, and you'll have lots of company. One alternative view that's been widely discussed is the *branching-universes* hypothesis, which really deals more with the problem of the discontinuous change in the wave function than with the difficulty of defining an observation. In this view, when an observation takes place the different options are *all* realized, and therefore none is excluded. Each option gives rise to an alternative universe, and all the alternatives coexist from that time on. The number of coexisting universes multiplies ferociously, it is clear, and many of them contain replicas of you leading your different lives depending on the options that have given rise to them. This approach is fun to play with but seems to be totally untestable. Furthermore, it doesn't really address the problem I've been discussing, because it doesn't help resolve the question of when an observation takes place. If there are no observations, then there's no branching. I think the only way to eliminate specific discrete observations from the branching-universes picture is to suppose that the branching takes place *continuously*, which I find difficult to conceive of.

Two further difficulties also come to mind:

1. If each branch goes its separate way, then there can be no superposition of alternative histories and no quantum interference—contrary to what we see. If the different histories *do* interfere, then no observations have taken place and we're back to the Superposition picture.

2.   The branching-universes picture is completely asymmetrical in time. Quantum theory as we know it works just as well if you follow the course of events backward in time, as I explained in §15.2, whereas if the universe is going to branch, it can do so only in one time direction and not in the other.

Another way to address the problem, proposed by Eugene Wigner, is to accept consciousness as being outside the framework of conventional physics but to suppose that there is an extended description, still *physical* in character, that includes additional degrees of freedom corresponding to conscious states. In order to avoid a reappearance of the same problems, the laws governing these degrees of freedom would have to be different in character from those of quantum physics. They could be deterministic, I believe, so long as the interaction with the quantum world involved the kind of random results I've described before. As far as I know, no one has tried to construct such a picture, and I think it would be extremely difficult to do so. The critical question is whether such a theory could ever be tested experimentally.

The watcher and freedom of choice

I feel myself that physics is a Good Thing and should be saved if at all possible; and I believe that the most economical way, and the one most directly consistent with our own experience, is simply to postulate a watcher, along the lines of the description I've given of the Observation hypothesis. This can be thought of as an interim assumption, one that provides a convenient language of discourse in the absence of a better understanding of quantum theory and human personality, and it does have some real merits. It provides an objective basis for the notion of personal identity and the human spirit that runs deeper than just an appeal to the complexity of the brain. It doesn't rule out conscious animals and even conscious machines if they are sufficiently complex (who knows where the "watcher" comes from or how the connection is arranged?), but it draws a sharp distinction between a pure machine—without consciousness—and a conscious individual. The assumption of a watcher—a real observer—also enables you to give meaning to the traditional values and characteristics associated with the human spirit, such as goodness, love, and the sense of beauty.

The assumption of a watcher also opens the way for a model of free will. It's notoriously difficult to prove either the existence or the nonexistence of any genuine freedom of choice, but just about all of us humans act as if we believed in it. For free will to have any real meaning, there must be choices that are made intentionally—that are not the outcome of either deterministic laws or blind chance. The most economical way to accomplish this (and I suspect the *only* way) is for the agent to have an existence outside the framework of physical laws and yet to have some mode of interaction with the physical world. And by happy coincidence I have a candidate ready to hand—the watcher. You need to add two things: first, that the watcher is not just a passive observer but is also capable of making decisions and, second, that the watcher, as agent, can influence the course of physical events. It's tempting to suppose that the watcher might control the outcome of some crucial quantum observation in the brain, since physics doesn't say how that outcome is determined, but I believe such an assumption gives rise to contradictions in a case where two observers, spatially separated, decide to see mutually contradictory things.* It would be quite sufficient, though, to suppose that the watcher chooses *what* to observe, because it's known that just the fact of making a large number of selected observations can have a strongly controlling effect on the quantum state of a system. I like this picture, believing as I do that human life without free will is essentially sterile, but I believe that the evidence for the agent is much less compelling logically than the evidence for the watcher. The latter rests on the observed fact of conscious knowledge and on the need to give substance to physics itself, while the former, agency via free will, rests only on the subjective feeling that we do indeed make choices.

15.5 QUANTUM INTERFERENCE AND THE CONSCIOUSNESS DETECTOR

I mentioned earlier the impracticality of observing interference effects for states of a macroscopic system such as a camera or a human being. The reason this is impractical is that it's necessary to have a system totally isolated in order that the wave function be well defined; it's especially important that the wave functions for the two alternative histories have a sharply defined relative phase so that we can determine whether the

---

*This difficulty is related to what is known as the *EPR paradox*. (EPR stands for Einstein, Podolsky, and Rosen, who introduced such a scenario in trying to find flaws in quantum theory.)

interference is constructive or destructive. This means there must be absolutely no interaction with the world external to the system during the process. Just one photon escaping or entering the system would totally destroy the necessary phase coherence. This degree of isolation is quite reasonable for a very simple system like one photon or one atom, but it is extremely unreasonable to hope for with a large system. Remember that macroscopic systems can have of the order of 10^{24} atoms, and all systems radiate unless they're at absolute zero temperature. There's a further difficulty, too, in that you have to bring the two histories together again in such a way that there's a possibility of cancellation. This means, essentially, that the two alternative wave functions have to be almost exactly identical except for a complex phase factor, so that when the phase factors cancel there is something like complete cancellation. The wave function, though, involves every one of the particles present, so you have to arrange things such that each of the two histories brings every single particle to the same location at the same time as the other. This is incredibly difficult even to imagine, when the number of particles is as large as it is for a laboratory-sized object. The degree of incredibility is comparable with that for spontaneous time reversal, which I talked about in Chapter 8.

Nevertheless, I'd like to imagine that we can construct an interferometer that can perform this task—a device similar in concept to the Michelson interferometer of fig. 1.3.5 or the two-slit interference set-up of figs. 10.5.1 and 10.5.2, though of course radically different in design. The experimental program that I'd propose with our new apparatus would go like this: I'd start by taking the system of photon plus camera shown in fig. 15.3.2 and demonstrate to everyone's satisfaction that the two scenarios, labeled A and B, can indeed be made to interfere with each other—in short, that our interferometer works. This would confirm the description of the state as a superposition, as in eq. 15.3.1. Then I'd ask for a volunteer and set up the experiment of human plus photon as shown in fig. 15.4.1 and see whether interference occurs. Thus we'd have a direct experimental test to tell whether conscious observation destroys the wave function—the superposition represented by eq. 15.3.1—and we could thereby determine which of the two difficult scenarios we actually have to deal with: a nonphysical Observer, on the one hand, or perpetual Superposition—a world without any observations at all—on the other.

If you find that consciousness does indeed have observable consequences, corresponding to the destruction of quantum interference, then you can use the interferometer as a consciousness detector. You can check out your best friend or your pet dog or cat—I feel sure you'll find they are conscious—and then work your way down to goldfish, insects, and more primitive life forms, such as amebas and supercomputers.

PROBLEM

1. Write a short story about a person who builds a consciousness detector and then discovers, for example, that some of his or her acquaintances pass the test and some don't—or devise your own plot line.

CHAPTER 16

Photons

Let there be light.

 WHERE WE'VE GOT TO

Let's review for a moment how far we've come in our effort to understand the quantum nature of things. We found that many of the puzzles boiled down to two: the particlelike behavior of light and the wavelike behavior of particles (electrons mostly). What I described in Chapter 14 is a legitimate wave theory of particles that goes a long way toward resolving these puzzles. The Schrödinger equation provides a solid basis for understanding not only the wavelike properties of electrons, as seen in interference phenomena, but also the discrete energy levels of atoms—both in the simple case of hydrogen and in that of more complicated atoms with two or more electrons.

However, two of the puzzles remain: the nature of the photon and the process of atomic transitions. Even the Schrödinger equation does not tell us how the atom gets from one state to another or how to figure out the transition probability, which is what determines the intensity—the brightness—of the different spectral lines. It turns out that these two puzzles go hand in hand: The atomic transitions fall into place when we've understood the photon, which is what I'll describe in this chapter. My description will be largely verbal, because the mathematical theory, known as *quantum electrodynamics,* is pretty abstruse, but the basic ideas are quite simple.

You might think that we could just use the Schrödinger approach and treat the photon like the electron—that is, construct a Schrödinger equation for the photon, viewed as a point particle, and generate the wave properties in that way. Unfortunately, the Schrödinger equation works only for a nonrelativistic particle, and the photon is hopelessly relativistic because it always travels *at* the speed of light. The major difficulty is that a single-particle wave equation like the Schrödinger equation is suitable only for a particle with a continuing existence—one that cannot be absorbed or created—while relativistic particles like photons can always be created and destroyed as a consequence of the equivalence of mass and energy, as you saw in Chapter 5. Paul Dirac, in 1928, devised an ingenious Schrödingerlike equation for a relativistic electron that deals with this problem elegantly—up to a point. Nevertheless, you find in the end that the single-particle wave equation approach of Schrödinger and Dirac is inappropriate for relativistic particles of any kind and that the methods of quantum field theory need to be used. I'll explain what quantum field theory is in §16.4.

16.2   QUANTIZING THE EM FIELD

In Chapter 14 I outlined a very general procedure for producing a quantum theory for any system you like. You start with the classical expression for the energy of the system in terms of the momentum and position variables and go through a procedure analogous to the one I described in §14.1, and you come out with a Schrödinger equation suitable for the system. This procedure automatically produces a reasonable quantum theory for the system in question and in the process guarantees (by a theorem beyond our range here) that the new theory will satisfy the correspondence principle—that it will behave in the proper classical fashion in the large-scale limit (§13.1).

Thus, if you want to understand photons, you have to apply this quantization procedure to Maxwell's theory of the EM field. This was first done by Dirac in 1927, very soon after the basic ideas of a mathematical quantum theory were developed by Schrödinger in 1925–1926. The logic of this procedure is quite straightforward up to a certain point, although the mathematical details are not. Some severe diseases break out at a later stage of the process, but they don't affect the basic ideas— and we've learned how to deal with them.

I've broken down the basic logic into six steps:

1. For convenience in understanding the steps, enclose the EM field in a large cubical box with reflecting walls, so that the EM waves are standing waves of the sort discussed in §11.4 and can be thought of as normal modes, labeled by some index ℓ. Each normal mode behaves independently of all the others and simply oscillates sinusoidally, exactly like a simple nonrelativistic oscillator, with frequency f_ℓ given by

$$f_\ell = \frac{c}{\lambda_\ell} \tag{1}$$

where λ_ℓ is the wavelength of the ℓth mode ($\lambda_\ell = 2\pi/k$, where k is the effective angular wave number introduced in prob. 11.4.2).

2. The quantum theory of a simple oscillator is well understood, and the allowed excitation energies are discrete and equally spaced:

$$E_N = Nhf_0, \qquad N = 0, 1, 2, \ldots \tag{2}$$

Here f_0 is the classical frequency of the oscillator, and N is called the excitation number.

3. Each normal mode ℓ of the EM field has a different natural frequency f_ℓ, so the quantum states of that mode are given by eq. 2 with f_0 replaced by f_ℓ. If the excitation level of the mode ℓ is $N = N_\ell$, then its energy is

$$E = N_\ell h f_\ell \tag{3}$$

4. The ground state of the whole system is the state of lowest possible energy, in which every mode is in its ground state, $N_\ell = 0$. The energy of the ground state is therefore zero. We call this state the vacuum state, because there's nothing there and the energy is zero.

5.   The simplest excited state is one in which just one mode is excited to its first excited state, so that the energy of the field is just hf_ℓ. This is what we call a *one-photon state. A photon isn't a particle*—the word *photon* is just a way of talking about these excited states of the EM field.

6.   If two modes are excited, we say that there are two photons present, and the energy is just the sum $hf_\ell + hf_{\ell'}$. If one mode is excited to its second excited state, the energy is $2hf_\ell$, and we say that there are two identical photons, each of energy hf_ℓ.

The mathematics works just as well if you remove the walls of the box and allow traveling-wave modes filling all of space (§11.2) instead of the standing-wave modes in the box. In this case the modes are described by a true wave vector $\underline{k}_\ell$, which points in the direction the wave is traveling and whose magnitude is $2\pi/\lambda_\ell$ (eq. 11.1.2). The classical EM field has momentum as well as energy, and the excited states of a traveling wave mode are in fact eigenstates (see §14.4) of the momentum as well as of the energy,

$$\underline{P} = N_\ell \hbar \underline{k}_\ell \tag{4}$$

just as if you had N_ℓ photons, each with momentum $\hbar \underline{k}_\ell$. (The state is an eigenstate of each of the three components of the momentum vector $\underline{P}$.)

I haven't mentioned the matter of polarization. Classical EM waves can be polarized, which means that there are different ways for the fields $\underline{E}$ and $\underline{B}$ to be oriented, as described in §11.5, for a particular direction of travel given by the wave vector $\underline{k}$. It is a consequence of Maxwell's equations that EM waves are *transverse,* which means that the electric and magnetic field vectors oscillate in a plane perpendicular to $\underline{k}$. As a result of all this, there are *two* normal modes for each value of the wave vector $\underline{k}$, and thus, in the language of particles, there are two possible states for a photon of given momentum $\underline{p}$, in the sense that either of the two modes could be excited.

What I've hoped to get across is the idea that a photon isn't really a particle. The quantized EM field has states of excitation that in some ways resemble particles but in other ways do not. Each excitation, as I've described it, has a definite energy and momentum, with the right relation for a relativistic massless particle; it's completely spread out over the whole volume, however, and thus bears little resemblance to the particlelike photons that I've referred to in earlier sections. This difficulty is answered in almost exactly the same way, in fact, that we used in §14.3

and §14.4 to reconcile the Schrödinger wave picture of an electron with the pointlike behavior that we see experimentally. In experimental terms, if you make an observation you'll find the photon at one location or another, and the extended wave of the normal mode reflects merely an extended probability distribution rather than a spreading out of the particle itself. Furthermore, the probability distribution itself can be made localized, in analogy to the wave pulse of §14.3, by forming a quantum superposition of different single-photon quantum states with slightly differing energies and momenta. Like an electron, a photon travels like a wave and hits like a particle.

These photons also have no identities: They are absolute clones. You can't say, "It's particle 1 that has momentum p and particle 2 that has momentum p'" as distinct from the other way around; all you're allowed to say is that there is one particle with momentum p and one particle with momentum p'. You can't specify which is which, because all you are really saying is that both of those two *modes* are excited. The language of excitations of normal modes is more precise for describing these states than the language of particles.

PROBLEM

1. (a) In a particular excited state of the EM field, the modes de-scribed by wave vectors $\underline{k}_1$, $\underline{k}_2$, and $\underline{k}_3$ are in their first excited states and the modes $\underline{k}_4$, $\underline{k}_5$, and $\underline{k}_6$ are in their second excited states. Describe the state of the system in terms of photons, and give expressions for the total energy and the total momentum of the state. Ignore polarization.
    (b) Two photons of momentum $\underline{p}_1$ and $\underline{p}_2$ scatter from each other (the interaction is extremely weak, but it can happen) so that their final momenta are $\underline{p}_3$ and $\underline{p}_4$. Give a verbal description of the initial and final states in terms of excitations of modes of the EM field.

16.3   INTERACTIONS

One of the unanswered questions that has showed up occasionally is, How do photons interact with other kinds of matter? In particular, does a photon bounce off an electron? Is there a force between them? How is a photon created when an atom drops from one energy level to another? Once you've understood the photon as a quantum excitation of the EM field, you can address these questions in terms of the interactions between charged particles and the EM field. We looked at these interactions in §11.5 and found that the interaction goes both ways: The EM field produces forces on charged particles, and charged particles produce changes in the EM field. In the quantum picture, as I'll explain, we know how to take these interactions into account and to give a complete mathematical description of processes involving photons, but I can't say that this picture does much for our intuitive understanding of what's happening.

In order to make a quantum theory of interactions between photons and matter, you need to generalize eq. 15.2.2 with a Hamiltonian operator that now includes the EM field itself as well as whatever charged particles you are looking at. Following a standard procedure, we construct the quantum Hamiltonian out of the classical expression for the energy of the combined system, which includes terms that describe the interaction of the field with the charged particles. Since the Hamiltonian governs the physics, these terms act in both ways, as required: They describe the effect of the field on the particles, and they describe the effect of the charged particles on the field distribution itself.

The point here is that these interaction terms, properly treated, tell you how the interaction works quantum mechanically—how the charged particle produces quantum excitations and de-excitations of the modes of the EM field—that is, how photons are produced and absorbed and how this process is associated with appropriate jumps of the charged particle from one atomic state to another. The whole system, atom plus EM field, is described by a set of complex probability amplitudes for all the different states the system could be in. Each of these states corresponds to some energy level of the atom together with a certain distribution of photons (a certain level of excitation for each mode of the EM field). If you start at $t = 0$ with a state of the whole system in which the atom is in one of its excited states and there are no photons present, then as time goes on the probability amplitude increases for the atom to be in a lower state and a photon of appropriate energy to be present. The squared amplitude then gives the probability for this transition to have taken place—that is, for the system to be found in the state that corresponds to the presence of a photon. The system thus gets from one state to the other, not by a process in the classical sense at all, but by a continuous increase in the probability that the system simply *is* in the second state.

What I've just described is the calculation of a *transition probability*, which is what determines the intensities of the different spectral lines of atoms and molecules. The earlier models of Bohr and Sommerfeld, which could make fairly good predictions of the energy levels of simple atoms, were totally incapable of determining transition probabilities, a problem that was satisfactorily solved only with the kind of calculation I've described: combining the Schrödinger equation for the atom with the quantized EM field.

Our final two puzzles are answered, after the fashion of quantum theory, which as usual tells you that you asked the wrong question in the first place. The question "What's a photon?" presupposes that there are *things* to which that name can be applied, and the answer is that they're not things at all. The word turns out to refer rather to a *state of affairs,* in which a quantum system, the EM field, is excited in a certain way, more nearly analogous to the excited energy levels of an atom than to the presence or absence of particles.

The wave properties of light are now seen to reside in the wave character of the different normal modes. If you look back at the two-slit experiment, you see that it's the normal modes of the field *in a geometry that includes the screen with the slits* that display the interference pattern, so that when one of these modes is excited (when there's a photon in the state with that wave pattern), the probability of an interaction with an atom of the detecting film is governed by the way that wave pattern is distributed. Quantum theory's answer to the question "How can light be both waves and particles?" is that it is neither.

In a similar fashion, the question "How does an atom emit a photon and perform the jump from one state to another?" doesn't get a direct answer either. The system, atom plus EM field, *sidles* from one state to another, as the probability of its being in one state decreases while the probability of its being in the other state increases. You never actually have a *process* of transition at all.

16.4 *ALL* PARTICLES ARE FIELD QUANTA

We seem to have ended up with an asymmetry between photons and electrons, because the Schrödinger theory of electrons treats them still very much as particles, whose probability amplitudes have wave characteristics, while the EM field seems to be the other way around. It's a funny sort of symmetry that says that if you quantize particles you get wavelike properties, and if you quantize waves you get particlelike properties. When you face up to the need for a *relativistic* quantum theory, however, you find, as I indicated earlier, that you can't really start with the particle picture at all, because in a relativistic world all the different kinds of particles can be created and annihilated, just like photons. As we now understand things, the only way to have a consistent relativistic theory is to treat *all* the particles of nature as the quanta of fields, like photons. Electrons and positrons are to be understand as the quanta of excitation of the *electron–positron field*, whose "classical" field equation, the analog of Maxwell's equations for the EM field, turns out to be the Dirac equation, which started life (§16.1) as a relativistic version of the single-particle Schrödinger equation. The electron–positron field equations then are quantized by a procedure very much like the one I gave in §16.2 for photons. The quanta, which now come in two types because of the nature of the field, are the electrons and positrons of nature, and they act just like particles for practical purposes.

This approach now gives a unified picture, known as *quantum field theory*, of all of nature, and as usual it works beautifully—up to a point. We've resolved all the bothersome puzzles and eliminated the apparent contradictions of the early days of quantum theory, but we run into a new generation of diseases, and a new generation of elegant unifying ideas, when we probe this new approach more deeply.*

---

*These matters are discussed in a supplement, "Part III, Particles and Fields" (see Preface).

APPENDIX A

Math Facts

A.1 TRIGONOMETRY

For the right triangle shown in fig. 1, the trig functions are as follows:

sine
$$\sin\theta = \frac{A}{C} \qquad A = C\sin\theta \tag{1}$$

cosine
$$\cos\theta = \frac{B}{C} \qquad B = C\cos\theta \tag{2}$$

tangent
$$\tan\theta = \frac{A}{B} \qquad A = B\tan\theta \tag{3}$$

cotangent
$$\cot\theta = \frac{B}{A} = \frac{1}{\tan\theta} \tag{4}$$

Fig. 1   Right triangle, for eqs. 1–6.

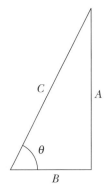

secant
$$\sec\theta = \frac{C}{B} = \frac{1}{\cos\theta} \tag{5}$$

cosecant
$$\csc\theta = \frac{C}{A} = \frac{1}{\sin\theta} \tag{6}$$

Inverse functions:

$$x = \sin\theta \rightleftarrows \theta = \arcsin x \tag{7}$$

$$x = \tan\theta \rightleftarrows \theta = \arctan x \tag{8}$$

and so forth. The inverse functions are also often written $\sin^{-1}x$, $\tan^{-1}x$, and so on. The notation $\sin^{-1}x$ is never used to represent the reciprocal of $\sin x$, which would be written $(\sin x)^{-1}$. Note that the inverse functions are not uniquely defined, because there are many values of θ that have the same value of the sine (or other trig function). Usually $\arcsin x$ and $\arctan x$ are taken to be in the first or fourth quadrant $(-\pi/2 \le \theta \le \pi/2)$, and $\arccos x$ is taken to be in the first or second quadrant $(0 \le \theta \le \pi)$. That's the way calculators and computers normally handle them.

Relationships:

$$\sin^2\theta + \cos^2\theta = 1 \tag{9}$$

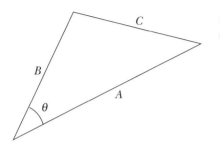

Fig. 2 Arbitrary triangle, for law of cosines.

$$\tan^2\theta + 1 = \sec^2\theta \tag{10}$$

Sum of two angles:

$$\sin(\theta_1 + \theta_2) = \sin\theta_1\cos\theta_2 + \cos\theta_1\sin\theta_2 \tag{11}$$

$$\cos(\theta_1 + \theta_2) = \cos\theta_1\cos\theta_2 - \sin\theta_1\sin\theta_2 \tag{12}$$

$$\tan(\theta_1 + \theta_2) = \frac{\tan\theta_1 + \tan\theta_2}{1 - \tan\theta_1\tan\theta_2} \tag{13}$$

Law of cosines: For an arbitrary triangle, as shown in fig. 2,

$$C^2 = A^2 + B^2 - 2AB\cos\theta \tag{14}$$

A.2   VECTORS

A *vector* $\underline{A}$ has both magnitude and direction. It can be expressed in terms of *components* with respect to any orthogonal coordinate system:

$$\underline{A} = A_x\hat{x} + A_y\hat{y} + A_z\hat{z} \tag{1}$$

where $\hat{x}$, $\hat{y}$, and $\hat{z}$ are unit vectors—vectors of length 1 (no units)—in the x, y, and z directions. This is shown in the two-dimensional case in fig. 1.

Fig. 1   A vector and its components.

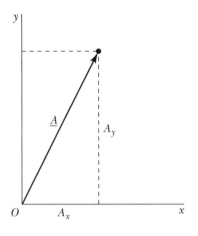

The *magnitude* of $\underline{A}$ is the length A of the arrow representing $\underline{A}$:

$$A = \sqrt{A_x^2 + A_y^2 + A_z^2} \tag{2}$$

Vector addition, graphical method: Place the arrows representing the vectors in sequence, with the tail of each arrow at the head of the previous arrow; the sum, or resultant, of all the vectors is then an arrow from the tail of the first arrow to the head of the last. Some examples are shown in fig. 2.

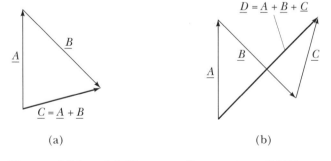

(a)   (b)

Fig. 2   Vector addition. (a) The sum of two vectors. (b) The sum of three vectors.

Vector addition, algebraic method: Simply add the corresponding components. If $\underline{C} = \underline{A} + \underline{B}$, as in fig. 2(a), then

$$C_x = A_x + B_x \tag{3}$$

$$C_y = A_y + B_y \tag{4}$$

$$C_z = A_z + B_z \tag{5}$$

and similarly for three or more vectors, as in fig. 2(b).

The negative of a vector: $-\underline{A}$, is a vector of the same magnitude as $\underline{A}$ and exactly opposite direction.

To subtract $\underline{A}$ from $\underline{B}$: Form the sum of $\underline{B}$ and $-\underline{A}$:

$$\underline{C} = \underline{B} - \underline{A} = \underline{B} + (-\underline{A}) \tag{6}$$

Alternatively, draw $\underline{A}$ and $\underline{B}$ with their tails touching, and then $\underline{B} - \underline{A}$ is the vector drawn from the head of $\underline{A}$ to the head of $\underline{B}$, as in fig. 3.

Algebraically,

$$C_x = B_x - A_x \tag{7}$$

$$C_y = B_y - A_y \tag{8}$$

$$C_z = B_z - A_z \tag{9}$$

Fig. 3   The difference of two vectors.

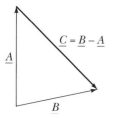

$\underline{C} = \underline{B} - \underline{A}$

$\underline{A}$

B

Fig. 4   The angle θ, for calculating the scalar and vector
products.

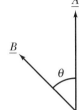

similarly to the sum (eqs. 3–5).

Note that

$$\underline{A} + \underline{C} = \underline{A} + (\underline{B} - \underline{A}) = \underline{B} \tag{10}$$

Scalar product (or dot product):

$$\underline{A} \cdot \underline{B} = AB \cos\theta = A_x B_x + A_y B_y + A_z B_z \tag{11}$$

where θ is the angle between the directions of A and $\underline{B}$ (fig. 4). The
scalar product can also be written as

$$\underline{A} \cdot \underline{B} = A_{\parallel} B = AB_{\parallel} \tag{12}$$

where $B_{\parallel}$ is the component of $\underline{B}$ parallel to $\underline{A}$, and $A_{\parallel}$ is the component of $\underline{A}$ parallel to $\underline{B}$:

$$A_{\parallel} = A \cos\theta \qquad B_{\parallel} = B \cos\theta \tag{13}$$

Vector product (or cross product):

$$\underline{C} = \underline{A} \times \underline{B} \tag{14}$$

$\underline{C}$ is perpendicular to the plane in which $\underline{A}$ and $\underline{B}$ lie and has magnitude

$$C = AB \sin\theta \qquad (15)$$

The three vectors $\underline{A}$, $\underline{B}$, and $\underline{C}$ form a right-handed triad, like the x, y, and z axes in a right-handed coordinate system, as shown in fig. 5.

In terms of components,

$$\underline{C} = (A_y B_z - A_z B_y)\hat{x} + (A_z B_x - A_x B_z)\hat{y} + (A_x B_y - A_y B_x)\hat{z} \qquad (16)$$

Note the *cyclic* permutation of components x, y, and z: in each successive term, $x \to y$, $y \to z$, and $z \to x$. The rule (eq. 16) is equivalent to writing the vector product $\underline{C}$ in the form of a determinant :

$$\underline{C} = \det \begin{vmatrix} \hat{x} & \hat{y} & \hat{z} \\ A_x & A_y & A_z \\ B_x & B_y & B_z \end{vmatrix} \qquad (17)$$

where the *determinant* of a square array of numbers, called a *matrix*, is the sum of all the products that can be formed with one factor from each row and one from each column of the array. Plus and minus signs are assigned in such a a way that two terms in the sum

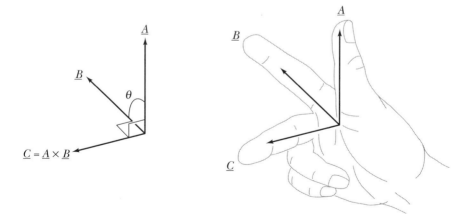

Fig. 5   The vector product and the right-hand rule.

that differ in the interchange of two rows or two columns have opposite signs. For a 3×3 array the rule gives six terms, corresponding to the six terms in eq. 16, the only difference being that only numbers are normally involved, rather than the unit vectors that appear in eqs. 16 and 17. Thus the general rule for a 3×3 determinant is

$$\det \begin{vmatrix} A_x & A_y & A_z \\ B_x & B_y & B_z \\ C_x & C_y & C_z \end{vmatrix} = \begin{aligned} & A_x B_y C_z - A_x B_z C_y + A_y B_z C_x \\ & - A_y B_x C_z + A_z B_x C_y - A_z B_y C_x \end{aligned} \tag{18}$$

 ANALYTIC GEOMETRY

Analytic geometry is a way of analyzing geometrical problems entirely in terms of the x and y coordinates of all the points involved. The coordinates are defined relative to a pair of *coordinate axes,* lines perpendicular to each other and referred to as the x axis and the y axis, as shown in fig. 1. The x axis is usually taken as horizontal and the y axis as vertical. The x coordinate of a point on the plane is equal to its distance from the y axis, with a sign convention to show whether it's to the right or left, and the y coordinate is the distance of the point from the x axis. A distinctive

Fig. 1   The x and y coordinates of the point P.

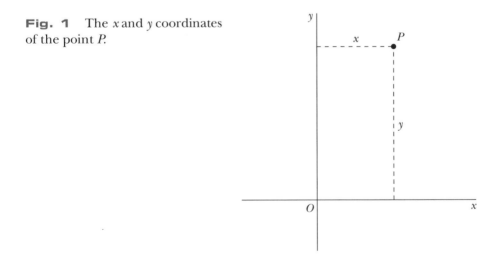

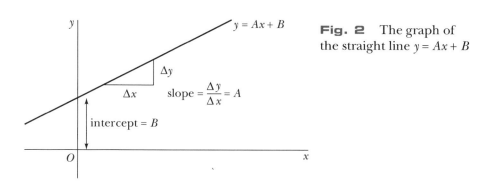

Fig. 2 The graph of the straight line $y = Ax + B$

feature of this approach is that any line or curve is represented by an *equation* that relates the x and y coordinates for each point of the line or curve. A straight line is represented by a linear equation; the line represented by

$$y = Ax + B \qquad (1)$$

is shown in fig. 2. A circle is represented by a quadratic equation that reflects the Pythagorean theorem; the circle represented by

$$(x - x_0)^2 + (y - y_0)^2 = R^2 \qquad (2)$$

is shown in fig. 3.

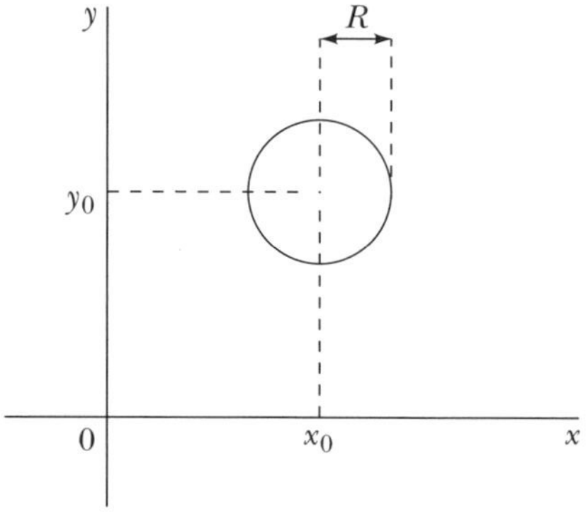

Fig. 3 The graph of a circle, eq. 2.

A general curve can be represented in the form

$$y = f(x) \tag{3}$$

where every pair of numbers x and y that satisfy eq. 3 corresponds to a point on the curve, called the *graph* of the function f. In eq. 3, x is referred to as the *independent variable* and y as the *dependent variable*, because the value of y depends on the value assigned to x. The variable x is also referred to as the *argument* of the function f.

The *slope* of a straight line is defined as the tangent of the angle that it makes with the x axis, measured counterclockwise from the +x direction. Thus the slope of a horizontal line, parallel to the x axis, is zero, and the slope of a vertical line, parallel to the y axis, is infinite. Eq. 1 represents a straight line of slope A that intercepts the y axis at the point $(0,B)$—that is, $x = 0$, $y = B$—as shown in fig. 2.

Mathematicians are usually dealing with pure numbers when they do analytic geometry, but physicists are almost always interested in physical quantities that carry units, so the variables x and y nearly always have the dimensions of length, time, energy, or other such physical quantity.

 **A.4**   DERIVATIVES (see table 1)

The idea of a derivative was introduced and explained in §4.2, which you should review. I'll discuss derivatives here in the context of functions of the time t, since that's the case that concerns us most, but the logical ideas apply to any independent variable. The derivative of any function $f(t)$ can be thought of in several ways: (i) as the rate at which f is changing as a function of t at any given moment, (ii) as the slope of the graph of $f(t)$ at any point on the graph, and (iii) as the quotient of df by dt, where df is the small change in f that takes place in the small time dt.

$$\frac{df}{dt} = \frac{f(t + dt) - f(t)}{dt} \tag{1}$$

The views (ii) and (iii) are illustrated in fig. 1. If the time increment dt is taken sufficiently small, then the quotient approximates the slope at that point to arbitrary accuracy, provided the curve is smooth there. Note that the slope is different at different points on the graph, so that the derivative is itself a function of t. When the independent variable is the

Table 1   Table of derivatives (The argument x can be any physical
variable or simply an ordinary number without units.)

$f(x)$	$\dfrac{df}{dx}$	
$u(x)\,v(x)$	$\left(\dfrac{du}{dx}\right)v + u\left(\dfrac{dv}{dx}\right)$	(differentiation by parts)
$\dfrac{u(x)}{v(x)}$	$\dfrac{(du/dx)v - u(dv/dx)}{v^2}$	
$v(u),\ u = u(x)$	$\dfrac{dv}{du}\dfrac{du}{dx}$	(chain rule)
x^λ	$\lambda x^{\lambda-1}$	
$\sin x$	$\cos x$	
$\cos x$	$-\sin x$	
$\tan x$	$\sec^2 x$	
$\sec x$	$\tan x \sec x$	
$\arcsin x$	$\dfrac{1}{\sqrt{1 - x^2}}$	
$\arccos x$	$-\dfrac{1}{\sqrt{1 - x^2}}$	
$\arctan x$	$\dfrac{1}{1 + x^2}$	
$\ln x$	$\dfrac{1}{x}$	
e^x	e^x	

time t, physicists commonly indicate the derivative by an overhead dot:
$\dot{f}(t)$ (introduced in §4.2).

Note that the definition of the derivative as a quotient, eq. 1, lets you
express df, the change in f during the interval dt, in terms of the derivative:

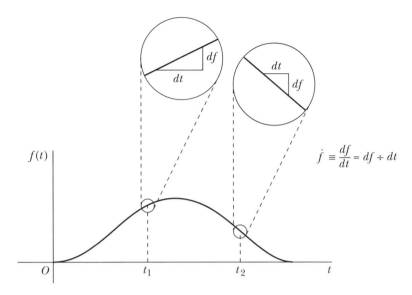

Fig. 1   The derivative of a function.

$$df = \frac{df}{dt}\, dt \tag{2}$$

as you can see also from fig. 1. As shown in the table, the sign of the derivative for the inverse trig functions is chosen to be correct when the angle (the value of $\arcsin x$, and so on) lies in the first quadrant ($0 \le \theta < 90°$). The exponential function e^x and the natural logarithm $\ell n x$ are introduced in §3.3.

   INTEGRALS

There are two meanings to the idea of an integral, which turn out to be very closely related for practical purposes. These are the *indefinite integral* and the *definite integral*.

The indefinite integral

The indefinite integral, written as $\int f(x)\, dx$, of a function $f(x)$ is the in-verse of the derivative: It is simply the function whose derivative is $f(x)$. It's not unique, because you can add an arbitrary constant to a function

Table 1 Table of indefinite integrals

$f(x)$	$\int f(x)\,dx$
x^{n-1}	$x^n/n + C$
$\cos x$	$\sin x + C$
$\sin x$	$-\cos x + C$
$\sec^2 x$	$\tan x + C$
$\tan x \sec x$	$\sec x + C$
$\dfrac{1}{\sqrt{1-x^2}}$	$\arcsin x + C = -\arccos x + C'$
$\dfrac{1}{1+x^2}$	$\arctan x + C$
$\dfrac{1}{x}$	$\ln x + C$
e^x	$e^x + C$

of x and the derivative is unchanged. Table 1 gives the indefinite integrals that correspond to the derivatives given in table A.4.1. The arbitrary constant is given in each case as C or C'.

The definite integral

The second kind of integral is the definite integral, which is the sum of a very large number of very small terms (or, more accurately, the limit of such a sum as the number of terms becomes infinite and the individual terms become infinitely small). A typical form of definite integral is the integral of a given function $f(x)$ between two limits x_1 and x_2, written as

$$I = \int_{x_1}^{x_2} f(x)\,dx \tag{1}$$

The limits are often omitted when it's clear from the context what the limits are. The meaning of the definite integral is this: The interval from

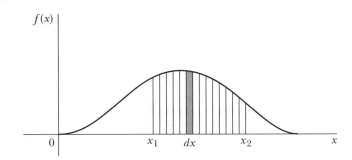

Fig. 1   The definite integral as an area.

x_1 to x_2 is broken up into a large number of very small intervals of length dx, each interval being located at a different value of the variable x, and the expression $f(x)\,dx$ is evaluated for each interval and then summed over all of them. The symbol $\int$ is nothing but an old-fashioned S, signifying a sum. The intervals are supposed to be so small that it doesn't matter where in each interval you take the value of $f(x)$. This is illustrated in fig. 1, where the area of each vertical slice is simply the product $f(x)\,dx$, and the integral, the sum of the areas of the slices, is the total area under the curve between the two limits x_1 and x_2. The error due to the jagged edge along the curve becomes negligible when dx is taken sufficiently small.

If $f(x)$ is the derivative of some other function $g(x)$,

$$f(x) = \frac{dg(x)}{dx} \tag{2}$$

then the expression to be summed, $f(x)\,dx$, is simply the change in g over the interval dx, according to eq. A.4.2:

$$f(x)dx = \frac{dg}{dx}\,dx = dg \tag{3}$$

When you now integrate $f(x)$, you find that you have just the sum of the changes dg, which is equal then to the net change of g over the entire interval from x_1 to x_2:

$$I = \int_{x_1}^{x_2} f(x)dx = \int dg = \Delta g = g(x_2) - g(x_1) \tag{4}$$

You'll see that the function $g(x)$ is the indefinite integral of $f(x)$—that is, the function whose derivative is $f(x)$—and eq. 4 then shows the relation between the definite integral and the indefinite integral. If you know the indefinite integral, you can do the definite integral. Note also that the arbitrary constant C in the indefinite integral (see table 1) doesn't affect the value of the definite integral, because it's only the *change* in $g(x)$ that's needed.

 ## A.6 APPROXIMATIONS

Here $|\varepsilon| \ll 1$, except in eq. 2 (see note below).

$$(1 + \varepsilon)^n \approx 1 + n\varepsilon \qquad (|n\varepsilon| \ll 1) \tag{1}$$

$$f(x + \varepsilon) \approx f(x) + \varepsilon \frac{df(x)}{dx} \qquad \text{(eq. A.4.2)} \tag{2}$$

$$\sin\varepsilon \approx \varepsilon \qquad (\varepsilon \text{ in radians}) \tag{3}$$

$$\cos\varepsilon \approx 1 - \frac{\varepsilon^2}{2} \qquad (\varepsilon \text{ in radians}) \tag{4}$$

$$e^\varepsilon \approx 1 + \varepsilon \tag{5}$$

$$\ln(1 + \varepsilon) \approx \varepsilon \tag{6}$$

In eq. 2 the validity depends on the character of $f(x)$ in a somewhat more complicated way than for the others, but in general the approximation is all right if f is smooth and ε is sufficiently small.

A.7 COMPLEX NUMBERS (see §14.1)

A *real number* is an ordinary number, rational or irrational, and positive, negative, or zero. It is called real simply to distinguish it from imaginary and complex numbers.

An *imaginary number* is any multiple of the imaginary unit i, defined as a number whose square is -1:

$$i^2 = -1 \tag{1}$$

The square of any imaginary number is negative.

A *complex number* is the sum of a real number and an imaginary number:

$$z = x + yi \tag{2}$$

where x and y are both real numbers. A complex number can be visualized as a point in a plane, called the *complex plane*, with coordinates x and y, as shown in fig. 1. Real and imaginary numbers are special cases of complex numbers (in eq. 2, $y = 0$ for real numbers and $x = 0$ for imaginary numbers). The distance ρ in the diagram is referred to as the *magnitude*, or *modulus*, of the complex number z, written as $|z|$:

$$\rho = |z| = \sqrt{x^2 + y^2} \tag{3}$$

and the orientation angle θ is called the *argument* of z:

$$\theta = \arg z = \arctan\left(\frac{y}{x}\right) \tag{4}$$

$$x = \rho \cos\theta, \ y = \rho \sin\theta \tag{5}$$

$$z = \rho(\cos\theta + i \sin\theta) \tag{6}$$

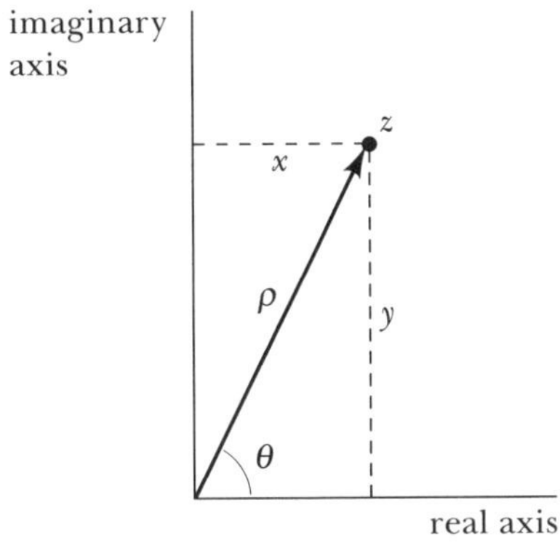

Fig. 1   The complex plane.

Complex arithmetic

Addition and multiplication obey the associative, commutative, and distributive laws of ordinary arithmetic, so that, for example, the sum and product of two complex numbers z_1 and z_2 are given by

$$z_1 + z_2 = (x_1 + iy_1) + (x_2 + iy_2) \tag{7}$$

$$= (x_1 + x_2) + (y_1 + y_2)i \tag{8}$$

$$z_1 z_2 = (x_1 + iy_1) \times (x_2 + iy_2) \tag{9}$$

$$= (x_1 x_2 - y_1 y_2) + (x_1 y_2 + y_1 x_2)i \tag{10}$$

The *complex conjugate* z^* of the complex number z is defined by

$$z^* = (x + iy)^* = x - iy \tag{11}$$

and is shown in fig. 2. The product of a number and its complex conjugate is always real and positive (or zero, but only if the number itself is zero):

$$zz^* = z^*z = (x + iy)(x - iy) = x^2 + y^2 \tag{12}$$

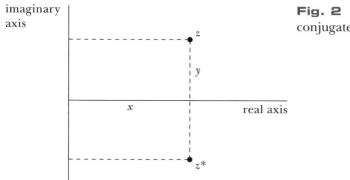

Fig. 2   The complex conjugate z^*.

This result lets you express the magnitude ρ of a complex number (eq. 3) in terms of explicit operations on z itself:

$$\rho^2 = zz* \tag{13}$$

It also opens the way to reciprocals and quotients, because it's easy to rescale $z*$ so that the product with z is unity instead of ρ^2:

$$z \frac{z*}{\rho^2} = 1 \tag{14}$$

We can thus identify the factor $z*/\rho^2$ as the reciprocal of z:

$$\frac{1}{z} = \frac{z*}{\rho^2} = \frac{z*}{x^2 + y^2} \tag{15}$$

Note that *every* complex number z except zero has a reciprocal, a fact that is very significant mathematically because it permits you to find the quotient of any two complex numbers and makes the arithmetic of complex numbers complete. To construct the quotient you just multiply by the reciprocal:

$$\frac{z_1}{z_2} = z_1 \left(\frac{1}{z_2} \right) = \frac{z_1 z_2*}{x_2{}^2 + y_2{}^2} \tag{16}$$

Complex exponentials

Multiplication by i is equivalent to a 90° rotation:

$$i(x + iy) = -y + ix \tag{17}$$

as shown in fig. 3. The fact that $i^2 = -1$ in this view corresponds simply to the fact that two 90° rotations make a 180° rotation, which thus takes z into $-z$.

It's natural to ask now how you can produce a rotation through an arbitrary angle θ. In fig. 4 the complex number z', which results from rotating the point z through an angle θ, can be broken down into two components: one parallel to the direction of z and one perpendicular to it, as shown. The parallel component can be expressed as z itself multi-

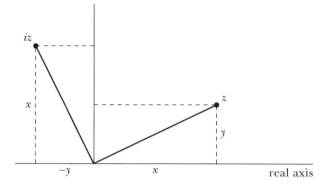

Fig. 3   Multiplication by i produces a 90° rotation.

plied by the real factor cosθ, which gets both the magnitude and the direction right, while the perpendicular component needs a factor sinθ to get the magnitude right and a factor i to provide the 90° rotation. Thus you can express z' as

$$z' = z \cos\theta + iz \sin\theta \tag{18}$$

$$= z (\cos\theta + i \sin\theta) \tag{19}$$

The factor that produces the arbitrary rotation θ is therefore the combination (see prob. 14.1.1)

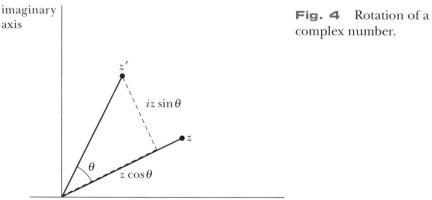

Fig. 4   Rotation of a complex number.

$$R(\theta) = \cos\theta + i\sin\theta \tag{20}$$

The complex number $R(\theta)$, a number of unit magnitude with orientation angle θ, is shown in fig. 5. The odd fact is that it's appropriate to represent $R(\theta)$ as a complex exponential $e^{i\theta}$, as I've done in the figure:

$$e^{i\theta} = \cos\theta + i\sin\theta \tag{21}$$

The combination $\cos\theta + i\sin\theta$ satisfies several of the properties of an exponential function, including (1) the chain rule for the derivative:

$$\frac{d(e^{i\theta})}{d\theta} = ie^{i\theta} \tag{22}$$

and (2) the product rule for exponentials:

$$e^{i\theta_1}\,e^{i\theta_2} = e^{i(\theta_1 + \theta_2)} \tag{23}$$

Another way to get a feeling for the exponential of a complex number is

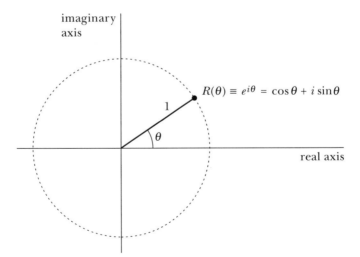

Fig. 5   The complex exponential $e^{i\theta}$.

to look at small arguments. For a small argument ε, the real exponential has the approximate value

$$e^\varepsilon \approx 1 + \varepsilon, \quad \varepsilon \ll 1 \tag{24}$$

(see eq. A.6.5 or A.4.2). The only reasonable way to define the complex exponential is to let the same relation apply for small complex arguments:

$$e^z \approx 1 + z, \quad |z| \ll 1 \tag{25}$$

In fig. 6 this relation is shown for several possible complex values of z. For z real and positive, $z = \varepsilon$, you get case 6a, where e^ε is greater than 1. For z real and negative, $z = -\varepsilon$, you get case 6b, where $e^{-\varepsilon}$ is less than 1. Case 6c is the case where z is pure imaginary, $z = i\varepsilon$, which is the limiting case for very small θ of fig. 5. Since e^z is greater than 1 for real positive z and less than 1 for real negative z, it's very reasonable that its magnitude should be neither greater nor less than 1 for the case where z is pure imaginary.

Case 6d extends the idea to complex z, in this case $z = (1 + i)\varepsilon$, which gives you a hint of the fact that as ε increases, e^z traces out a spiral. You can see this by writing it as

$$e^z = e^{(1 + i)x} = e^x e^{ix} \tag{26}$$

the product of the rotating function e^{ix} (fig. 5) and the ever-increasing real exponential e^x.

Fig. 6   The exponential $e^z \approx 1 + z$ for small z, real or complex.

The use of the factor $e^{i\theta}$ to produce a rotation provides an alternative—and very useful—way of representing a general complex number. If ρ and θ are the magnitude and orientation angle of z (eqs. 3 and 4), then z can be obtained by rotating the real number ρ through the angle θ:

$$z = \rho e^{i\theta} \tag{27}$$

corresponding exactly to eq. 6, which was derived by expressing the real and imaginary parts of z in terms of ρ and θ.

Notice too that $e^{i\theta}$ is a periodic function; that is, it repeats itself:

$$e^{2\pi i} = \cos(2\pi) + i \sin(2\pi) = 1 \tag{28}$$

so that

$$e^{i(\theta + 2\pi)} = e^{i\theta} e^{2\pi i} = e^{i\theta} \tag{29}$$

A rotation through $360°$ is equivalent to no rotation at all.

The device used in eq. 26 to get the exponential of a complex argument can be used for any complex z, written as $z = x + iy$:

$$e^z = e^{(x + iy)} = e^x e^{iy} = e^x(\cos y + i \sin y) \tag{30}$$

Conversely, you can use the same kind of factorization to get the natural logarithm of a complex number. Representing z as

$$z = \rho e^{i\theta} \qquad \text{(eq. 27)} \tag{31}$$

you get, using the standard rules for logs,

$$\ln z = \ln \rho + \ln(e^{i\theta}) = \ln \rho + i\theta \tag{32}$$

Note that $\ln z$ isn't uniquely defined, since any integer multiple of 2π can be added to θ without altering z, as in eq. 29.

Newtonian Mechanics

I'll discuss only *particle mechanics,* which tells how Newton's laws work for *point particles.* Point particles ideally are infinitesimally small objects with finite mass, but in practice they include any objects small enough that their size and internal motion can be neglected. Particle mechanics can be used to derive the behavior of extended objects as well, because big objects can be thought of as made of large numbers of little ones, each of which obeys the rules for point particles.

 KINEMATICS

The word *kinematics* refers to the description of the motion of particles without regard to how that motion is influenced by forces. Kinematics involves only positions, velocities, and accelerations as they vary with time; it does not involve quantities like force, mass, momentum, and energy.

The position $\underline{r}$ of a particle, viewed as a function of the time t, is fundamental to this description. From it you can derive the velocity $\underline{v}$, the first derivative of the position variable, and the acceleration $\underline{a}$, the derivative of the velocity (or the second derivative of position). The standard problem is to reconstruct the position as a function of time, given the acceleration and appropriate initial conditions. This is because it's the acceleration that is determined by the applied forces and that therefore governs how the velocity and, in turn, the position evolve in time.

Table 1   Basic formulas of kinematics (The integrals are definite, from t_0 to t.)

One-dimensional motion	Three-dimensional motion	Rotation about a fixed axis	
General			
$v = \dot{x} \equiv \dfrac{dx}{dt}$	$\underline{v} = \dot{\underline{r}} \equiv \dfrac{d\underline{r}}{dt}$	$\omega = \dot{\theta} \equiv \dfrac{d\theta}{dt}$	(1)
$a = \dot{v} = \ddot{x} \equiv \dfrac{d^2x}{dt^2}$	$\underline{a} = \dot{\underline{v}} = \ddot{\underline{r}}$	$\alpha = \dot{\omega} = \ddot{\theta}$	(2)
$v(t) = v_0 + \int a\,dt$	$\underline{v}(t) = \underline{v}_0 + \int \underline{a}\,dt$	$\omega(t) = \omega_0 + \int \alpha\,dt$	(3)
$\Delta x = \int v(t)\,dt$	$\Delta \underline{r} = \int \underline{v}(t)\,dt$	$\Delta\theta = \int \omega(t)\,dt$	(4)
Constant acceleration			
$a = $ constant	$\underline{a} = $ constant	$\alpha = $ constant	(5)
$v(t) = v_0 + at$	$\underline{v}(t) = \underline{v}_0 + \underline{a}t$	$\omega(t) = \omega_0 + \alpha t$	(6)
$\Delta x = v_0 t + \dfrac{1}{2}at^2$	$\Delta \underline{r} = \underline{v}_0 t + \dfrac{1}{2}\underline{a}t^2$	$\Delta\theta = \omega_0 t + \dfrac{1}{2}\alpha t^2$	(7)
$v^2 = v_0^2 + 2a\Delta x$	$v^2 = v_0^2 + 2\underline{a} \cdot \Delta\underline{r}$	$\omega^2 = \omega_0^2 + 2\alpha\Delta\theta$	(8)

The basic formulas of kinematics are displayed in table 1 for both one-dimensional and three-dimensional motion. These all follow from basic properties of derivatives and, occasionally, integrals. The equations describing rotation of a rigid object about a fixed axis are included also, because they are so often useful and the mathematics is identical to that for one-dimensional particle motion. The orientation is given by an angle θ, the *angular velocity* is ω, and the *angular acceleration* is α. An overhead dot represents differentiation with respect to time (see §4.2).

There are four lines in the first section of table 1; the first two lines are the definitions of velocity and acceleration as derivatives, and the second two lines are the reconstruction of velocity and position, given the acceleration and initial conditions.

Note that in the equations for three-dimensional motion, with the exception of the formula for v^2, the different components work independently of each other, so that, for example, the behavior of x and v_x are entirely governed by the x component of the acceleration, a_x. Thus $\Delta x = v_{0_x} + a_x t^2/2$ for the case of constant acceleration, and $\Delta y = v_{0_y} + a_y t^2/2$, involving only a_y. The formula for v^2—the last line in the table— can be seen by noting (for the one-dimensional case) that

$$d(v^2) = 2v\, dv = 2va\, dt = 2a(v\, dt) = 2a\, dx \qquad (9)$$

from which (see eq. A.5.4)

$$\Delta(v^2) = v^2 - v_0{}^2 = 2a\Delta x \qquad (10)$$

The formula for v^2 in three dimensions can be thought of as the sum of three expressions like eq. 10, one for each dimension because v^2 is the sum of squares of the components of $\underline{v}$, and the dot product $\underline{a} \cdot \Delta\underline{r}$ is the sum of products, component by component (eq. A.2.11).

Note that for three-dimensional motion a distinction must be drawn between *velocity* $\underline{v}$, a vector quantity, and *speed* v, defined as the magnitude of the velocity:

$$v = |\underline{v}| \qquad (11)$$

To specify the velocity, you have to give both the speed and the direction of the motion.

Motion of a point particle on a circle

θ, ω, and α behave as above.

s is the distance traveled (arc length).

$a_{\parallel}$ is the tangential component of $\underline{a}$, parallel to $\underline{v}$.

$a_{\perp}$ is the radial component of $\underline{a}$, perpendicular to $\underline{v}$ and pointing radially inward.

$$s = r\theta \tag{12}$$

$$v = |\underline{v}| = r\omega \tag{13}$$

$$a_{\parallel} = r\alpha = \frac{dv}{dt} = \frac{d|\underline{v}|}{dt} \tag{14}$$

$$a_{\perp} = \omega^2 r = \frac{v^2}{r} \qquad \text{(eq. 12.3.4)} \tag{15}$$

Note that $a_{\parallel}$ is just the rate of change of speed—a conventional sort of acceleration—while $a_{\perp}$ doesn't involve dv/dt at all; it is related entirely to the way in which the *direction* of $\underline{v}$ is changing with time.

B.2   DYNAMICS—MOMENTUM AND FORCE

The word *dynamics* refers to the way forces act to influence the motion of particles and other kinds of matter. Since momentum p and energy E are fundamental conserved quantities of nature, it's natural to use them as the starting point for describing the dynamics of a particle or set of particles. *Force,* in this approach, is defined as that external influence that causes the momentum and energy of a particle to change, and we use the rate of change of the momentum of a particle as a measure of the force acting on it:

$$\underline{F} = \underline{\dot{p}} \equiv \frac{d\underline{p}}{dt} \tag{1}$$

(The rate of change of the energy is also a natural quantity to define, but it turns out to depend in a direct way on the force as defined here, so it doesn't add anything.) Eq. 1 is Newton's second law of motion, more or less in the form in which Newton conceived it.

In Newtonian—nonrelativistic—dynamics, the momentum is simply proportional to the velocity $\underline{v}$ and points in the same direction. The constant of proportionality, called the *mass m*, is an intrinsic property of the particle and is a fixed quantity for each given particle:

$$\underline{p} = m\underline{v} \tag{2}$$

The standard unit of mass is the *kilogram* (kg), defined as the mass of an actual standard object (kept in a vault, in France), so the units for momentum are kg · m/s. The standard unit of force is called the *newton* (N), and by eqs. 1 and 2 it is given by

$$1 \text{ N} = 1 \text{ kg} \cdot \text{m/s}^2 \tag{3}$$

From eqs. 1 and 2 you find that the acceleration, the rate of change of $\underline{v}$, is determined by the force:

$$\underline{F} = \frac{d(m\underline{v})}{dt} = m\underline{a} \tag{4}$$

or

$$\underline{a} = \frac{\underline{F}}{m} \tag{5}$$

It's more common to give Newton's law of motion as $\underline{F} = m\underline{a}$, but I have a preference for the form shown in eq. 5 because force is the cause and acceleration is the effect, and I think it's appropriate to express the effect in terms of the cause instead of the other way around. It's not Newton's law that determines the force: That has to be given in terms of some *force law* that describes an actual physical force of nature, such as the gravitational force, the electromagnetic force, or some other force. It's common also to speak of a *contact force*—a force due to actual contact with some other object ("If Alice pushes on Harry with a force of 53 N, what is Harry's acceleration?"). Such a contact force represents in an idealized form the electromagnetic forces between the surface molecules of the two objects, which actually never come in contact with each other at all.

In summary, then, the force determines the acceleration (eq. 5) and the acceleration in turn determines the motion (eqs. B.1.3 and B.1.4). If the force is constant—a common simplification—then the acceleration is constant and the simpler eqs. B.1.6 and B.1.7 apply.

Newton's *first* law of motion, important historically because it represents Newton's first breakthrough in understanding, is in fact just a special case of the second law—namely, that when the net force on an object is zero it's velocity is constant. In a standard kind of statement, the first law says that *a body at rest tends to remain at rest and a body in motion tends to remain in motion with the same velocity, unless acted on by external forces.*

Newton's *third* law (§4.6), treated by Newton as separate from the others, also ties in with the basic idea of conservation of momentum that I used in arriving at a definition of force. It says, in brief, that *the forces that two objects exert on each other are equal and opposite.* The forces here are vector quantities. This law holds because in the absence of external forces, the net momentum of the two objects must be constant, which means that the rate of momentum change of the second object, $\dot{p}_2$, must be equal and opposite to the rate of momentum change of the first, $\dot{p}_1$. By eq. 1, then, the forces also must be equal and opposite. Newton's third law ceases to be valid if the force fields that are ultimately responsible for the forces carry significant momentum also, as is the case with electromagnetic forces when magnetic effects are important.

The role of work and kinetic energy in nonrelativistic mechanics is sketched in §5.1 and §5.2, and the idea of *potential energy,* a most fruitful and versatile concept in classical physics, is discussed in some detail in §5.3. You should review those sections.

APPENDIX C

Miscellaneous Information

 C.1 THE GREEK ALPHABET

Capital	Lower case	Name in English	Capital	Lower case	Name in English
A	α	alpha	E	ε	epsilon (EPsuhlon)
B	β	beta (BAYtuh)	Z	ζ	zeta (ZAYtuh)
Γ	γ	gamma	H	η	eta (AYtuh)
Δ	δ	delta	Θ	θ	theta (THAYtuh)

(*Continued*)

Capital	Lower case	Name in English	Capital	Lower case	Name in English
I	ι	iota (ieOHtuh)[1]	P	ρ	rho
K	κ	kappa	Σ	σ	sigma
Λ	λ	lambda	T	τ	tau (rhymes with how or haw)
M	μ	mu (mew)	Y	υ	upsilon (OOPsuhlon)[2]
N	ν	nu	Φ	φ	phi (fie or fee)
Ξ	ξ	xi (ksie or ksee)	X	χ	chi (kie or kee)
O	o	omicron (OHmuhcron)	Ψ	ψ	psi (psie or psee—or sie)
Π	π	pi (pie—no one says pee)	Ω	ω	omega (ohMAYguh or OHmuhguh)

[1]The "ie" is a long *i*, to rhyme with "pie."
[2]The pronunciation of *upsilon* varies from YOOPsuhlon to UPsuhlon. My pronunciation has "oo" as in *book*.)

 **UNITS**

The SI (Système Internationale) is the standard metric system of units. It's an MKS system, the <u>m</u>eter, <u>k</u>ilogram, and <u>s</u>econd being basic, and includes also the coulomb as the unit of electric charge. Many physicists continue to use a CGS—<u>c</u>entimeter, <u>g</u>ram, <u>s</u>econd—system, while normally adhering to the coulomb as the unit of charge. The basic units of the SI system are given in the following tables.

Standard units

Physical quantity	Symbol	Unit	Abbreviation	Equivalence
length	ℓ, x, etc.	meter	m	
mass	m	kilogram	kg	
time	t	second	s	
force	$\underline{F}$	newton	N	$kg \cdot m/s^2$
energy	E, K, U, etc	joule	J	$N \cdot m = kg \cdot m^2/s^2$
power	P	watt	W	J/s
momentum	$\underline{p}$			$N \cdot s = kg \cdot m/s$
temperature	T	kelvin	K	
entropy	S			J/K
charge	Q	coulomb	C	
electric current	I	ampere	A	C/s
potential, electromotive force	V, etc.	volt	V	J/C
electric field	$\underline{E}$			$N/C = V/m$
magnetic field	$\underline{B}$	tesla	T	$N/A \cdot m = W/m^2$
magnetic flux	Φ	weber	W	$T \cdot m^2$
resistance	R	ohm	Ω	V/A
capacitance	C	farad	F	C/V
inductance	L	henry	H	$V \cdot s/A$
frequency	f	hertz	Hz	cycle/s
angular frequency	ω			rad/s

Nonstandard units

Physical quantity	Unit	Abbre- viation	Equivalence
length	fermi $\equiv$ femtometer	fm	10^{-15} m
	angstrom	Å	10^{-10} m
	micron $\equiv$ micrometer[1]	$\mu \equiv \mu$m	10^{-6} m
	inch	in	2.54 cm (exact)
	foot	ft	30.48 cm
	mile	mi	1.610 km
	light-year[2]	lt-yr	9.4605×10^{15} m
	parsec[3]	pc	3.262 lt-yr
mass	eV over c squared	eV/c^2	1.78266×10^{-36} kg
	atomic mass unit	u	1.66054×10^{-27} kg = 931.49 MeV/c
energy	electron-volt[4]	eV	1.60218×10^{-19} J
	calorie[5]	cal	4.186 J (exact)
charge	electron charge	e	1.60218×10^{-19} C

[1]Pronounced MIEkruhMEEter.

[2]Equal to the distance light travels in 1 year.

[3]The distance from which the radius of the earth's orbit about the sun subtends an angle of 1 second of arc.

[4]Equal to the electron charge e times 1 volt, the energy acquired by a charge e when accelerated through a potential difference of 1 volt.

[5]Approximately equal to the energy needed to raise 1 gram of water by 1 K (1 K is the same as 1 centigrade degree). The "kilocalorie," which is equal to 1000 cal and abbreviated Cal, is rarely used in physics but is the standard unit of food energy used in dietetics.

PREFIXES FOR METRIC UNITS

This table is not complete but includes the most commonly encountered prefixes.

Prefix	Abbreviation	Equivalence
femto-	f	$\times 10^{-15}$
pico- (PIEco)	p	$\times 10^{-12}$
nano-	n	$\times 10^{-9}$
micro-	μ	$\times 10^{-6}$
milli-	m	$\times 10^{-3}$
centi-	c	$\times 10^{-2}$
deci-	d	$\times 10^{-1}$
kilo-	k	$\times 10^{3}$
mega-	M	$\times 10^{6}$
giga- (JIGuh)	G	$\times 10^{9}$
tera-	T	$\times 10^{12}$

PHYSICAL CONSTANTS

Constant	Symbol	Numerical value
speed of light[1]	c	2.9979×10^{8} m/s
Planck's constant	h	6.6261×10^{-34} J · s
Dirac h-bar[2]	$\hbar = \dfrac{h}{2\pi}$	1.05457×10^{-34} J · s = 197.327 nm · eV/c
gravitational constant	G	6.6726×10^{-11} N · m²/kg²

(*Continued*)

419

Constant	Symbol	Numerical value
Boltzmann constant	k_B	1.38065×10^{-23} J/K = (1 eV)/(11604.5 K)
Avogadro's number[3]	N_A	6.0221×10^{23} mol^{-1} = 1 g/1 u
gas constant	R	8.3145 J/mol $\cdot$ K = $N_A \cdot k_B$
dielectric constant of empty space[4]	ε_0	8.8542×10^{-12} C^2/N $\cdot$ m^2
Coulomb's law constant[4]	$k = \dfrac{1}{4\pi\varepsilon_0}$	8.9876×10^9 N $\cdot$ m^2/C^2
permeability of empty space[4]	μ_0	$4\pi \times 10^{-7}$ N/A^2
electron charge	e	1.60218×10^{-19} C
electron mass	m_e	9.1094×10^{-31} kg = 5.4858×10^{-4} u = 0.51100 MeV/c^2
proton mass	m_p	1.67262×10^{-27} kg = 1.00728 u = 938.27 MeV/c^2
neutron mass	m_n	1.67493×10^{-27} kg = 1.00866 u = 939.57 MeV/c^2

[1]The unit of length, the meter, has recently been redefined to make the speed of light equal to *exactly* 299,792,458 m/s.

[2]Dirac's modified Planck's constant, $\hbar$, proves to be much more useful for many purposes than Planck's original definition, because it reduces substantially the number of times the factor 2π has to appear in formulas. The combination $\hbar c$, equal to 197 eV $\cdot$ nm (or 197 MeV $\cdot$ fm), is particularly useful in particle physics.

[3]Avogadro's number, the ratio of the laboratory unit of mass and the atomic unit of mass, provides the link between the atomic scale and the laboratory scale. A mole of a substance is defined to be just this many atoms or molecules, so the mass of a mole in grams is just equal to the mass of an atom or molecule in atomic mass units u.

[4]These constants are artifacts, resulting from the arbitrary choice of the coulomb as the unit of electric charge. The coefficient $k = 1/4\pi\varepsilon_0$, which is numerically equal to $c^2 \times 10^{-7}$ in these units, appears in eq. 12.3.2 for the Coulomb force, and the coefficient $\mu_0/4\pi$, numerically equal to 10^{-7}, appears in a similar way in the basic formulas for magnetic forces. The velocity of electromagnetic waves deduced from Maxwell's equation is $1/\sqrt{\varepsilon_0\mu_0}$ and is thus exactly equal to c.

When you're calculating to a certain accuracy, always keep one or two more figures of accuracy than you need throughout the calculation and then round off to the desired accuracy only at the end. It's good practice to keep an additional figure if the initial digit is 1; note, for example, that the relative error in rounding 10.4 to 10 is almost ten times greater than the relative error in rounding 99.4 to 99.

 **ASTRONOMICAL FACTS**

mass of earth	5.976×10^{24} kg
mass of sun	1.989×10^{30} kg
mass of moon	7.347×10^{22} kg
equatorial radius of earth	6.378×10^{6} m
equatorial radius of sun	6.960×10^{8} m
equatorial radius of moon	1.738×10^{6} m
mean radius of earth's orbit	1.496×10^{11} m
mean radius of moon's orbit	3.84×10^{8} m

Glossary of Physical Terms

Words in boldface within entries (or closely related words) are also listed in the glossary. A section reference is given for terms that are discussed in the book. Where appropriate, the SI (Système Internationale), or standard metric, unit is given (see Appendix C, §2).

absolute temperature (§9.1)   The **temperature** of an object or **system,** measured in degrees above **absolute zero,** and related in a direct way to the amount of thermal **energy** in the **system.** The absolute temperature and the **entropy** are defined together, making use of the **second law of thermodynamics,** in such a way that when a small amount of **heat** dQ is added to a system at temperature T, the change in entropy of the system is given by $dS = dQ/T$. Symbol: T. SI unit: kelvin (K).

absolute value (§14.1)   The magnitude of a **complex number.** If $z = x + iy$ is the complex number, then the absolute value, written as $|z|$, is given by $|z| = (x^2 + y^2)^{1/2}$. Also known as the modulus.

absolute zero (§9.1)   The temperature at which the thermal **energy** of a **system** is exactly zero. (Such a **quantum state** may have a finite **zero-point energy,** but this energy is nonrandom, and therefore nonthermal, in character.)

acceleration (§4.2)   A **vector** equal to the rate of change of **velocity** of an object: $\underline{a} = d\underline{v}/dt$. The acceleration vector may represent an increase or decrease in **speed** or a change in the direction of motion. Symbol: $\underline{a}$. SI unit: meter/(second)^2 (m/s^2).

acceleration of gravity (§4.4)   In **Newtonian** gravity, the **acceleration** of an arbitrary mass caused by a **gravitational field** g. It is exactly equal to the field g itself, the units of gravitational field, N/kg, being identical to the units of acceleration, m/s^2.

alpha particle (§5.6)   A nuclear particle, the **nucleus** of a ^{4}He atom, consisting of two **protons** and two **neutrons,** tightly bound. Symbol: α.

amplitude (§1.3)   For a **sine wave** of the general form $y(x) = y_0 + A \sin(cx + \varphi)$, the amplitude A is the magnitude of the maximum displacement of the wave from its central value y_0 (y_0 is usually taken as zero).

angular frequency (§11.2)   For a **sine wave oscillation** in time of the general form $y(t) = A \sin(\omega t + \varphi)$, the angular frequency ω is equal to the time rate of change, in radians per unit time, of the **argument.** The argument $\omega t + \varphi$ represents the instantaneous **phase** of the motion at the time t. The angular frequency is related to the ordinary **frequency** f by $\omega = 2\pi f$. *Symbol:* ω. SI unit: radians per second (rad/s).

angular momentum (§12.3) A conserved **vector** quantity representing the amount of rotational motion of an object or a general physical **system** about some chosen origin. For a set of **particles** labeled by $i = 1, 2, \ldots, N$, with instantaneous positions and **momenta** $\underline{r}_i$ and $\underline{p}_i$, respectively, the angular momentum due to the orbital motion at that moment is given by the sum of contributions due to the separate particles: $\underline{L} = \sum_i \underline{r}_i \times \underline{p}_i$. Each term involves the **vector product** of the vectors $\underline{r}_i$ and $\underline{p}_i$. In addition, each particle may have a **spin** angular momentum $\underline{S}_i$ that must be added in to give the total angular momentum. The **conservation** of angular momentum is related by **Noether's theorem** to the rotational **symmetry** of physical laws. Symbol: $\underline{L}$ (for orbital motion), $\underline{S}$ (for spin angular momentum), $\underline{J}$ (for total angular momentum). SI units: kilogram · (meter)^2/second   (kg · m^2/s).

angular momentum quantum numbers (§14.2)   In **quantum theory,** parameters labeling the different possible **eigenstates** of **angular momentum.** For orbital motion of a **particle** (such as an **electron** in an atom), the **quantum numbers** are ℓ and m_ℓ, with ℓ taking the integer values $\ell = 0, 1, 2, \ldots$, without limit, and m_ℓ taking the values $0, \pm 1, \pm 2, \ldots, \pm \ell$, a total of $2\ell + 1$ possible values. The quantum number ℓ determines the magnitude of the angular momentum $\underline{L}$: $|\underline{L}|^2 = \ell(\ell + 1)\hbar^2$, and m_ℓ determines the z component of $\underline{L}$: $L_z = m_\ell \hbar$. For **spin** angular momentum, the intrinsic angular momentum of an elementary particle, the angular momentum quantum numbers are normally written s and m_s, with the allowed values $s = 0, \pm 1/2, \pm 1, \pm 3/2, \ldots$ (any given elementary particle has a single fixed value of s—see **spin-statistics theorem**) and $m_s = s$, $s - 1, s - 2, \ldots, -s$, a total of $2s + 1$ possible values ($2s + 1$ may be even or odd, depending on s).

angular wave number (§11.1)   For a **sine wave** of the form $y(x) = A \sin(kx + \varphi)$, the quantity $k = 2\pi/\lambda$ (equal to number of radians of **phase** per unit length), where λ is the **wavelength.** The term *angular wave number* is nonstandard; I've adopted it in this book for clarity. See **wave number.** SI unit: radian/meter (m^{-1}, since radians are dimensionless).

anticommute   In the algebra of quantum **operators,** two operators $\hat{A}$ and $\hat{B}$ anticommute if $\hat{A}\hat{B} = -\hat{B}\hat{A}$. See **commute.**

antimatter (§5.6)   A theoretically allowed form of matter composed entirely of the **antiparticles** of the **particles** that make up ordinary matter. Since ordinary matter is made of **quarks** and **electrons,** antimatter is made of antiquarks and **positrons,** which form exact antiparticle analogs of all the known elements and chemical compounds. In conventional terms, **protons** are replaced by **antiprotons** (negatively charged), and **neutrons** by **antineutrons.** The **photons** and **gluons** that hold matter together are unaltered.

antineutron (§5.6)   The **antiparticle** of the **neutron,** with exactly the same **mass** and, like the neutron, uncharged. Symbol: $\overline{n}$.

antiparticle   The **charge conjugate** twin of any **fermion** in nature, with identical **mass** and opposite **electric charge.** A **particle** and its antiparticle can annihilate each other, with the entire conversion of their combined masses into **energy.** Symbol: $\bar{\text{P}}$, where P is the symbol for the particle.

antiproton (§5.6)   The **antiparticle** of the **proton,** with exactly the same **mass** and negative **electric charge.** Symbol: $\bar{\text{p}}$.

antisymmetric   A function of two or more variables, such as a **wave function,** is antisymmetric if the interchange of any two variables causes the function to change sign. The function $f(x)g(y) - g(x)f(y)$ is antisymmetric in the variables x and y. The wave function for **identical fermions** is antisymmetric in the coordinates of all the different **particles.** See **symmetric.**

antisymmetrize   To construct an **antisymmetric wave function** out of one that is not antisymmetric by combining terms with different **permutations** of the **arguments** (sense 1). The function $f(x)g(y)$ is antisymmetrized by forming the combination $f(x)g(y) - g(x)f(y)$. See **symmetrize.**

Argand diagram (§14.1)   A graphical method (fig. 14.1.2) of displaying **complex numbers** as points on a two-dimensional plane, with the **real** and **imaginary parts** of each complex number taken as the x and y coordinates, respectively, of the corresponding point on the plane. That is, the complex number $z = x + iy$ is represented by the point (x,y) on the plane.

argument   **1.** (§A.3)   The independent variable on which a given function depends. The variable x is the argument of the function $f(x)$.   **2.** (§A.7) For a **complex number,** the angle in the **Argand diagram** made with the x axis by the line from the origin to the point representing the complex number (see fig. 14.1.2). The argument θ of the complex number $z = x + iy$ is determined by the relations $x = |z|\cos\theta$ and $y = |z|\sin\theta$, where $|z|$ is the **absolute value** of z.

Avogadro's number (§C.4)   A number equal to 6.02×10^{23}, relating the **macroscopic** scale to the atomic scale. Avogadro's number is equal to the ratio of two **mass** units, one large-scale and one small-scale: the gram and the atomic mass unit u (see Appendix C, §2). (The atomic mass unit is defined as $1/12$ the mass of 1 atom of carbon twelve (^{12}C) and is thus approximately equal to the mass of a hydrogen atom.) Symbol: N_A.

Balmer series (§12.1)   A well-known series of **spectral lines** in the **spectrum** of atomic hydrogen, with **wavelengths** given by the **Rydberg formula:** $1/\lambda = R(1/m^2 - 1/n^2)$, with $m = 2$ and $n = 3, 4, \ldots$. Other values of m also give series of spectral lines. The constant R is the **Rydberg constant.**

baryon (§5.6)   Any of the **strongly interacting fermions** consisting of three **quarks,** or one of their **antiparticles.** The two types of baryon that are prolifically common in ordinary matter are the **proton** and the **neutron.** All the other varieties of baryon are extremely short-lived and have been observed only in high-energy physics experiments.

baryon number (§5.6)   A **conserved particle number,** equal to the number of **baryons** in the universe (or in an isolated physical **system**) minus the number of antibaryons. See **antiparticle.**

beauty   One of the six **flavors** of **quark,** more commonly referred to as the **bottom quark.** Symbol: b.

beta decay   The radioactive decay of a **neutron,** or of a **nucleus** containing a neutron, in which the neutron becomes a **proton** (of slightly lower mass) by emitting an **electron** and an antineutrino of the electron family: $n \rightarrow p^+ + e^- + \bar{v}_e$. See **antiparticle** and **electron neutrino.**

Big Bang (§7.2)   The assumed cataclysmic explosion in which the universe began, around 15 billion years ago. The structure of **spacetime** would have been so violently distorted and discontinuous at that point that it would be impossible to track the history of our universe back to any earlier times and, in fact, meaningless even to suppose that there was any earlier time.

Big Crunch (§8.4)   A possible cataclysmic implosion in which the universe will end if the gravitational forces are great enough to bring its expansion to a halt and cause it to collapse. The general belief among astrophysicists is that the average **mass** density of the universe is not great enough for this to happen.

binding energy (§5.6)   The amount of **energy** that would have to be supplied in order to separate particles that are bound together by forces of any sort. The binding energy of a **system** represents not energy available for use but a *deficit* of energy.

binomial coefficient (prob. 10.7.1)   A mathematical expression for the number of different ways in which m objects can be chosen from a collection of n objects, where $m \leq n$. The binomial coefficient is also the coefficient of $x^m y^{n-m}$ in the expansion of $(x + y)^n$. It is given by $\binom{n}{m} = n!/m! \, (n-m)!$, where $n!$, called n-factorial, is the product of the first n integers: $n! = 1 \cdot 2 \cdot 3 \cdots n$.

black-body radiation (§10.3)   The characteristic distribution of **wavelengths** in the **electromagnetic radiation** emitted by an idealized hot object (a *black body,* which is a perfect absorber and a perfect emitter of radiation) at a given **absolute temperature** T. The peculiar character of black-body radiation gave the first clue to the **quantum** nature of electromagnetic radiation.

black hole (§7.1)   The final result of the gravitational collapse of a massive star into a gravitational distortion so severe that no **particles** or radiation can escape, except by very tiny **quantum tunneling** effects. The **event horizon,** surrounding the black hole at roughly the **Schwarzschild radius,** represents the limit within which no object can be seen from outside.

Bohr radius (§12.3)   The radius of the circular orbit of the **electron** in the lowest **energy eigenstate** of the hydrogen atom, in Bohr's semiclassical, pre-Schrödinger model. The Bohr radius is given by $a_0 = \hbar^2/kme^2 = 0.528$ Å and is a common reference length for distances at the atomic scale. The factor k is the **Coulomb's law** coefficient (eq. 12.3.2).

Boltzmann constant (§9.2)   The constant $k_B = 1.38 \times 10^{-23}$ J/K, relating **absolute temperature** to **energy.** The average energy of a particle, when the ambient temperature is T, is of order of magnitude $k_B T$.

Boltzmann factor (§9.2)   The statistical factor, $e^{-E/kT}$, that represents the relative **probability** for a **system** to be in a state of **energy** E when it is in **thermal equilibrium** at **absolute temperature** T. The factor k is the **Boltzmann constant,** usually written k_B in this book.

Bose–Einstein condensation   The **quantum** phenomenon displayed by a **system** consisting of a fixed number of noninteracting **bosons** at very low **absolute temperature,** in which a finite fraction of the **particles** are all in the same **single-particle state.** The effect may also occur for interacting particles, depending on the circumstances, and is an important part of the story of **superfluidity** in liquid **helium four** (^{4}He).

Bose–Einstein statistics   A phrase referring to the way you count **quantum states** for a **system** of **identical bosons.** These statistical rules result from the indistinguishability of the bosons and from the fact that **wave functions** for identical bosons must be completely **symmetric.** Bosons are said to "obey Bose–Einstein statistics." See **Fermi–Dirac statistics.**

boson   In **quantum theory,** one of the two broad classes of **particles,** bosons and **fermions,** distinguished by their **symmetry** properties under **permutations.** The **wave function** for a set of **identical** bosons must be completely **symmetric** under **permutations** of the particles. **Photons, gluons, gravitons,** and **mesons** are examples of bosons.

bottom quark   One of the six **flavors** of **quark.** It has **electric charge** $-(1/3)e$, like the **down** and **strange quarks,** but is much more massive. It is closely paired with the **top quark** (assumed to exist but not yet observed). The lightest observed **particle** containing a bottom quark is the bottom **meson,** with a **mass** over five times that of a **proton.**

broken symmetry   The common occurrence of a **symmetry** of nature that appears to be violated. The general belief is that such a symmetry is a true symmetry of the laws of physics and that the symmetry violation is due to **spontaneous symmetry breaking.**

central field approximation   The approximation, widely used in atomic and nuclear physics, that each **particle** of a many-particle **system,** such as an atom or **nucleus,** moves freely in a spherically symmetrical **central force field** that is due to the averaged-out effects of all the other particles (together with the **electric field** due to the **nucleus** in the case of atoms).

central force   A spherically **symmetrical** central **force field,** producing a force on a **particle** that depends only on its distance from the center and is always directed radially inward or outward. Examples include the **gravitational field** of the sun and the **electric field** due to a charged **point particle,** such as an atomic **nucleus.**

charge conjugate   The partner, of equal mass and opposite electric charge, related by a fundamental **symmetry** to each elementary particle of nature (or quasi-elementary, like the **hadrons**). In the case of **fermions,** the charge conjugate particle is called the **antiparticle**; in the case of bosons, the charge conjugate particle may be a different particle ($\pi^+ \to \pi^-$), or it may be the particle itself (the π^0 or the photon).

charge quantum number   A rational number—positive, negative, or zero—giving the **electric charge** of a **particle** in units of e, the magnitude of the **electron** charge. All the particles that can be seen experimentally have integer values of the charge **quantum number,** while **quarks** have values that are integer multiples of $1/3$.

charm quark   One of the six **flavors** of **quark.** It has **electric charge** $+(2/3)e$, like the **up quark,** and is closely paired with the **strange quark,** though quite a bit more massive. The lightest observed **particle** containing a charm quark is the D **meson,** with a **mass** about twice that of a **proton.** Symbol: c.

circular polarization (§11.5)   A wave pattern for **electromagnetic waves** in which the **electric** and **magnetic field vectors** rotate, with constant magnitude, in a plane perpendicular to the direction of propagation. See **polarization.**

closed shell   A completely **occupied shell** of **single-particle states** of approximately equal **energy** in a many-**particle system,** such as the **electrons** of an atom or the **protons** or **neutrons** of a **nucleus.** A closed-shell system tends to be more stable—less likely to be influenced by interactions—than others.

closed universe (§7.3)   A universe that is of finite spatial size. It has no boundaries but is closed on itself like a circle or the surface of a sphere. A closed universe is expected by **general relativity** to collapse after a finite lifetime and end in a **big crunch.**

color, or color charge   The **conserved** analog of **electric charge** that interacts, according to the **Standard Model,** via the **color gauge field** (or **gluon field**). This interaction is much stronger than the electromagnetic interaction. The color charge is an eight-component **vector** in an abstract color space, but owing to a color version of the **uncertainty principle,** only two of the eight components can be well specified at the same time. Thus the color of a **quark** or **gluon** (the particles that carry color) is specified by a vector in a two-dimensional space. The three possible colors for a quark are conventionally called red, blue, and green (or R, B, and G) in analogy with the primary colors. The conservation of color charge is related by **Noether's theorem** to an abstract **symmetry group** known as color SU(3).

color confinement   The mechanism responsible for the fact that we never see free **quarks** or **gluons.** It is usually supposed that **color charges** interact so strongly, even at large distances, that only color-neutral objects can exist in isolation. Only objects with quark number (see **particle number**) equal to a multiple of 3 (namely the **baryons** and **mesons** of nature) can be color-neutral.

color gauge field   The **gauge field,** analogous to the **electromagnetic field,** that is responsible for the **strong interactions.** Its **quanta** are called **gluons.** The color gauge field, often referred to as the **gluon field,** is responsible for interactions among **color charges,** and these interactions are responsible in turn, though somewhat indirectly, for what we see phenomenologically as the **strong forces** that hold **nuclei** together and generate nuclear energy of all sorts.

commutator   In the algebra of quantum **operators,** the commutator of two operators $\hat{A}$ and $\hat{B}$, written $[\hat{A}, \hat{B}]$, is equal to the difference $\hat{A}\hat{B} - \hat{B}\hat{A}$

commute   Two **operators** $\hat{A}$ and $\hat{B}$ commute if $\hat{A}\hat{B} = \hat{B}\hat{A}$. Note that operator multiplication is not normally commutative. In the **group** of ordinary rotations in three dimensions, for example, a 90° rotation about the x axis followed by a 90° rotation about the y axis yields a different net rotation than the same operations in reverse order.

complementarity (§13.3)   The idea, in quantum physics, that certain kinds of information, or experiments, *complement* each other. See **complementary variables.**

complementary variables (§13.2)   Any pair of variables in a **quantum system** that can't both be accurately known at the same time, although information about both is needed for a complete picture of the state. Examples of pairs of complementary variables include position and **momentum, energy** and time, and **electric field** strength and **photon** number.

complex conjugate (§A.7)   A kind of mirror image of a given complex number. If $z = x + iy$ is the given number, then its complex conjugate, z^*, is the complex number $x - iy$. Two useful properties: (i) $(z^*)^* = z$; (ii) $z^*z = |z|^2$, where $|z|$ is the **absolute value** of z. Note that z^*z is always **real** and positive.

complex numbers (§14.1, §A.7)   An artificial but highly useful set of mathematical objects, constructed as combinations of a **real number** and a real multiple of the **imaginary** unit i, defined as $\sqrt{-1}$. A general complex number z has the form $x + iy$, where x and y are ordinary real numbers.

composition of velocities (§3.2)   A rule for combining relative **velocities.** In **special relativity,** the velocity of object C relative to object A is not simply the sum of its velocity relative to an object B and the velocity of B relative to A, but is given by eq. 3.2.4.

configuration   In general, a particular arrangement in space of the various components of a physical **system.** More specifically, in describing the **quantum state** of a many-**particle** system like an atom or **nucleus,** the configuration is a specification of all the **occupied single-particle states.** You can talk about a configuration in this way only if the **independent-particle model** is at least approximately valid.

confinement   See **color confinement.**

conservation law (§I.2, §4.1)   Any law by which a physical quantity, normally some sum or integral over the whole universe, is constant in time. Every conservation law is related by **Noether's theorem** to some **symmetry** property of the laws of physics. Standard examples are **energy, momentum, angular momentum, electric charge,** and **color charge.**

Cooper pair   A very loosely bound pair of **electrons** in a simple **superconductor** or of atoms in **superfluid helium three.** The Cooper pairs are **bosons,** and the superconducting **state** can be viewed in simplified terms as a **Bose–Einstein condensation** of these bosons.

coordinate axes (§2.1)   In any flat space or **spacetime,** the imaginary lines that provide the framework for the coordinate system. Each axis is associated with one of the coordinates and is the set of points for which all the *other* coordinates are zero. The intersection of all the coordinate axes is the origin.

correspondence principle (§13.1)   The very general principle that when a new theory is proposed to take the place of a previously accepted theory, the predictions of the new theory must agree with those of the old one in all situations where the old theory has been tested and found accurate. The correspondence principle was first stated as a principle in connection with **quantum theory,** where the conceptual changes proposed were quite radical and the constraints imposed by the correspondence principle were very severe—and very necessary.

cosmology   The study of the size, shape, and behavior of the entire universe, the *cosmos,* especially the study of its origins.

Coulomb's law (§4.3, §12.3)   The historically important **electrostatic force** law for two **nonrelativistic point charges.** The magnitude of the force is $F = kQ_1Q_2/r^2$, where k is a standard factor (eq. 12.3.2), given in SI units by $k \equiv 1/4\pi\varepsilon_0 = 8.99 \times 10^9$ N $\cdot$ m^2/C^2. The Coulomb force is repulsive for charges of the same sign and attractive for charges of opposite sign. The Coulomb force can be understood as the force on each particle that is due to the **electric field** produced by the other.

coupling constant   In a **field** theory, the parameter that represents the strength of interaction between the different fields.

covariant derivative   In a **gauge theory** (including **general relativity**), a modification of the **partial derivative** that tranforms in a standard way under the **transformations** of the **local symmetry group.** The simplest example is **electro-**

magnetic theory, where the covariant derivative of a **complex field** ψ is given by $D_\alpha \psi = (\partial_\alpha + ineA_\alpha)\psi$. Here ∂_α represents the ordinary partial derivative $\partial/\partial x^\alpha$, n is the **charge quantum number** associated with the field ψ, and A_α is a component of the gauge field.

cross product (§11.5, §A.2)   **Vector product**

curvature tensor (§6.4)   A **tensor** whose components give complete information about the curvature of a curved space, such as the **spacetime** of **general relativity.** The curvature tensor is exactly zero for a flat space, such as the spacetime of special relativity. Symbol (in this book): $\underline{\underline{R}}$.

curvilinear coordinates (§6.4)   An arbitrary coordinate system on a flat or curved space in which the coordinate surfaces (surfaces for which one of the coordinates is constant) may be curved in any way.

daughter nucleus   The nucleus that remains after a radioactive decay process such as **beta decay.** See **parent nucleus.**

definite metric   The **metric** of a space such as a Euclidean space that has only one type of **vector,** like the **spacelike vectors** in **spacetime,** and no **null vectors.** The metric on a curved surface in a three-dimensional Euclidean space is definite.

degree of freedom   A **system** is said to have one degree of freedom for each of the real variables required to specify completely the **state** of the system at a given time. A single **particle** has six degrees of freedom, corresponding to the three position coordinates and the three components of its **velocity.**

derivative (§4.2, §A.4)   A measure of the rate at which some function varies as a function of one of its arguments. The derivative of a function of a single variable, say $f(x)$, with respect to x is given by the quotient $df(x)/dy = [f(x+dx) - f(x)]/dx$, where dx is an **infinitesimal** increment in x and $df(x)$ is the corresponding increment in $f(x)$. For the case of a function of several variables, see **partial derivative.**

deterministic (§10.7)   A word describing classical physical laws, which completely determine the future behavior of a **system** if you're given precise and complete information about the **initial conditions** of the system—positions and **velocities** of all particles, and **field** values at every point in space. The laws of **quantum** physics, in contrast, are not deterministic and can give only **probabilistic** predictions about the future behavior of a physical system.

differential equation   Any relationship among the derivatives of an unknown function. If the function is of just one variable, the equation is an **ordinary differential equation,** and if the function is of more than one variable, it is a **partial differential equation.** A simple example is the equation of motion of a single **nonrelativistic particle** in one space dimension, whose position at time t is the function $x(t)$. The equation of motion is an ordinary differential equation that is equivalent to **Newton's second law** of motion: $d^2x/dt^2 = F(t)/m$, where $F(t)$ is the force acting on the particle at time t.

diffraction (§10.5) The spreading of a beam of light (or other kind of **wave**) that results from the tendency of any wave motion to fill all available space. Diffraction is often accompanied by **interference** effects. The spreading is small if the **wavelength** is small compared to the width of the beam.

Dirac $\hbar$ (§12.3) **Planck's constant** divided by 2π: $\hbar = h/2\pi$, a much more useful combination in modern physics than Planck's constant itself. Called "h-bar." Value in SI units: 1.055×10^{-34} J · s.

discrete (§12.1) Distinct and unconnected, describing, for example, a variable whose possible values have finite spacings, like the integers or the energy levels of an atom.

displacement A shift in the value of some quantity. Specifically, the relocation of an object or a **reference frame** in **spacetime** (§4.1). If the shift involves only the time coordinate it's a *time displacement,* and if it involves only spatial coordinates it's a *spatial displacement.*

Doppler shift (§7.2) The **frequency** shift of a wave that is due to the motion of either the source or the observer. The relative shift in frequency is approximately equal, for low **velocity** v, to the ratio v/c, where c is the wave speed.

double Hubble (§8.2) My invented term for a time period consisting of the age of the universe multiplied by the number of seconds in the age of the universe, which comes out to around 10^{28} years. See **Hubble time.**

down quark One of the six **flavors** of **quark.** It has **electric charge** $-(1/3)$ e. The **up** and the down quarks, which are closely related to each other, are the constituents of **protons** and **neutrons** and thus make up some 99.98% (by mass) of all ordinary matter.

dynamical symmetry A **symmetry** associated with **transformations** of the coordinates of **spacetime,** as distinct from an **internal symmetry.**

dynamics (§4.1) The way in which the motion of a mechanical **system** is governed by the **forces** acting. More generally, the rules of behavior of any physical system, as in **electrodynamics** or **quantum chromodynamics.**

eigenstate (§14.4) A **quantum state** in which some physical variable is known to have a precisely predictable value, the **eigenvalue,** rather than just a **probability** distribution over possible values.

eigenvalue (§14.4) In **quantum theory,** the precise value that a physical variable has when the **system** is in an **eigenstate** of that variable. In an arbitrary **state** of the system, which is always a superposition of those eigenstates, a measurement of the physical variable must yield one or another of the eigenvalues.

eigenvalue spectrum (§14.4) The set of possible values that a physical variable may take for some given **quantum system.** The eigenvalue spectrum may be **discrete,** like the **energy eigenvalues** of an atom, or continuous, like those of a free **particle.**

Einstein field equations (§6.5) The **field equations** devised by Einstein to describe the **dynamics** of curved **spacetime** in **general relativity.**

electric charge (§11.5) The **conserved** physical quantity attached to certain **particles** of nature and acted on by **electric** and **magnetic fields.** Electric charges also produce electric and magnetic fields. The fundamental unit of electric charge is the magnitude e of the charge on the electron: $e = 1.60 \times 10^{-19}$ C. (A more accurate view is that it is the **charge quantum number** that is conserved and that the quantity e should be regarded as a **coupling constant,** giving the strength of the electromagnetic interaction.) Symbol: q or Q. SI unit: coulomb (C).

electric field (§11.5) That part of the **electromagnetic field** that produces a **force** on **electrically charged** particles that's independent of their **velocity.** It's a vector quantity, with three components. Symbol: $\underline{E}$. SI unit: volt/meter (V/m) $\equiv$ newton/coulomb (N/C).

electromagnetic (EM) field (§11.5) The force **field** that acts on **electric charge.** It is a combination of the **electric** and **magnetic fields,** viewed as a single field of six related components and often written as a **tensor.** The electromagnetic field is derived from the **gauge field** A^{μ}, whose components are the **scalar** and **vector potentials** of **electromagnetic theory.**

electromagnetic (EM) radiation (§10.3) **Electromagnetic waves,** with particular reference to their being emitted from some source.

electromagnetic (EM) theory (§11.5) The theory of **electromagnetic fields** and their interactions. Electromagnetic theory is the prototype of all **gauge theories.**

electromagnetic (EM) waves (§11.5) **Traveling-wave** or **standing-wave** patterns in the **electric** and **magnetic fields.**

electron (§5.5, §10.4, §12.1) An elementary **particle,** one of the class of **leptons.** It is negatively charged and is the lightest of all known charged particles. The arrangement of electrons in atoms and molecules is responsible for all the chemical properties of matter. The magnitude e of the electron charge is the fundamental unit of **electric charge.** Symbol: e$^-$.

electron neutrino (§5.5) One of the three known types of **neutrino.** It is a **lepton** and, together with the **electron,** is subject to a **conservation law** that requires the electron number plus the electron neutrino number (see **particle number**) to be strictly constant in time. Symbol: ν_e.

electron–positron field In **quantum field theory,** the **field** whose **quanta** are **electrons** and **positrons.** The **field equation** is called the Dirac equation.

electron-volt (§5.5)   A unit of **energy** commonly used by most physicists. It is equal to the amount of **kinetic energy** gained by a **particle** of **charge** e (the **electron** charge) when it is accelerated through a **potential** difference of 1 volt, so it is equal to the electron charge e times 1 volt. Symbol: eV. Value in SI units: 1.602×10^{-19} J.

electrostatic   A term referring to **electric charges** at rest or to effects that are not related to the motion of electric charges.

electrostatic potential   Also called **scalar potential.** The **potential energy** per unit charge of an **electrically charged particle** in an **electric field.** Because the **force** on a charged particle is proportional to the charge, the potential depends only on the electric field distribution and not on the charge on the particle. The electrostatic potential is the time component ($\mu = 0$) of the **four-vector field** A^μ, the **gauge field** of **electromagnetic theory.** Symbol: V or φ. SI unit: volt (V) = joule/coulomb (J/C).

electroweak interaction   The partially **unified** theory of **electromagnetic** and **weak forces,** consisting of two **gauge theories** with very nearly equal interaction strengths, strongly intermixed. The physical **photon** is understood as a **superposition** of the charge-zero **quanta** of the two theories. See **intermediate vector bosons.**

EM   Electromagnetic

energy (§4.1, §5.1)   A very fundamental **conserved** quantity that is a measure of the amount of **work** that a given **system** is capable of doing. The **conservation of energy** is related by **Noether's theorem** to the invariance of physical laws under time **displacements.** See **kinetic energy, potential energy, rest energy,** and **binding energy.** The energy density of the **electromagnetic field** is $(\varepsilon_0/2)(E^2 + c^2 B^2)$, where ε_0 is related to the **Coulomb's law** factor k (eq. 12.3.2) by $k = 1/4\pi\varepsilon_0$. Symbol: E or U. SI unit: joule (J) = newton $\cdot$ meter (N $\cdot$ m) $\equiv$ kilogram $\cdot$ (meter)^2/(second)^2 (kg $\cdot$ m^2/s^2).

energy eigenvalue (§14.4)   An **eigenvalue** of the **energy** of a **quantum system.** The energy **eigenvalue spectrum** of a bound system is **discrete,** while that of an unbound system is continuous.

energy level (§12.2, §12.3)   Any of the distinct **energy eigenvalues** of a bound **system** such as an atom, molecule, or **nucleus.** An energy level is said to be *degenerate* if there are two or more states with the same energy eigenvalue.

energy–momentum four–vector (§5.7)   The **relativistic four-vector** whose components are the **energy**/c (including **rest energy**) and **momentum** (three components) of a **particle** or of a general physical **system.** Symbol: $p^\mu, \mu = 0, 1, 2, 3$. SI unit: kilogram $\cdot$ meter/second (kg $\cdot$ m/s).

energy operator   The **quantum operator** corresponding to the **energy** of a physical **system.** It is more commonly referred to as the **Hamiltonian.**

entropy (§9.1, §9.2)   A **thermodynamic** measure of the unavailability of **energy** in a **system** for doing useful **work,** or, at the microscopic level, a measure of the thermal disorder. According to the **second law of thermodynamics** the entropy of the universe, or of any isolated system, always increases with time. See **absolute temperature.** Symbol: S. SI unit: joule/kelvin (J/K).

equations of motion (§4.3)   The equations that describe the motion of a physical **system** from one moment to the next. Together with the **initial conditions,** the equations of motion completely determine the behavior of the system at all times. The equations of motion are **ordinary differential equations,** involving time derivatives, for a system with a finite number of **degrees of freedom,** and they are typically **partial differential equations** for systems with infinitely many degrees of freedom, such as **fields.** The equation of motion of any system, including for example the whole universe, can be written symbolically as $dx/dt = f(x)$, where the letter x represents a complete set of variables describing the state of the system at any given time.

equilibrium (§9.1)   The **state** of a **system** that is not changing with time in some important respect. A **macroscopic** system can be in **thermal equilibrium** at constant **temperature** even though there is random thermal motion of all the molecules.

equivalence principle (§6.3)   The principle that **inertial forces,** the apparent forces associated with **acceleration,** are indistinguishable from gravitational forces. This principle was a driving force in the development of **general relativity.**

ether (§1.1, §1.4)   The extremely tenuous material that was supposed by an earlier generation of physicists to fill all of space and to provide the necessary **medium** for the propagation of **electromagnetic waves.** The idea has been completely abandoned.

even permutation   See **permutation.**

event horizon (§7.1)   A surface in the curved **spacetime** of **general relativity** from beyond which, for various reasons, it's impossible to make any kind of observation of events inside. The standard example is the event horizon surrounding a **black hole,** at approximately the **Schwarzschild radius.**

event point (§2.2)   A point in **spacetime,** or on a **spacetime diagram,** representing an event at a definite location and a definite time.

exclusion principle   The law that no two **identical fermions** can be in the same **single-particle quantum state.** The exclusion principle is a consequence of the requirement that the **wave function** for identical fermions must be an **antisymmetric** function of the positions of the **particles.**

exponential function (§3.2)   The function e^x, which increases by a factor $e = 2.71828$ each time the argument x increases by 1. The base e (which has *nothing* to do with the **electron** charge) is determined by the property that the derivative of the function e^x is just equal to the function itself. Also, any function of the form a^x, for some constant a.

Fermi–Dirac statistics   A phrase referring to the way you count **quantum states** for a **system** of identical **fermions.** These statistical rules result from the indistinguishability of the fermions and from the fact that **wave functions** for identical fermions must be completely **antisymmetric.** Fermions are said to "obey Fermi–Dirac statistics." See **Bose–Einstein statistics.**

Fermi energy   The **energy** of the highest **occupied single-particle state** in a gas of noninteracting **fermions** at **absolute zero temperature.**

fermion   In **quantum theory,** one of the two broad classes of **particles, bosons** and fermions, distinguished by their **symmetry** properties under **permutations.** The **wave function** for a set of **identical** fermions must be completely **antisymmetric** under **permutations** of the particles, with the consequence that no two identical fermions can occupy the same single-particle state (the **exclusion principle**). **Electrons, neutrinos, protons, neutrons,** and **quarks** are examples of fermions.

field (§6.2, §10.1)   A physical **system** consisting of one or more variables, the **field variables,** that take a value at every point in space. The variations of the field variables form a **wave** pattern that fills all of space and changes with time according to some **partial differential equation** known as the **wave equation.** Examples are the **gravitational field,** the **electromagnetic field,** and, more generally, **gauge fields.**

field equation   A **partial differential equation** describing the **dynamical** behavior of a **field.** The field equations for the **electromagnetic field** are **Maxwell's equations,** and other standard field equations are the Klein–Gordon equation (with **quanta** of **spin** zero) and the Dirac equation (with quanta of spin one half).

field variable (§10.1)   One of the physical variables that make up a **field.** The **electric** and **magnetic vectors** are the field variables that make up the **electromagnetic field.**

fine structure (§12.3)   The delicate splitting of **spectral lines** of the hydrogen atom that is due to **spin** interactions and **relativistic** effects. The magnitude of the splitting, relative to the spacing of lines given by the **Rydberg formula,** is of the order of α^2, where α is the **fine-structure constant.**

fine-structure constant (§12.3)   A dimensionless constant, written as α and equal to $ke^2/\hbar c$, characterizing the strength of the **electromagnetic** interaction in fundamental terms. [The constant k is the **Coulomb's law** factor (eq. 12.3.2).] The historical significance of the fine-structure constant comes from the **fine structure** of the **spectral lines** of hydrogen. Value: $1/137.04$

first law of thermodynamics (§9.1)   The law of the **conservation of energy** in the context of **thermodynamics** and with special reference to the role of thermal energy.

flavor   A term used to describe the six known (or presumed) varieties of **quark.** These flavors come in three "generations" of two flavors each, with the two quarks in each generation having **electric charge** –1/3 and +2/3, in units of *e*. The three generations are **down** (d) and **up** (u), **strange** (s) and **charm** (c), **bottom** (b) and **top** (t). Top flavor has not been observed, but for strong theoretical reasons it is generally believed to exist.

force (§4.3, §B.2)   An external influence on an object that causes (or tends to cause) a change in its **momentum,** or state of motion. It is a **vector** quantity. Symbol: $\underline{F}$. SI unit: newton (N) = kilogram · meter/(second)^2 (kg · m/s^2).

Fourier analysis (§11.3)   The procedure for expressing an arbitrary **wave** pattern as a **superposition** of simple **sine waves.** In one simple form, a function $f(x)$ that vanishes at $x = 0$ and $x = L$ can be expressed as the sum $f(x) = \sum_n A_n \sin(n\pi x/L)$, where the sum runs from 1 to ∞. There is a theorem that says that, for any reasonable function $f(x)$, coefficients A_n exist such that the sum of sine waves duplicates $f(x)$ completely.

four-vector (§3.4, §5.7)   A set of four quantities, the *components* of the four-vector, related to the four dimensions of space and time in **special relativity.** The position and time of an event constitute a four-vector, as do the **momentum** and **energy** of a moving particle. The components of a four-vector relative to two different frames of reference (see **inertial reference frame**) are related to each other by the **Lorentz transformation.** Symbol: $\underline{V}$, with components V^μ, $\mu = 0$, 1, 2, 3, where V is replaced by x for the **spacetime** position four-vector, by p for the **energy–momentum four-vector,** and so on.

frame of reference (§2.1)   **Reference frame**

frequency (§1.3, §11.2)   For any simple **oscillation** or repetitive motion, the number of cycles or repetitions per unit time. The frequency is equal to $1/T$, where T is the **period.** Symbol: f. SI unit: hertz (Hz) $\equiv$ cycle/second.

fundamental mode (§11.4)   That **normal mode** of a confined **wave** that has the lowest **frequency.**

Galilean invariance (§1.1, §3.1)   The invariance of **nonrelativistic** physics under **Galilean transformations.**

Galilean transformation (§3.1)   The change from the description of events in one **inertial reference frame** to the description of the same events in a second frame moving at constant **velocity** relative to the first, in **nonrelativistic** physics. The time is not affected by a Galilean transformation, as it is by a **Lorentz transformation.**

gamma   1. (§1.4) The dimensionless **relativistic** factor γ that describes the contraction of moving objects and the slowing of moving clocks. It is given by $\gamma = (1 - v^2/c^2)^{-1/2}$, where v is the **speed** of the motion and c is the speed of light. **2.** (§5.5) A highly energetic **photon** (**energy** of the order MeV or greater). In this energy range the **wave**like characteristics of the radiation are almost impossible to observe, and **particle**like characteristics predominate. Also referred to as a gamma ray. Symbol: γ.

gauge field   The fundamental **field** in a **gauge theory,** related in a special way to a **symmetry** of nature and its associated **conserved** quantity. The transformation law for the gauge field is designed to allow the symmetry in question to be a **local symmetry,** known as the *gauge symmetry.* See **SCF principle.** So far as we now know, all the forces of nature are mediated by four fundamental gauge fields. The gauge field of **electromagnetic theory** is the **four-vector** field A^μ, whose components are the **scalar** and **vector potentials.**

gauge invariance   The invariance of a physical theory, known as a **gauge theory,** under a particular **group** of local **transformations,** or **gauge transformations.** See **local symmetry.**

gauge symmetry   Gauge invariance

gauge theory   A theory of fundamental interactions that relates a basic **conservation law,** a related **group** of **local symmetries,** and a fundamental **field,** the **gauge field.** The interrelations among these three parts are described by **Noether's theorem** and the properties of the **symmetry group,** and the resulting theory, uniquely determined by the local symmetry requirement, gives a detailed account of the structure and **dynamics** of the gauge field and the nature of its interactions. The only adjustable parameter in a normal gauge theory is a single **coupling constant,** which determines the strength of the interaction. Sometimes referred to as Yang–Mills theory. See **SCF principle.**

gauge transformation   One of the local **transformations** associated with a particular **gauge theory.** See **local symmetry.**

general relativity   Einstein's theory of gravity as a manifestation of the curvature and other **dynamical** properties of **spacetime** itself. See **Einstein field equations, equivalence principle,** and **curvature tensor.**

geodesic (§6.5)   The nearest thing to a straight line that can be constructed in a curved space (or **spacetime**). A geodesic has no more curvature than that required by the curvature of the space itself, and it is typically, but not necessarily, the shortest distance between two points in that space.

global symmetry   The invariance of physical laws under a global **transformation,** in which the same transformation is performed at every point in **spacetime**—in contrast to a **local symmetry.**

gluon   The **quantum** of the **color gauge field.** Gluons themselves carry **color charge** and therefore interact with each other in a way that's radically different from the behavior of **photons,** which are electrically neutral.

gluon field   The **color gauge field,** which interacts with **color charge.** The **quanta** of the color gauge field are known as **gluons.**

grand unified theory (GUT)   A theory that would **unify** all the **particles** and all the interactions that make up the **Standard Model.** It would presumably be a **gauge theory** with a single **symmetry group** of which the symmetries of the Standard Model would all be subgroups.

gravitational field (§6.2) The **force field** that acts on **mass** or, more generally, **energy.** In **nonrelativistic** physics the gravitational field acting on a mass produces an **acceleration** exactly equal to the strength of the field itself; the field is therefore often referred to as the **acceleration of gravity.** Symbol: g. SI unit: newton/kilogram (N/kg) $\equiv$ meter/second^2 (m/s^2).

gravitational mass (§5.6) The **mass** of an object as measured by the gravitational **force** on it. It is a consequence of the **equivalence principle** that gravitational mass is identical to **inertial mass.**

gravitational red shift (§6.5) The lowering of the **frequency** of light as it leaves a massive object like a star. You can think of the **photons** as losing **energy,** like a ball thrown straight upward from the surface of the earth, so that the frequency of the light, being proportional to the photon energy by the **Planck relationship** $E = hf$, must drop also.

gravitino The **fermion** partner of the **graviton** in a **supersymmetric** theory of gravity. The gravitino would have a **spin** of $3/2$, in contrast to the graviton, with spin two. Such a "supergravity" theory is still quite speculative.

graviton The spin-two **quantum** of the **gravitational field.** No one has yet figured out how to **quantize** the gravitational field, and gravitons interact so weakly with matter of any sort that we are unlikely ever to detect them, but physicists nevertheless have very little doubt that gravitons exist.

ground state The lowest **energy eigenstate** of any **quantum system.**

group An interrelated set of operations or **transformations** with a multiplication rule and three basic properties, called the group properties. In physics the operations or transformations act on some space or set of objects, the *operands;* in mathematics the operands are typically dispensed with, and only the relationships among the group elements are considered. Objects that are invariant under the operations of a group are said to have the **symmetry** of that group. The group multiplication rule gives the "product" of any two of the operations—that is, the result of applying them in succession—which must also be an operation that belongs to the set. The group properties are as follows:
(i)   One member of the set is the identity operation *I,* which doesn't change the operand.
(ii)   Every member of the set has an *inverse,* also a member, such that the product of the operation and its inverse is *I,* the identity.
(iii)   The multiplication rule must be *associative,* which means that $a(bc) = (ab)c$, where *a, b,* and *c* are members, or *elements,* of the group.
A group whose elements all **commute** is called *Abelian;* otherwise, it is **non-Abelian.**

group velocity (§14.3) The **speed** of a localized **wave pulse** in a **traveling wave** motion of any kind. The group velocity is often different from the **phase velocity,** which is the speed of the individual crests and troughs.

GUT Grand unified theory

hadron Any of the quasi-fundamental strongly interacting **particles** of nature. There are two distinct families of hadrons, the **baryons,** which are **fermions,** and the **mesons,** which are **bosons.** Each baryon is composed of three **quarks** or three antiquarks, and each meson is composed of a quark and an antiquark.

Hamiltonian In **quantum theory,** the **operator** associated with the **energy** of a given **system**; it is also known as the **energy operator.** The Hamiltonian is particularly important in quantum theory because it is the operator that generates time **displacements.** This means that the form of the Hamiltonian determines the equations of motion of the system, so that the physical laws are incorporated in the Hamiltonian.

heat (§5.1, §9.1) A broad term that refers mainly to the energy of random molecular motion, or *thermal energy,* in **macroscopic systems** such as solids, liquids, and gases. Symbol: Q. SI unit: joule (J).

heat reservoir (§9.1) In **thermodynamics,** a large body of matter kept at a fixed **temperature,** supplementary to the **system** under study and used to supply or absorb **heat** at that fixed temperature. The heat reservoir is assumed to be large enough that the heat transfers considered cause only negligible changes in its temperature.

helium four The **isotope** of helium ^{4}He, of atomic mass four. Its **nucleus** is made of two **protons** and two **neutrons,** and both the nucleus and the atom as a whole are **bosons.** The **superfluidity** of helium four is a **macroscopic quantum** phenomenon that is essentially a **Bose–Einstein condensation.**

helium three The **isotope** of helium ^{3}He, of atomic mass three. Its **nucleus** is made of two **protons** and one **neutron,** and both the nucleus and the atom as a whole are **fermions.** The **superfluidity** of helium three is a **macroscopic quantum** phenomenon closely analogous to **superconductivity** and distinctively different from the more familiar superfluidity of **helium four,** whose atoms are **bosons.**

hidden variable (§10.7) Submicroscopic physical variables, introduced in an effort to avoid the nondeterministic, or **probabilistic,** character of **quantum theory.** Hidden variables are imagined as behaving in a complicated but totally **deterministic** way, and since they would be impossible to observe directly, the larger-scale objects for which they provide a substructure would appear to behave in a random way. The possibility of hidden variables has been almost completely ruled out by a class of theorems known as *Bell's theorems.*

Higgs boson The **quantum** of the **Higgs field.** In view of the elusive character of the Higgs field—hypothetical, yet necessary to the success of the **gauge theory**

of **weak interactions**—the experimental search for the Higgs, as it's usually called, is seen as crucially important. Its **mass** is not well fixed by the theory, which makes it that much harder to look for.

Higgs field   The **boson field** devised to provide **spontaneous symmetry breaking** in the **gauge theory** of the **electroweak interactions.** The Higgs field, a **vector** quantity in what's called **weak isospin** space, is constructed to have a lower **energy** at *finite* field strength than at zero field, so that what we see as the **vacuum state**—empty space—is really a state in which the Higgs field has a nonzero value and points in a particular nonsymmetrical direction in weak isospin space, thus breaking the weak isospin **symmetry** for all practical purposes.

Hubble constant (§7.2)   The constant of proportionality H in **Hubble's law,** which gives the **speed** of separation of galaxies in terms of their distance apart. It is probably between 5 and 10 km/s/Mpc, where Mpc represents a distance of 10^6 **parsecs.** Note that H has the dimensions of 1/time.

Hubble's law (§7.2)   The law that all the galaxies we can see are moving away from each other with a **speed** proportional to their separation. The constant of proportionality is called the **Hubble constant** H, which probably falls between 5 and 10 km/s/Mpc. That is, the speed in kilometers per second is equal to from 5 to 10 times the separation in megaparsecs. (See **parsec.**) The age of the universe is roughly equal to $1/H$, which comes out to be between 10 and 20 billion years.

Hubble time (§7.2)   The inverse of the **Hubble constant,** representing the approximate age of the universe and falling somewhere between 10 and 20 billion years.

i (§14.1)   The symbol for the **imaginary** unit used in constructing **complex numbers;** $i = \sqrt{-1}$

identical particles   **Particles** that are so completely indistinguishable that in **quantum theory** they cannot be treated as having separate identities in any sense. All the elementary particles of nature fall into a small number of types of identical particle—the **electrons,** the **photons,** etc.—and each type of identical particle falls into one of the two broad classes called **fermions** and **bosons,** which are distinguished by the way in which many-particle **quantum states** can be formed.

imaginary axis (§14.1)   One of the two axes in an **Argand diagram,** used in making a graphical representation of **complex numbers.** The imaginary axis is the y axis if a general complex number is written as $x + iy$ and displayed as the point with coordinates (x,y) on the Argand diagram. See **real axis.**

imaginary number (§14.1)   A **complex number** equal to a multiple of the imaginary unit i. On an **Argand diagram** an imaginary number is represented by a point on the **imaginary axis.** See **real number.**

imaginary part (§14.1)   The coefficient of the imaginary unit i in a **complex number.** If the complex number is written as $z = x + iy$, then the imaginary part of z, written Im z, is the **real number** y. See **real part.**

indefinite metric The **metric** of a space such as **spacetime** that has two in-equivalent types of **vector,** corresponding to **spacelike** and **timelike** vectors in spacetime. The quantity ds^2 (see **metric**), which gives the **Lorentz invariant** separation between two neighboring points in spacetime, may be either positive or negative if the metric is indefinite. See **definite metric.**

independent-particle model A model of a many-**particle system** in which the particles do not interact with each other. In such a model each particle may move in an effective fixed force field that represents the average effect of all the other particles. The independent-particle model makes the extremely difficult many-body problem much more tractable, and it works remarkably well for the **electrons** in an atom or molecule and even for the **protons** and **neutrons** in a **nucleus.**

inertia (§4.1) The resistance of an object to **acceleration** of any sort, including deceleration and change of direction of motion. The inertia of a **nonrelativistic** object is described by its **inertial mass,** which is identified simply as its **mass.**

inertial force (§6.3) An apparent **force** associated with accelerated motion. Examples include the force that seems to be pulling you down when you are in an elevator accelerating upward and the "centrifugal force" that seems to be pushing you to the right when you're in a car that's turning sharply to the left.

inertial mass (§5.6) The **mass** of an object as measured by its resistance to **acceleration. Nonrelativistically,** the acceleration of an object is given by **Newton's second law** of motion: $a = F/m$, where m is the inertial mass. It is a consequence of the **equivalence principle** that inertial mass is identical to **gravitational mass.**

inertial reference frame (§1.5) In **Newtonian** physics and **special relativity,** a **reference frame** that is moving at constant **velocity** relative to all other inertial reference frames. To avoid circularity, you can say that an inertial frame is one of a class of reference frames, all moving at constant velocity relative to each other, such that in each of them **momentum** is conserved, that is, objects travel at constant velocity unless acted on by real **forces.**

infinitesimal (§3.3, §A.4) A small change in some variable, which you allow to approach zero as you look at some limiting behavior such as, for example, the ratio of two infinitesimals. Thus the **derivative** df/dx of a function $f(x)$ is the limit of the quotient of df by dx, where df is the change in f induced by the small change dx in the variable x. In physical situations it's appropriate to think of an infinitesimal as being simply a very small increment and not to worry about taking limits.

infinities Apparently meaningless infinite quantities that appear in almost all nontrivial calculations in **quantum field theory.** Infinities generally appear as a bad oversupply of high-energy contributions from **virtual processes,** somewhat like the **ultraviolet catastrophe** in the classical theory of **electromagnetic radiation.** In many cases we have learned to control these infinities by the process of **renormalization**—a formal rescaling of the parameters of the theory to absorb the infinite factors and produce finite results.

initial conditions (§8.2)   The conditions on the behavior of a **system** at some initial time (such as the initial position and **velocity** of an object) that, together with the **equations of motion,** determine all of the subsequent behavior of the system.

integral   **1.** (§A.5)   A function related to a given function by the requirement that its derivative be equal to the given function. The function $g(x)$ is the *indefinite integral* of $f(x)$, written $g(x) = \int f(x)dx$, if $f(x) = dg(x)/dx$. This requirement determines $g(x)$ only up to an additive constant called a *constant of integration.* **2.** (§3.3, §A.5)   A number, called the *definite integral,* associated with a given function $f(x)$ and an interval, from $x = a$ to $x = b$, say, over which $f(x)$ is defined. The definite integral, written $\int_a^b f(x)dx$, can be thought of as the length of the interval $b - a$, multiplied by the average value of $f(x)$ over the interval.

interference (§1.3, §10.5)   The physical effect in which two or more **waves** arriving at the same point in space may reinforce or cancel each other depending on whether they arrive in **phase** (in step) or out of phase. It is *constructive interference* if they tend to reinforce each other, *destructive interference* if they tend to cancel each other. See **superposition.**

interference fringes (§1.3, §10.5)   The dark and bright areas appearing on an illuminated surface as a result of **interference.** The dark areas correspond to destructive interference, the bright areas to constructive interference.

interferometer (§1.3)   An optical instrument that makes use of the **interference** of monochromatic light to measure distances with high precision in terms of the **wavelength** of the light, or to measure the wavelength itself if that isn't known.

intermediate vector bosons   The **quanta** of the **gauge field** that mediates the **weak interaction.** There are three of these **bosons,** the W^+, the W^-, and the Z^0, related to each other and to the **photon** through the **unified gauge theory** of the **electroweak interaction.**

internal energy (§9.1)   In **thermodynamics,** the **energy** associated with the random thermal motion and interactions of all the **particles** and **fields** that make up a physical **system.** Symbol: U. SI unit: joule (J).

internal symmetry   A **symmetry** under **transformations** involving non**dynamical** parameters of a system, such as **complex phase** angle, **isospin,** or **color charge.**   In contrast to **dynamical symmetries,** an internal symmetry is not associated with **transformations** of the coordinates of **spacetime.**

invariant length (§3.4)   The magnitude of a **vector** of any sort, invariant under the **transformations** that describe the **symmetry** of the vector space in question. Specifically, invariant length commonly refers to the magnitude of a **four-vector** in **special relativity,** with different definitions for **timelike** and **spacelike** vectors.

For vectors representing **displacements** in **spacetime,** the invariant length is equal to the **proper time** or **proper distance** associated with the displacement. For the **energy–momentum four-vector** of a particle, the invariant length is equal to the **rest mass** of the particle multiplied by the **speed** of light c.

isospin   A three-component **vector,** mathematically analogous to an **angular-momentum** vector, that displays the approximate **internal symmetry** relating the **proton** and the **neutron,** the three kinds of **pions,** and other similar **multiplets** of **hadrons.** The magnitude I of the isospin tells how many **particles** make up the multiplet (namely, $2I + 1$), and the z component of the isospin, I_z (which has nothing at all to do with the z direction in real space), labels the different members of the multiplet just as the z component of the **spin** angular momentum specifies the different orientations of a particle with spin. The **nucleon** (**neutron** or **proton**) has isospin one half, and the triplet of pions has isospin one.

isotope   One of several kinds of atom that have the same number of **protons** in the **nucleus** (and hence the same number of **electrons** and the same chemical properties) but different numbers of **neutrons.** The two isotopes of helium, for example, are **helium three** (^{3}He) and **helium four** (^{4}He).

Kelvin temperature scale (§9.2)   The **absolute temperature** scale that uses kelvins (or degrees Kelvin) as **temperature** units. A kelvin is the same as a centigrade degree, but a temperature given in kelvins is always measured from **absolute zero.**

kinematics (§4.1, §B.1)   The description of motions in **spacetime,** including the relationships among the positions, **velocities,** and **accelerations** of **particles,** without regard to the **forces** that may control the motion. Contrast **dynamics.**

kinetic energy (§5.2)   The **energy** associated with the motion of an object. For a point particle of **rest mass** m and **speed** v, the **relativistic kinetic energy** is given by $K = mc^2(\gamma - 1)$, where c is the speed of light and γ is the relativistic factor **gamma.** For **nonrelativistic** particles this becomes $K = mv^2/2$. A rotating object also has kinetic energy of rotation. Symbol: K. SI unit: joule (J).

lab (laboratory) frame (§1.6)   A conventional phrase for the **reference frame** in which the experimental apparatus is at rest.

lambda point   The **temperature** at which **helium four** (^{4}He) undergoes a **phase transition** to its **superfluid** state. The name *lambda point* refers to the shape of the graph of specific heat as a function of temperature, which shows a singular peak that looks somewhat like the Greek letter lambda (λ).

laws of motion (§I.1, §4.3)   The laws describing the behavior in time of a physical **system.** See **equations of motion.**

lepton   One of the six fundamental **fermions** that do not participate in **strong interactions.** There are three generations, or families, of leptons, very different in **mass** but very similar in other properties: the **electron** (e$^-$) and its associated **neutrino** (ν_e), the **muon** (μ^-) and its neutrino (ν_μ), and the **tau** and its neutrino

(v_τ). Each of the six leptons has a distinct **antiparticle,** and each generation corresponds to a conserved **particle number.**

lifetime   The average life of an unstable **particle** or **system** of particles. Typically, the **probability** of decay after time t is governed by a **probability density** proportional to an **exponential** factor $e^{-t/\tau}$, where τ is the lifetime, or *mean life.*

lightlike vector (§3.4)   A **four-vector** whose **invariant length** is zero. A lightlike vector is neither **spacelike** nor **timelike.** The position vector of a light signal in **spacetime,** relative to its initial position, and the **energy–momentum four-vector** for a massless **particle** such as a **photon,** are examples of lightlike vectors.

light-year (§2.2)   A unit of length, equal to the distance that a light signal travels in one year, 9.4605×1015 m. Symbol: lt-yr.

local symmetry   The invariance of the laws of physics under a **group** of local **transformations,** where a local transformation consists of a different transformation at each point in **spacetime.**

logarithmic scale (§11.5)   A graphical representation of a physical variable in terms of its logarithm. Values of the variable equal to different powers of 10 are thus equally spaced and so labeled on such a graph. For an example of a logarithmic scale, see fig. 11.5.1, which shows the **electromagnetic spectrum.**

longitudinal wave   A **wave** motion in which the **displacements** that create the wave pattern are parallel to the direction of propagation of the wave. An example is a sound wave in air. See also **transverse wave.**

Lorentz contraction (§1.5)   According to **special relativity,** the **relativistic** contraction of moving objects along the direction of motion, given by the relativistic factor **gamma.**

Lorentz invariance (§1.1)   The invariance of a physical quantity, or a theory, under **Lorentz transformations,** reflecting the equivalence of all **inertial reference frames.** See **relativity principle.**

Lorentz transformation (§3.1)   In **special relativity,** the **transformation** from one **inertial reference frame** to any other frame moving with fixed **velocity** relative to the first. The moving frame may also have a different orientation than the first frame. The theory of relativity gives formulas for the values of physical variables in the new frame in terms of their values in the original frame. The three most familiar characteristics of a Lorentz transformation are **Lorentz contraction, time dilation,** and the **simultaneity shift.**

macroscopic (Part II Introduction)   Large-scale, referring to physical phenomena at the scale of our everyday experience or larger. The relation between the macroscopic and atomic scales is characterized by **Avogadro's number,** which is about equal to the number of hydrogen atoms in a gram of hydrogen, around 6×1023. **Quantum** effects are normally not observed at the macroscopic level.

macroscopically occupied state   A **single-particle quantum state** that is occupied by a substantial fraction of all the **particles** in a **macroscopic system,** as in **Bose–Einstein condensation.**

magnetic field (§11.5)   That part of the **electromagnetic field** that produces a **force** on **electrically charged particles** that depends on their **velocity.** It's a vector quantity with three components. Symbol: $\underline{B}$. SI unit: tesla (T) ≡ newton/ampere · meter (N/A · m).

magnitude (of a complex number)   **Absolute value**

mass (§4.1, §5.5, §5.6)   A physical quantity representing the amount of matter in an object. In **special relativity** mass is equivalent to **energy,** the two being related by the constant factor c^2, where c is the **speed** of light. In modern physics usage, the term *mass,* with its symbol m, refers to the **rest mass** of a particle, but people sometimes also talk about the **relativistic mass,** $m_{rel} = m\gamma$, where γ is the relativistic factor **gamma.** In **nonrelativistic** physics there is no such distinction, and mass and energy simply become distinct concepts. Symbol: m. SI unit: kilogram (kg).

mass number   The total number of **protons** plus **neutrons** in the **nucleus** of a given type of atom.

matter wave (§10.2, §10.4)   In the early days of **quantum theory,** the **wave** motion that must be associated with **particles** of matter to explain their wave-like properties. Matter waves are now simply identified with the **Schrödinger wave function.**

Maxwell's baby (§8.2)   My invented analog of **Maxwell's demon,** illustrating the overwhelming tendency for **macroscopic systems** to become disordered if they're not disordered in the first place.

Maxwell's demon (§8.2)   A hypothetical imp (invented by James Clerk Maxwell) who produces a violation of the **second law of thermodynamics** by tidying up the molecular chaos of thermal motion one molecule at a time. The concept of such an imp doesn't violate any accepted principle of physics.

Maxwell's equations (§1.1)   The **wave equations** of **electromagnetic theory,** describing the **dynamical** behavior of **electric** and **magnetic fields.**

mean life   **Lifetime**

medium (§1.3)   An extended distribution of matter through which a **wave** can propagate by mechanical vibration. Thus a gas, liquid, or solid can be the medium for propagation of sound waves, and a liquid, with a surface defined by gravity, can be the medium for surface waves like those on a body of water. In the early days of **electromagnetic theory** it was believed that such a mechanical medium (the **ether**) was needed for the propagation of **electromagnetic waves.**

Meissner effect   The complete exclusion of the **magnetic field** from the interior of a simple **superconductor.**

metric (§6.4)   In any space, flat or curved, the rule that gives actual distances along a path in terms of the coordinates used to label the points of the space (see **curvilinear coordinates**). In its usual form, the metric specifies the **infinitesimal** distance ds associated with an infinitesimal **displacement** by means of a formula giving ds^2 (the square of ds) as a quadratic expression in the increments dx^μ in each of the four coordinates x^μ: $ds^2 = \sum_{\mu,\nu} g_{\mu\nu}\, dx^\mu\, dx^\nu$. The coefficients $g_{\mu\nu}$ are known collectively as the metric **tensor** $\underline{\underline{g}}$ and are generally functions of the coordinates x^μ. In the geometry of **spacetime,** in both **special** and **general relativity,** the expression for ds^2 may be either positive or negative, depending on whether the displacement is **spacelike** or **timelike.** The metric in this case is called an **indefinite metric.** See also **definite metric.**

metric tensor (§6.4)   In **special** and **general relativity,** the set of 16 parameters (constituting a **tensor**) that specify the **metric** at each point in **spacetime.** In special relativity it takes a simple form with just four nonzero components, each equal to +1 or −1, reflecting the geometrical properties of spacetime. Symbol: $\underline{\underline{g}}$, with components $g_{\mu\nu}$ (μ, $\nu = 0, 1, 2, 3$).

Michelson interferometer (§1.3)   A kind of **interferometer** devised by Albert Michelson and used by him in an attempt to observe the effect of the earth's motion through space on the behavior of light **waves.**

Minkowski diagram   Spacetime diagram

Minkowski space   The **spacetime** of **special relativity,** with emphasis on its geometrical properties.

modulus (§14.1, §A.7)   The **absolute value** of a **complex number.**

momentum (§4.1, §B.2)   A fundamental **conserved vector** quantity of nature, describing the amount of motion of a **particle** or a general physical **system.** For a **relativistic** particle, the momentum vector is given by $\underline{p} = m\gamma\underline{v}$, where m is the **rest mass,** $\underline{v}$ is the **velocity** vector, and γ is the relativistic factor **gamma.** In the **nonrelativistic** limit, the momentum of a particle becomes simply $\underline{p} = m\underline{v}$. The momentum density of the **electromagnetic field** is given by $\varepsilon_0 \underline{E} \times \underline{B}$, where ε_0 is related to the **Coulomb's law** factor k (eq. 12.3.2) by $k = 1/4\pi\varepsilon_0$. The **conservation** of momentum is related by **Noether's theorem** to the spatial **displacement symmetry** of physical laws—that is, to the invariance of physical laws under a spatial displacement of the **reference frame.** Symbol: $\underline{p}$. SI unit: kilogram · meter/second (kg · m/s) ≡ newton · second (N · s).

momentum space   The space of **momentum vectors.** The coordinates of a point in momentum space are equal to the components of the corresponding momentum vector.

multiplet   A set of **particles** of identical or very similar properties, such as the doublet of **nucleons** or the triplet of **pions.**

muon (§5.5)   A **lepton** very similar in its properties to the **electron** but 207 times more massive. It is subject, together with the **muon neutrino,** to a conservation law that requires the muon number plus the muon neutrino number (see **particle number**) to be strictly constant in time. The μ^- has lepton number +1, like the **electron** (e–), while the μ^+ is treated as an **antiparticle,** with lepton number –1 like the **positron.** (The symbol $\bar{\mu}$ for the antiparticle is not normally used.) Symbol: $\mu^{\pm}$.

muon neutrino (§5.5)   One of the three known types of **neutrino.** It is a **lepton,** associated particularly with the **muon.** Symbol: ν_μ.

natural logarithm (§9.2)   The logarithm of a number taken to the base e (see **exponential function**). If $x = e^y$, then $y = \ell nx$, where ℓn is the symbol for the natural logarithm. The **derivative** of the logarithm of x to any base is proportional to the function $1/x$; a special property of the natural logarithm is that the proportionality constant is 1: $d(\ell nx)/dx = 1/x$. In terms of logarithms to base 10, $\ell nx = \log_{10}x/\log_{10}e = \log_{10}x/0.43429$.

neutrino   Any of the three varieties of neutral (zero **electric charge**) **lepton.** There is one variety of neutrino associated with each of the charged leptons: the **electron,** the **muon,** and the **tau.** It is possible, but not yet known, that all three types of neutrino may turn out to have a **mass** exactly equal to zero, like the **photon**; the **electron neutrino,** with which we are most familiar, is known to have zero mass to within 0.004% of the electron mass. Symbol: ν.

neutron (§5.5)   One of the two types of **particle,** the neutron and the **proton,** that make up the **nuclei** of all atoms of ordinary matter. These two are the most common varieties of **baryon,** a **strongly interacting** particle made of three **quarks.** The neutron is normally stable inside a nucleus but is unstable when it's isolated, with an average **lifetime** of about 15 minutes. It has zero **electric charge**. Symbol: n.

neutron star (§7.1)   A star consisting almost entirely of **neutrons,** packed together to a density a little greater than the density of atomic **nuclei,** around 1018 kg/m3 (a million tons per cubic millimeter).

Newton's law of gravity (§6.2)   The **nonrelativistic** law of gravitational attraction, almost identical in form to **Coulomb's law** of **electrostatic force** between **electric charges.** It states that the **force** between two **masses** is proportional to each of the masses and inversely proportional to the square of the distance between them: $F = -Gm1\,m_2/r^2$, where the minus sign indicates that the force is always attractive and the proportionality constant G is the *gravitational constant,* equal to 6.67×10^{-11} N · m^2/kg^2.

Newton's second law (§4.3, §B.2)   One of Newton's three **laws of motion** describing the behavior of **nonrelativistic** objects and their interactions by means of **forces.** The second law states that the rate of change of the **momentum** of a **particle** is equal to the net force acting on it: $dp/dt = \underline{F}$. In this form it is true also for **relativistic** particles, though in its more conventional form, $\underline{F} = m\underline{a}$, it is valid only for nonrelativistic particles. Newton's first law is simply equivalent to the second law applied to the case where the net force is zero, in which case the **acceleration** is zero and the **velocity** of the object remains constant. See **Newton's third law.**

Newton's third law (§4.6)   One of Newton's three **laws of motion** describing the behavior of **nonrelativistic** objects and their interactions by means of **forces.** The third law describes the forces of mutual interaction between two objects and states that the forces they exert on each other are exactly equal and opposite. The third law is a simple consequence of the **conservation** of **momentum** and applies only if changes in the momentum carried by the force **fields** can be neglected.

node (§11.2)   In a **standing wave,** a point where the **amplitude** of **oscillation** is zero. Two- and three-dimensional standing waves have nodal lines and nodal surfaces, respectively.

Noether's theorem   A very basic theorem of both classical and **quantum** physics that relates **symmetries**—invariance properties—to **conservation laws.** The theorem states that for every symmetry of the **laws of motion** of the universe (that is, an invariance of those laws under some **group** of **transformations**), there is an associated **conservation law**—the constancy in time of the total value of some physical quantity. The reverse is also true, that every conservation law is associated with a symmetry property of the physical laws. (There is one well-understood class of exceptions: any symmetry involving time inversion. Here the conditions of the theorem are not met and the theorem doesn't apply.)

non-Abelian   Referring to a **group** whose group multiplication law is noncommutative: $ab \neq ba$. See **commute.**

nonrelativistic   A term used to refer to the physics of objects moving much slower than the **speed** of light. Nonrelativistic physics can also be thought of as the limit of **special relativity** when the speed of light c is treated as infinite. Note that the **relativistic** factor **gamma** becomes equal to 1 in that limit, so relativistic effects like the **Lorentz contraction, time dilation,** and **simultaneity shift** all disappear in the nonrelativistic limit.

normal freqency (§11.4)   The natural **frequency** of one of the **normal modes** of an **oscillating system.** The normal frequencies of a vibrating string are integer multiples of the lowest one, the **fundamental** frequency.

normal mode (§11.4)   One of the distinct **standing-wave** patterns of an extended **oscillating system** such as a string or membrane under tension, or the **electromagnetic field** in a cavity with reflecting walls. Each normal mode has a characteristic **frequency,** the **normal frequency.**

nucleon   A general term for a **proton** or **neutron.** Thus the **mass number** of an atom is equal to the nucleon number (see **particle number**) of its nucleus.

nucleus (§5.6) The very small, massive core of every atom, around 10,000 times smaller in diameter than the atom itself and comprising around 99.98% of its **mass.** It is made up of **neutrons** and **protons** and is held together by **strong interactions.**

observable In **quantum theory,** a physical variable capable of being measured experimentally—a necessary category because the theory includes a lot of mathematical objects that can't be identified with observable physical quantities.

observation (§15.2, §15.3) In **quantum theory,** the actual act of making a measurement. Such acts play an important role in the logic because of their relation to the probabilistic predictions of quantum theory and because of the way in which observations fix the **quantum state** of the **system.**

occupied state A **single-particle quantum state** that is occupied by a **fermion.** The **exclusion principle** does not allow there to be two or more **identical** fermions in any one single-particle state, so each single-particle state may simply be either unoccupied or occupied.

odd permutation See **permutation.**

open universe (§7.3) A universe that is of infinite spatial size. It has no boundaries and extends infinitely far in all directions. An open universe is expected by **general relativity** to keep expanding forever.

operator In **quantum theory,** one of the operations that can be performed on a **wave function** to make a different wave function. Those operators that have what's called the *hermitian property* play a particularly important role in the theory: each hermitian operator corresponds to an **observable,** or measurable, physical variable and at the same time defines a class of **transformations,** such as those that represent **symmetries** of the theory. Thus the physical **momentum** of a particle (one-dimensional, say) corresponds to an operator $\hat{p}$ that takes the **derivative** of the wave function, $\hat{p}\psi(x) = -i\hbar d\psi(x)/dx$, and that can thus be used to generate spatial **displacements.** These two roles of the operator are related by **Noether's theorem**: In this example, the invariance of physical laws under spatial displacements is related by Noether's theorem to the **conservation** of momentum.

ordinary differential equation A **differential equation** describing a function of a single variable and therefore involving only ordinary derivatives, not **partial derivatives. Newton's second law,** the equation of motion for a particle (eq. 4.3.2 or eq. 4.3.10), is an ordinary differential equation.

oscillation (§11.2) The repetitive motion of a physical **system** at a particular **frequency.** The oscillations of an extended system may take the form of **standing waves** in any of a number of possible **normal modes** of oscillation.

parallax (§7.2) The angular shift in the apparent location of an object produced by shifting the point from which you observe it. Parallax can be used to measure the distance of faraway objects. The distance of nearby stars can be measured (in **parsecs**) by observing the parallax due to the earth's motion about the sun. The distance in parsecs is equal to half the number of seconds of parallax observed as the earth shifts by the diameter of its orbit.

parent nucleus The original nucleus that decays in a radioactive decay process such as **beta decay.** See **daughter nucleus.**

parsec (§7.2) An astronomical distance of 3.262 **light-years,** defined as the distance from which the mean radius of the earth's orbit about the sun would subtend an angle of 1 second. See **parallax.** Symbol: pc.

partial derivative The **derivative** of a function of several variables with respect to just one of them, the other variables being treated as constants for the purpose. Thus the partial derivative of $f(x,y)$ with respect to x is given by $\partial f(x,y)/\partial x = [f(x+dx,y) - f(x,y)]/dx$, where dx is an **infinitesimal** increment in x. The partial derivative $\partial f(x,y)/\partial y$ is similarly defined.

partial differential equation (§14.1) A **differential equation** describing a function of several variables and therefore involving only **partial derivatives,** not ordinary **derivatives.**

particle (§10.1) Classically, an idealized object of finite mass and **infinitesimal** size (a *point particle*), whose internal motion can be ignored. In the **quantum** world, the word *particle* refers normally either to the elementary "particle-like" constituents of matter (all of which are apparently **fermions**) or to the particlelike **quanta** of the fundamental force **fields** (all **bosons**). There may be exceptions to this pattern in the **Higgs boson** or in the fermionic quanta of a possible **supersymmetric gauge field.**

particle number The total number of a given class of **particles** in the universe (or in an isolated **system** under study). The phrase *particle number* is generally used in connection with one of the particle-number **conservation laws** that apply to the different classes of **fermions.** There are exact or approximate particle-number conservation laws for **baryons, quarks,** and each of the three types of **lepton.** In each case, the particle number is defined to be the number of particles *minus* the number of **antiparticles.**

period (§1.3, §11.2) The repetition time of an **oscillation,** equal to $1/f$, where f is the **frequency.** Symbol: T. SI unit: second (s).

permutation Any reordering of a set of objects that has a definite order. Any permutation can be achieved by making successive interchanges of pairs of objects. If the number of pair interchanges needed is even, then it's an *even permutation,* and if the number of pair interchanges is odd, it's an *odd permutation.* The total number of possible permutations of n objects is $n!$ (n = factorial), equal to the product of the first n integers: $n! = 1 \cdot 2 \cdot 3 \cdot \cdots \cdot n$

phase **1.** (§10.5, §11.1) An angle used to describe different stages in the cycle of an **oscillatory** motion and to give the degree to which different oscillations of the same **frequency** are out of step with each other. If $y_1 = A_1 \sin(\omega t + \varphi_1)$ and

$y_2 = A_2 \sin(\omega t + \varphi_2)$, for example, then the instantaneous phases of the oscillations y_1 and y_2 are $\omega t + \varphi_1$ and $\omega t + \varphi_2$, respectively. We also say that the oscillation y_1 *leads* the oscillation y_2 by a phase angle $\varphi_1 - \varphi_2$. **2.** (§14.1) The instantaneous angle of orientation of a rotating two-dimensional **vector**, called a **phasor**, or the **argument** of a **complex number** used to represent such a phasor. **3.** One of several possible states of an extended **macroscopic system**, such as the liquid, solid, and gaseous phases of water.

phase factor A **complex number** of **absolute value** 1 and specified **phase** (definition 2), of the form $R(\theta) = e^{i\theta}$ (see prob. 14.1.1), which has the effect, when it's a factor multiplying some given complex number, of changing the **argument** of that given number by an amount θ. The phase factor may be time-dependent, typically (in **quantum theory**) in the form $e^{-i\omega t}$. See **stationary state.**

phase transition The change in **phase** (sense 3) of an extended **macroscopic system**, such as the freezing of water.

phase velocity (§14.3) The **speed** with which a **wave** pattern (the crests and troughs in a water wave, for example) travels. The phase velocity may be different from the **group velocity**, the speed with which a localized **wave pulse** travels. Often called the *wave speed.*

phasor (§14.1) A rotating two-dimensional **vector** used to represent a simple **oscillation.** Both the x and y components of such a vector carry out simple **sine wave** oscillations with **angular frequency** ω if the phasor is rotating with constant **angular velocity** ω. Using **complex numbers** to represent an oscillatory motion is exactly equivalent to using phasors (see phase factor).

photon (§5.5, §10.3, §16.2) The **particle**like **quantum** of the **electromagnetic field.** Historically, the first quantum effects to be studied all had to do with photons. High-energy photons are usually referred to as **gammas.** Symbol: γ.

pion (§5.5) The lightest and most common of the **mesons,** made of **quark**-antiquark pairs in different combinations that involve only **up quarks** and **down quarks.** Also (and historically) called the *pi meson.* There are three types of pion, with charges $\pm e$ and 0. Symbol: π^+, π^-, π^0.

Planck relationship (§10.3) In **quantum theory,** the relationship $E = hf$ between the **energy** E of a **photon** and the **frequency** f of the corresponding **electromagnetic waves.** The parameter h is **Planck's constant.** The relationship is often written in the equivalent form $E = \hbar\omega$, where $\hbar$ is the **Dirac $\hbar$** and ω is the **angular frequency** of the waves.

Planck's constant (§10.3) The parameter $h = 6.626 \times 10^{-34}$ J · s, which is characteristic of all **quantum** phenomena. See **Dirac $\hbar$.**

plane polarization (§11.5) A **wave** pattern for **electromagnetic waves** in which the **electric** and **magnetic field vectors** oscillate in fixed directions, perpendicular to each other and also perpendicular to the direction of propagation. The plane common to the electric field vector and the direction of propagation is called the *plane of polarization.* See **polarization.**

plane wave (§11.2) A uniform **traveling wave,** with definite **wavelength** and **frequency,** traveling in a single well-defined direction. The wave fronts (surfaces of constant **phase**) form planes perpendicular to the direction of propagation— hence the name *plane wave.*

point particle (§10.1, Appendix B) See **particle.**

polarization (§11.5) A particular state of alignment of **vector field** variables in a **plane wave,** as in **plane** or **circular polarization** of **electromagnetic waves.** Also, an analogous alignment of the **spins** in a beam of **particles.** The **photon** beam corresponding to circularly polarized light is polarized in this sense, with all the spins aligned parallel to the direction of the beam.

positron (§5.5) The **antiparticle** of the **electron,** with **electric charge** $+e$. Symbol: e^+ (it is not usually represented by $\bar{e}$, which would be the standard symbol for such an antiparticle).

potential A measure of the **potential energy** associated with any kind of **field,** such as gravitational potential or **electrostatic potential.**

potential energy (§5.3) In **nonrelativistic** physics, **energy** stored in a **system** as a result of its configuration and the forces among its different parts. The phrase *potential energy* refers to the system's *potential* for doing work in the future, as distinct from **kinetic energy,** which is the energy of present motion of the system. In fact, the energy referred to as potential energy is in every case energy stored in the **fields** responsible for the forces in the system. Symbol: U or V. SI unit: joule (J).

probability (§10.7, §14.4) A number between zero and one, representing the likelihood of a given event. If the occurrence of the event is counted as a "success" and its nonoccurrence as a "failure," and if the trial is repeated with identical preparation a large number of times, then you can expect the fraction of successes (the number of successes divided by the total number of trials) to be equal on the average to the probability. It is a distinctive feature of **quantum theory** that it is impossible to predict future events with certainty and that it's possible only to predict probabilities. Symbol: P.

probability amplitude (§14.4) A **complex number** that comes out of the **quantum theory** of a given **system** and whose **absolute value** squared is a predicted **probability.** The **Schrödinger wave function** $\psi(x,t)$ is a probability amplitude.

probability density (§14.4) A way of describing the **probability** distribution for a variable that can take on a continuous range of values. The function $P(x)$ is a probability density for the variable x if $P(x)\,dx$ is equal to the probability that x shall be found within the **infinitesimal** range dx. Equivalently, the probability that x lies between two values a and b is given by the definite **integral** $\int_a^b P(x)\,dx$.

proper distance (3.4)   In **special relativity,** the **Lorentz invariant** distance between two events that have a **spacelike separation**—that is, two events that can
be seen as simultaneous in some **reference frame.** The proper distance is equal
to the actual distance between the two events in that reference frame in which
they are simultaneous.

proper time (§3.3)   In **special relativity,** the **Lorentz invariant** time along the
world line of a **particle,** corresponding to the reading on a clock that moves with
the particle, its **gamma** factor (definition 1) varying as the **speed** of the particle
varies. The proper time is given by the indefinite **integral** $\tau = \int dt/\gamma$, with an
arbitrary constant of integration and with γ determined as a function of t by the
instantaneous speed of the particle. The term *proper time* can also refer to the
invariant time between two specified points along the world line, in which case
it is given by the definite integral from t_1 to t_2, the times associated with the two
specified points. If the **velocity** of the particle is constant, then the proper time
is equal to the actual time between the events in a **reference frame** in which the
events occur at the same point in space.

proton (§5.5)   One of the two types of **particle,** the proton and the **neutron,** that
make up the nuclei of all atoms of ordinary matter. These two are the most
common varieties of **baryon,** a **strongly interacting** particle made of three **quarks.**
The proton has positive **electric charge**. Symbol: p.

QCD   Quantum chromodynamics

QED   Quantum electrodynamics

QFT   Quantum field theory

quanta   Plural of **quantum** (definition 2)

quantization   The procedure of setting up the **quantum theory** for a **system**
whose classical behavior is known. Typically, the procedure involves finding the
classical Hamiltonian, which is the **energy** expressed in a standard way in terms
of coordinates and **momenta,** and then taking the quantum **Hamiltonian** as the
same function of the position **operators** and momentum operators. The Hamiltonian operator then determines the physical behavior of the quantum system.
This procedure is not unique for a given system but is guaranteed to satisfy the
correspondence principle.

quantize   1. (§10.3) To assign discrete values to a physical variable.   **2.** (§16.2)
To construct a **quantum** version of a classical theory by **quantization.** Note that
quantizing is not something you do to the physical **system** itself but is rather a
way of devising a new theory out of an old one.

quantum   1. (Part II Introduction) An adjective referring to the various
effects and characteristics of **quantum theory.   2.** (§16.4)   A noun referring
to one of the **particle**like quantum excitations of a **field.** The quanta of the
electromagnetic field are called **photons,** those of the **gravitational field**
are called **gravitons,** and those of the **color gauge field** are called **gluons.** Ele-

mentary **fermions,** such as **quarks** and **leptons,** can also be understood as the quanta of appropriately defined fermion fields. See **quantum field theory.**

quantum chromodynamics (QCD) The **quantum theory** of the **color gauge field.**

quantum electrodynamics (QED) (§16.2) The amazingly successful **quantum** theory of the **electromagnetic field.** The term normally includes a full **quantum field theory** treatment of **electrons** and **positrons.** QED provides the only proper treatment of electromagnetic processes involving **photons** or variable numbers of electrons and positrons.

quantum field theory (QFT) (§16.4) The basic formulation of physics that treats all particles, **bosons** and **fermions** alike, as **quanta** of appropriate **fields.** Quantum field theory is the logical way, and the only way we know of, to combine **special relativity** with **quantum theory**—and it is still beset with difficulties, both mathematical and conceptual.

quantum number (§14.2) A parameter, typically an integer, used to label the **quantum states** of a physical **system.**

quantum state (§14.1, §14.4) The **state** of a physical **system** in **quantum theory,** represented by a **wave function** or a **state vector.** The mathematical information contained in the wave function or state vector enables you to predict the **probabilities** for the different possible outcomes of any given kind of **observation.**

quantum theory (Part II Introduction) The theory that describes many of the important characteristics of physics at the atomic and subatomic scale: **interference** among all the different possible paths that a **system** might follow; **probabilistic** rather than **deterministic** predictions; and **discrete** values of **energy** and some other physical variables, which are then said to be **quantized** (definition 1). The mathematics of quantum theory requires the handling of infinitely many **degrees of freedom** for every classical degree of freedom, because it's necessary to keep track not just of the one value that a variable classically has at any given time, but of all of the possible values that variable *might* have.

quantum tunneling The ability of a **particle** or a **system** in **quantum theory** to get from one place to another even when classically there is no possible route for it to follow without violating **energy conservation.** This kind of tunneling is a consequence of the **wave** nature of the theory and is in fact a feature of classical wave theories also.

quark The fundamental constituent of **hadrons,** which, together with leptons, make up all the matter we are familiar with. Quarks come in six **flavors,** which are observable, and in three **colors,** which are hidden from our view by the phenomenon of **color confinement.** The color charge of quarks is responsible for their interactions, via the **color gauge field.**

real axis (§14.1) One of the two axes in an **Argand diagram,** used in making a graphical representation of **complex numbers.** The real axis is the x axis if a general complex number is written as $x + iy$ and displayed as the point with coordinates (x,y) on the Argand diagram. See **imaginary axis.**

real number (§14.1)   An ordinary number, in the range from $-\infty$ to $+\infty$, often referred to in the context of **complex numbers** generally. A real number is a complex number with zero imaginary part. The term *real* has become necessary only in order to distinguish these numbers from complex numbers with non-zero imaginary parts. On an **Argand diagram** a real number is represented by a point on the **real axis.** See **imaginary number.**

real part (§14.1)   The part of a **complex number** that doesn't involve the imaginary unit *i*. If the complex number is written as $z = x + iy$, then the real part of z is the real number x, written Re z. See **imaginary part.**

red shift (§7.2)   The observed **Doppler shift** of light from distant galaxies that is due to their motion relative to our own galaxy. The shift is toward the lower-**frequency,** or red, end of the **spectrum** (definition 1) because the distant galaxies are moving away from our own, according to **Hubble's law.** See also **gravitational red shift.**

reference frame (§2.1)   A **coordinate** system representing measurements made by a particular observer, who may be moving in an arbitrary way relative to other observers. In classical physics and **special relativity** we are especially interested in **inertial reference frames,** and the term *reference frame* usually refers to an inertial frame.

relativistic   A term used to refer to the physics of objects moving at **speeds** comparable to the speed of light, at which relativistic effects like the **Lorentz contraction, time dilation,** and **simultaneity shift** cannot be neglected. See **non-relativistic.**

relativistic mass (§4.5, §5.5)   The effective **mass** of a moving **particle** in **special relativity,** given by $m_{rel} = m\gamma$, where m is the **rest mass** and γ is the **relativistic** factor **gamma.** The relativistic mass, which becomes arbitrarily large as the particle approaches the **speed** of light c, describes the **inertia** of the particle—its resistance to **acceleration**—and also governs the gravitational **force** on the particle. It is exactly proportional to the total **energy**—**rest energy** plus **kinetic energy**—of the particle, by the relation $E_{tot} = m_{rel}c^2$. This is one of several meanings that can be given to the famous relation $E = mc^2$; another involves the rest energy and the rest mass. The term *mass,* used alone, almost never refers to the relativistic mass.

relativity, principle of (§1.5)   The principle, fundamental to **special relativity,** that the laws of physics, including especially the behavior of light, are just the same in every **inertial reference frame.** The principle of relativity is equivalent to **Lorentz invariance.**

renormalizability   The property of a **relativistic quantum field theory** that the nonphysical **infinities** (characteristic of all such theories) can be eliminated, in every order of approximation, by a formal rescaling of a finite number of parameters of the theory, such as **masses** and **coupling constants.** Only certain types of theory, such as **gauge theories** in particular, are renormalizable.

renormalization   In a **renormalizable quantum field theory,** the formal procedure for rescaling the parameters of the theory, such as **masses** and **coupling constants,** to obtain a theory that's finite in every order of approximation.

resonance 1.   The strong response of an **oscillating system** when it is driven by external forces at a **frequency** close to one of its own natural, or **normal, frequencies.   2.**   The enhancement of a scattering process involving nuclear or subnuclear **particles** that is due to the momentary creation of an unstable particle whose **lifetime** is too short for it to be observed directly.

rest energy (§5.5, §5.6)   In **special relativity,** the **energy** E_0 inherent in the **mass** of a particle when it is at rest, given by $E_0 = mc^2$, where m is the **rest mass.** The total energy is equal to the rest energy plus the **kinetic energy.**

rest frame (§1.6)   The **reference frame** in which a given observer or object is at zero **velocity.** The rest frame of someone performing an experiment is referred to as the **lab frame.**

rest mass (§4.5)   A **Lorentz invariant** parameter that describes the **mass** of a **relativistic particle** and is equal to the mass as measured by conventional means when the particle is at rest. The rest mass is related to the **rest energy** by the formula $E_0 = m_0 c^2$.

right-hand rule   Any of several rules for getting the right relations between different directions in space. Specifically: **1.** (§A.2)   The rule for the direction of the **vector product** of two **vectors.** If $\underline{C} = \underline{A} \times \underline{B}$ is the vector product of $\underline{A}$ and $\underline{B}$, then the direction of $\underline{C}$ is perpendicular to both $\underline{A}$ and $\underline{B}$, and the orientation of $\underline{A}$, $\underline{B}$, and $\underline{C}$ is like the orientation of the thumb and first two fingers of the right hand with the thumb and forefinger outstretched and the middle finger bent (fig. A.2.5).   **2.** (§11.5)   The rule for the **magnetic force** on a moving charged **particle:** $\underline{F} = Q\underline{v} \times \underline{B}$. You can use rule 1 for the vector product, or you can use a more visual rule in which the fingers of the outstretched right hand are in the direction of the **magnetic field** lines, the thumb is in the direction of the **vector** $Q\underline{v}$ (opposite to $\underline{v}$ if Q is negative), and the force vector points out from the palm of the hand—the natural direction for pushing.   **3.**   The rule for the direction of the magnetic field lines around an electric current or a moving charged particle. The field lines form circles about the line of the current or particle **velocity,** and their orientation is like the direction the fingers of the right hand point when they are curled up and the thumb is outstretched in the direction of the current or of the vector $Q\underline{v}$.

Rydberg constant (§12.1)   The constant R appearing in the **Rydberg formula** for the **spectral lines** of atomic hydrogen. The formula for R, by **quantum theory,** is $R = ke^2/2hca_0$, where k is the **Coulomb's law** constant (eq. 12.3.2) and a_0 is the **Bohr radius** for the **ground state** of hydrogen. The numerical value of the Rydberg constant, known to one part in a billion, is 1.097×10^7 m^{-1}. (This is for the case of infinite nuclear mass; a well-understood correction is needed to allow for the motion of the nucleus.)

Rydberg formula (§12.1) The formula for the **spectral lines** of atomic hydrogen, first devised empirically and later derived from **quantum theory**: $1/\lambda = R(1/m^2 - 1/n^2)$, where m and n are any positive integers with $n > m$. The constant R is the **Rydberg constant.**

scalar Referring to a single physical quantity that is invariant under a **symmetry group,** especially rotational symmetry in three dimensions or the symmetry under **Lorentz transformations** in **special relativity.** A scalar variable is usually referred to as a scalar.

scalar potential The **electrostatic potential** V (or φ), which is a **scalar** (that is, invariant) under three-dimensional rotations, as distinguished from the **vector potential** $\underline{A}$. The scalar potential is related to the **electric field** distribution, while the vector potential is related primarily to the **magnetic field.** The scalar potential is the time component of the **four-vector field** A^μ, the **gauge field** of **electromagnetic theory.**

scalar product (§5.1) A kind of product of two **vectors** that yields a single number—an invariant **scalar.** If $\underline{A}$ and $\underline{B}$ are two three-dimensional vectors, their scalar product is written $\underline{A} \cdot \underline{B}$ and is given by $AB\cos\theta$, where A and B are the magnitudes of $\underline{A}$ and $\underline{B}$, and θ is the angle between them. The scalar product of two vectors is also equal, importantly, to the sum of products of their components: $\underline{A} \cdot \underline{B} = A_xB_x + A_yB_y + A_zB_z$. If $\underline{\underline{A}}$ and $\underline{\underline{B}}$ are two **four-vectors,** their scalar product is $\underline{\underline{A}} \cdot \underline{\underline{B}} = A^0B^0 - A^1B^1 - A^2B^2 - A^3B^3$.

SCF principle A fundamental principle of physics, as I conceive it, relating the continuous **symmetries,** the **conservation laws,** and the fundamental **forces** of nature in an exact correspondence. The continuous symmetries are **local symmetries,** and the forces are then due to the corresponding **gauge fields,** interacting with the associated conserved quantities. See **Noether's theorem.**

Schrödinger equation (§14.1) The **wave equation** that describes the **wave function** for **nonrelativistic particles** or, more generally, the time development of any **quantum system,** nonrelativistic or **relativistic.** Its general form is $i\hbar\partial\psi/\partial t = \hat{H}\psi$, where ψ is the wave function of the whole system and $\hat{H}$ is the **Hamiltonian,** or **energy operator.** The symbol $\hbar$ is the **Dirac $\hbar$.** For a free nonrelativistic particle, the Schrödinger equation takes the form shown in fig. 14.1.1.

Schwarzschild radius (§7.1) In **general relativity,** the radius of the spherical surface at which the **Schwarzschild solution,** the static, spherically symmetrical solution to the **Einstein field equations,** becomes singular—that is, its mathematical properties change discontinuously. This singular behavior is generally held to be responsible for the peculiarities of **black holes.** Both the nature of the singularity and the meaning of a distance are obscured by the distortions of the coordinate system itself in a region where **spacetime** is so radically curved. The conventional value for the Schwarzschild radius is $R_{SCH} = 2Gm/c^2$, where G is the gravitational constant (see **Newton's law of gravity**), m is the **mass** of the black hole (as determined from the **gravitational field** at a distance), and c is the **speed** of light.

Schwarzschild solution (§6.5)   The solution of the **Einstein field equations** in **general relativity,** spherically symmetrical and unvarying in time, that represents the **gravitational field** around a small spherical **mass** distribution. This solution describes the curvature of **spacetime** around a star, and it enables you to predict deviations from the predictions of **Newtonian** gravity such as the bending of light, the **gravitational red shift,** and the slow shifting—precession—of planetary orbits. The Schwarzschild solution also gives the extremely distorted behavior at a very small radius—the **Schwarzschild radius**—that constitutes a **black hole**; this singular behavior occurs only if the entire mass is concentrated inside that small radius.

second law of thermodynamics (§8.1, §9.1)   The law of irreversibility, which says that **macroscopic** processes can never reverse themselves. Stated and discussed in many different forms, such as (1) **Heat** never flows spontaneously from a cold to a hot object. (2) You can't take heat **energy** from a **heat reservoir** at a single **temperature** and convert it to mechanical **work** with 100% efficiency. (3) The total **entropy,** or microscopic disorder, of the universe can only increase as a result of real physical processes.

shell   A set of **single-particle states** of approximately equal **energy** in a many-particle system, such as the **electrons** of an atom or the **protons** or **neutrons** of a **nucleus.** A shell often consists of several **subshells.**

simultaneity shift (§2.4)   In **special relativity,** the degree to which two **spacetime** events are seen as nonsimultaneous in one **reference frame** when they are simultaneous in another frame. The formula (eq. 2.4.10) is $\Delta t = v\Delta x/c^2$, where Δx is the spatial separation of the events in frame S, Δt is the time difference according to frame S, and v is the **velocity** relative to S of the frame S' in which the events appear simultaneous.

sine wave (§11.1, §11.2)   A sine or cosine function, as applied either to a spatial **wave** pattern or a simple **oscillation**—$y = A \sin(kx + \varphi)$ or $y = A \sin(\omega t + \varphi)$—or to a **traveling** or **standing wave** in space and time—$y = A \sin(kx - \omega t)$ or $y = A \sin kx \sin \omega t$.

single-particle state   A **quantum state** for a single **particle.** The term is used most often in connection with the **independent-particle model,** in which the many-particle **wave function** is assumed to be a **symmetrized** or **antisymmetrized** product of single-particle functions—that is, each particle is assigned to a particular single-particle state.

spacelike separation (§3.4)   Describing two events in **spacetime** that are simultaneous in some **reference frame.** The **four-vector** representing a spacelike separation is a **spacelike vector,** whose **invariant length** is the actual distance between the events in a **reference frame** in which the events are **simultaneous** (see also **proper time**). It's impossible to send information between two space-time points whose separation is spacelike.

spacelike vector (§3.4)   In **special relativity,** any **four-vector** that in some frame of reference has its time component equal to zero.

spacetime (§2.1)   Space and time taken together as a single four-dimensional space. Points in spacetime represent **events** and are labeled by the four position and time coordinates x, y, z, and t needed to locate an event. Both in **special relativity** and **general relativity,** spacetime has its own characteristic geometry.

spacetime diagram (§2.2)   A graphical representation of the **spacetime** of **special relativity,** typically showing the time dimension and just one of the three space dimensions. A point on the spacetime diagram, which I've called an **event point,** represents an event in spacetime, and its coordinates give both the position and the time of the event. The t axis is the set of points that might represent events that occur at different times at the spatial origin, $x = 0$, and the x axis is the set of points that might represent events that all occur at $t = 0$ but at different locations along the x direction.

special (theory of) relativity   The theory of space, time, and **dynamics** generated by the **relativity principle,** namely the requirement that the laws of physics, including the properties of light, be exactly the same in every **inertial reference frame.**

spectral line (§12.1)   Any **discrete wavelength** in the **spectrum** (sense 1) of an atom or molecule; so called because it's normally observed as a narrow line in standard ways of visually observing spectra.

spectrum   **1.** (§11.5, §12.1)   The distribution of wavelengths in the **electromagnetic radiation** emitted by atoms or molecules. The spectrum may be **discrete** or continuous, depending on the circumstances; an atom or molecule at rest has a discrete spectrum, while the thermal radiation from a hot object or gas has a continuous spectrum.   **2.** (§14.4)   **Eigenvalue spectrum**

speed (§B.1)   The magnitude of the **velocity** of an object or a **wave,** without reference to direction of travel. The word *velocity* is often used to mean speed. Symbol: v. SI unit: meter/second (m/s).

spin (§14.4)   The intrinsic **angular momentum** of a **particle,** whose magnitude cannot change and which cannot properly be associated with an angular rotation of the particle. Note that the spin of the **photon** is the **quantum** reflection of **circular polarization** in classical **electromagnetic waves.** The spin **quantum number** s of a particle (see **angular momentum quantum numbers**) can take the values $0, 1/2, 1, 3/2, \ldots$.

spin-statistics theorem   A theorem in **quantum field theory** that **bosons** can have only integer **spin** and **fermions** can have only half-odd-integer spin ($1/2$, $3/2, \ldots$). This rule is completely confirmed experimentally.

spontaneous symmetry breaking   The mechanism that is generally believed to be reponsible for **broken symmetries** in physics. In this view, the **symmetry** that appears to be broken is in fact a true symmetry of the laws of physics that is spoiled by an asymmetry in the environment—that is, in what we call the **vacuum state,** where no **particles** or **fields** are present. For this to happen, the vacuum state of our universe must be one of a family of equivalent vacuum states, all nonsymmetrical and all related to one another through the symmetry in question. The best-known and most strongly supported example of spontaneous symmetry breaking is the *Higgs mechanism.* See **Higgs field.**

Standard Model   A widely accepted approximation to the physics of **strong, electromagnetic** and **weak interactions** (not including gravity). The ingredients of the Standard Model are **quarks** and **leptons** as the fundamental **particles; color, electric charge,** and **weak isospin** as the basic **conservation laws** (besides **energy, momentum,** and **angular momentum**); and the **color gauge field** and the **electroweak gauge fields** as the fundamental gauge fields responsible for all the **forces.** The **Higgs field,** needed to break the weak isospin **symmetry** and to give **mass** to the W and Z bosons, is also part of the Standard Model.

standing wave (§11.2)   A **wave** motion, normally in a confined region with fixed boundaries, in which the zeros of the spatial pattern remain fixed, at points called **nodes,** and the **amplitude** of the spatial pattern **oscillates** in time. A typical standing wave in one space dimension has the form $y = A \sin kx \sin \omega t$, with fixed nodes at the points $x = n\pi/k$ (the points where $\sin kx = 0$). See **traveling wave.**

state (§14.1, §14.4)   The exact condition of a **system** at a given moment of time. Classically, the state of a system of **particles** and **fields** is specified by giving the positions and **velocities** of all the particles and the values of the field variables at every point in space. In **quantum theory,** the state of a system is specified by giving its **wave function** or **state vector,** which determines only a **probability** distribution for each of the variables that would have been exactly specified classically.

state vector   A **vector** in an infinite-dimensional vector space, used in formal **quantum theory,** especially in **quantum field theory,** to specify the **quantum state** of a **system.** It is accurate to think of the **Schrödinger wave function** of a system of particles as a vector of this kind in a vector space consisting of all possible wave functions. [The function $\psi(x)$ is analogous to a vector with components ψ_n, but with the **discrete** index n replaced by the continuous parameter x.]

stationary state (§14.1)   A **quantum state** in which the **probability** distributions for the different physical variables don't change with time. The **wave function** for a stationary state has a time-dependent **phase factor** $e^{-i\omega t}$ but is otherwise constant in time. A stationary state is necessarily an **eigenstate** of the **energy.**

statistical mechanics (§9.1)   The study of the thermal properties of a **system** by analyzing the random statistical behavior of the atomic or subatomic **particles** that make it up.

strangeness A **quantum number** associated with the **strange quark.** In pre-quark days, **conservation** of strangeness was postulated as a device for understanding the odd characteristics of the recently discovered **strange particles.** An s quark has strangeness −1.

strange particle One of several types of **hadron,** whose odd behavior when they were first produced in high-energy particle accelerators gave rise to their name. They are now known to contain **strange quarks.** See **strangeness.**

strange quark One of the six **flavors** of **quark.** It has **electric charge** $-(1/3)\,e$, like the **down quark,** and is closely paired with the **charm quark,** though closer in mass to the u and d quarks. The lightest **particle** containing a strange quark is the K **meson,** which has a **mass** about half that of a **proton.** Symbol: s.

strong forces The forces among **hadrons—protons, neutrons, pions,** and so forth—that hold the **nucleus** together and are responsible for the various kinds of nuclear **energy.** See **color gauge field.**

strong interactions The fundamental interactions that are reponsible for the **strong forces. Color gauge theory** is the theory of strong interactions in the **Standard Model.**

subshell A set of **symmetrically** related **single-particle states** of approximately equal **energy** in a many-**particle system,** such as the **electrons** of an atom or the **protons** or **neutrons** of a **nucleus.** Several subshells whose energies are close together may be grouped into a single **shell.**

superconductivity The total vanishing of electrical resistance in certain materials at **temperatures** below a well-defined transition temperature T_c (see **phase transition**). The superconducting state is a **macroscopic quantum state** that, in a simple superconductor, can be viewed as a **Bose–Einstein condensation** of **Cooper pairs,** loosely bound pairs of electrons that act as **bosons.**

superfluidity The total vanishing of viscosity in both isotopes of helium (**helium four** and **helium three**) below a certain transition temperature. In both cases the system is in a **macroscopic quantum state,** though the mechanisms are quite different. In the case of ^{4}He (where the atoms are **bosons**), it is essentially a **Bose–Einstein condensation** of the atoms. In the case of ^{3}He (where the atoms are **fermions**), it is a **superconductivity**like condensation of bound **Cooper pairs.** The transition temperatures are also very different—about 2.2 K for ^{4}He and about 1 mK (10^{-3} K) for ^{3}He (see **phase transition**).

superposition (§10.5, §11.3, §13.4) The property of **waves** that permits **interference.** The superposition of two or more waves is the algebraic sum of the functions describing the separate waves, and the sum may be greater or less than the separate functions at any given point in space and time. Superposition is also a property of **wave functions** in **quantum theory,** and of **quantum states** generally, and is thus the cause of all kinds of quantum interference phenomena. See **superposition principle.**

superposition principle (§13.4) A guiding principle in the early days of **quantum theory,** saying that two or more **quantum states** can be combined by **superposition** to yield a resultant state whose properties are related to those of the constituent states and which can exhibit **interference** of various sorts.

supersymmetry A proposed type of **symmetry** scheme in which invariance is required under a new type of **transformation** that mixes **boson** and **fermion** **states.** The transformations involve both ordinary **complex** parameters and Grassman variables (**anticommuting** analogs of ordinary numbers). Supersymmetric theories predict the existence of related pairs of **gauge** bosons and gauge fermions (see **gravitino**).

symmetric Exhibiting some **symmetry.** Specifically, a many-particle **wave function** is symmetric if it is invariant under all possible **permutations** of the particles. The function $f(x)g(y) + g(x)f(y)$ is symmetric in x and y. The wave function for a system of **identical bosons** is required to be symmetric in the coordinates of all the different particles. See **antisymmetric.**

symmetrize To construct a **symmetric wave function** out of one that is not symmetric by combining terms with different **permutations** of the **arguments.** The function $f(x)g(y)$ is symmetrized by forming the combination $f(x)g(y) + g(x)f(y)$. See **antisymmetrize.**

symmetry (§I.2) The invariance of some **system** or theory under a **group** of **transformations,** known as the **symmetry group.** A sphere, for example, is invariant under all possible three-dimensional rotations, and so are the laws of physics. Any symmetry of the laws of physics is related by **Noether's theorem** to some definite **conservation law.** See also **local symmetry** and **global symmetry.**

symmetry group A **group** of **transformations** that leave the laws of physics (or some theoretical candidate for the laws of physics) invariant—that is, exactly the same. See **Noether's theorem.**

system (§15.2) A word used to represent some set of physical objects under study, with all interactions with the rest of the universe neglected. A system might consist of a single atom, two colliding **particles** and the junk that results, all the **electrons** in a **macroscopic** piece of metal, and so on. The ultimate system is the whole universe.

temperature (§9.1) A measure of how much thermal **energy** there is in an object or a **system,** defined in such a way that if two systems at the same temperature are in thermal contact, then no **heat** will flow between them. The **absolute temperature** scale, defined in connection with the **second law of thermodynamics,** is a particularly important example of a temperature scale.

tensor (§6.4, §6.5) A generalization of the idea of a **vector** (whose components are labeled by a single index) to an object whose components are labeled by two or more indices. Thus p^μ represents the four components of the **energy–momentum vector** in **special relativity,** while $g_{\mu\nu}$, the **metric tensor** (for **special**

or **general relativity**), has 16 components. The **transformation** laws for a tensor are a natural generalization of the transformation laws for a vector.

term values (§12.1) In pre-quantum days, and in practical spectroscopy, a set of numbers characterizing in a compact way all the **wavelengths** in the **spectrum** of any atom or molecule. The **wave number** $1/\lambda$ for every **spectral line** is equal to the difference of two term values (see **Rydberg formula**). It turns out from **quantum theory** that the term values are simply the **energy levels** of the atom or molecule, apart from a scale factor $1/hc$, where h is **Planck's constant** and c is the **speed** of light.

Theory of Everything A hoped-for completely **unified theory** of all physical laws, including gravity, that would be uniquely determined by such requirements as freedom from **infinities** and freedom from ambiguities.

thermal equilibrium (§9.1) A **macroscopic system** is in **thermal equilibrium** when its macroscopic properties, such as **temperature,** pressure, and volume, are static even though there is random thermal motion of all the molecules.

thermodynamics (§9.1) The study of thermal properties of **macroscopic** systems. The laws of thermodynamics (see **first law, second law**), originally treated as postulates for a self-contained logical structure, are now derivable as theorems about very large numbers of constituent particles, using the methods of **statistical mechanics.** The proofs become rigorous in the *thermodynamic limit,* the limit in which the number of particles becomes infinite.

time dilation (§1.6) In **special relativity,** the slowing down of moving clocks and physical **systems,** given by the relativistic factor γ (**gamma,** definition 1).

timelike vector (§3.4) In **special relativity,** any **four-vector** that in some frame of reference has spatial components all equal to zero. The **energy–momentum four-vector** for a **particle** with nonzero **mass** is timelike, since in the **rest frame** of the particle the **momentum** components are all zero.

top quark One of the six **flavors** of **quark,** assumed to exist but not yet observed. It has **electric charge** $+(2/3)e$, like the **up** and **charm quarks,** but is evidently much more massive. It is closely paired with the **bottom quark.** Symbol: t.

transformation (§3.1) A mathematical operation on something, which may be an object, a **reference frame,** or the **state** of a physical **system,** that converts it into a different thing of the same class. The operation might be, for example, a rotation, a **displacement** in space or time, or a replacement of certain particles by different particles of the same general type. The transformations of particular interest in this book are those related to a **symmetry,** especially a symmetry of the laws of physics themselves. The laws display the symmetry if they are are invariant under a family of related transformations, called the **symmetry group,** and the laws are said to be invariant if the transformed states develop in time according to the same rules as the untransformed states.

transition probability (§16.3)   The **probability**, under the rules of **quantum theory**, that a **system** that starts off in one **quantum state** will at a later time be found, by an appropriate **observation**, to be in some other specified state. The transition probability is the **absolute value** squared of a **complex probability amplitude**, called the *transition amplitude*, which can be expressed as the sum of contributions from all the possible **virtual processes** that might take the system from the given initial state to the given final state.

transverse wave (§16.2)   A **wave** motion in which the **displacements** that create the wave pattern are perpendicular to the direction of propagation of the wave. Examples include vibrational waves in a stretched wire and **electromagnetic waves**. See also **longitudinal wave.**

traveling wave (§11.2)   A **wave** motion in which the spatial pattern remains fixed in shape while it travels at a constant **velocity** through space. A typical traveling wave in one space dimension has the form $y = A \sin(kx - \omega t)$, with zeros at the points where $kx - \omega t = n\pi$. These zeros therefore (like the entire wave pattern) move with speed $v = \omega/k$, the **wave speed.** See **standing wave.**

truth   One of the six **flavors** of **quark,** more commonly referred to as the **top quark.** Symbol: t.

tunnel   A verb referring to **quantum tunneling.**

twin paradox (§1.6, §2.6)   The odd fact, in **special relativity,** that if one of two people (invariably chosen to be twins) takes a trip and returns while the other remains at rest in a single **inertial reference frame,** then the person who took the trip ages less (by **time dilation**) than the one who stays at home. The scenario doesn't contradict the **principle of relativity,** because the traveler doesn't stay in a single inertial reference frame, while the person at home does.

ultraviolet catastrophe (§10.3)   The prediction of classical **electromagnetic theory** that any object not at **absolute zero** would emit **electromagnetic radiation** of infinite intensity in the short **wavelength**—ultraviolet—part of the **spectrum.** The discovery that **quantizing** (sense 1) the energy of radiation avoids the catastrophe marked the first step in the development of **quantum theory.**

uncertainty principle (§13.2)   The characteristic of **quantum theory** that certain related variables, known as **complementary variables,** cannot be known with arbitrarily high precision at the same time. Stated as one of the fundamental guiding principles in the early days of quantum theory, the uncertainty principle became a provable theorem, applicable to a wide range of situations, after a mathematically coherent quantum theory was developed. The position and **momentum** of a one-dimensional **particle** provide the most familiar example of the uncertainty principle. Time and **energy** are similarly related, although the interpretation is less straightforward.

unified theory   A theory of the fundamental **particles** and their interactions that combines two or more apparently unconnected parts of earlier theories into a single logically coherent one. The hope, still unrealized, is to unify all the parts of the **Standard Model,** together with gravity, through some unifying principle that relates them all and at the same time shows why they appear so different experimentally. See **grand unified theory, Theory of Everything.**

unit vector (§6.2)   A **vector** of length 1, in any vector space, whose function is to specify a direction. A radial **force,** for example, pointing outward from the origin, can be written $\underline{F} = F\hat{r}$, where F is the magnitude of the force and $\hat{r}$ is a unit vector in the same direction as the radius vector $\underline{r}$.

up quark   One of the six **flavors** of **quark.** It has **electric charge** $+(2/3)$e. The up and the **down** quarks, which are closely related to each other, are the constituents of **protons** and **neutrons** and thus make up some 99.98% (by mass) of all ordinary matter.

vacuum   Empty space; a **state** of space in which there is no matter present and no **quanta** of any **fields.** The term is used in **quantum field theory** to represent the **ground state** of a **system** of fields.

valence   An integer describing the kinds of chemical bonds that an atom can form. In terms of the **shell** structure of an atom, the valence is normally equal to the number of **electrons** in excess of a **closed shell** (positive valence) or short of a closed shell (negative valence).

vector   1. (§3.4, §A.2)   In ordinary three-dimensional space, a quantity that has both magnitude and direction, as, for example, **velocity** or **force.** Such a vector can be represented by its three components in some coordinate system— the values of its projections onto the three **coordinate** axes. Symbol: $\underline{A}$; magnitude A; components $A_x, A_y,$ and A_z, or A_i, $i = 1, 2, 3$.   **2.** (§3.4)   A generalization of a three-dimensional vector to a vector space of any dimension, with a number of components equal to the number of dimensions of the vector space, such as **four-vectors** in spacetime.   **3.**   A further mathematical generalization, to what is often called an *abstract vector,* defined as one of a family of objects that have the essential properties common to all vectors, such as addition, multiplication by an ordinary **real** or **complex number,** and **scalar product.** The concept of a **state vector** in **quantum theory** falls in this class.

vector boson   The **quantum** of a **vector field** in **quantum field theory**—that is, a field whose field variable is a **relativistic four-vector.** Specifically, one of the **intermediate vector bosons,** W^+, W^- and Z^0, that are the quanta of the **gauge field** that mediates the **weak interaction.** The photon is also a vector boson, though it is not usually referred to that way.

vector potential   In **electromagnetic theory,** the three spatial components of the basic **gauge field** A^μ from which the **electric** and **magnetic fields** are derived. These three components form a three-dimensional **vector** (definition 1) $\underline{A}$, and the zero[th] component A^0, a **scalar** in the three-dimensional space, is equal to the **scalar potential** φ (or V).

vector product (§11.5, §A.2)   A kind of product of two **vectors** in three dimensions that gives a third vector, written as $\underline{C} = \underline{A} \times \underline{B}$, where $\underline{A}$ and $\underline{B}$ are the two given vectors. The vector product $\underline{C}$ has magnitude $C = AB \sin\theta$ (where θ is the angle between $\underline{A}$ and $\underline{B}$) and direction given by the **right-hand rule** (definition 1). Also called cross product.

velocity (§B.1)   A three-dimensional **vector** quantity $\underline{v}$ giving the **speed** and direction of motion of a **particle.** Its components are the time **derivatives** of the three coordinates of the particle: $v_x = dx/dt$, $v_y = dy/dt$, and $v_z = dz/dt$. In vector notation, $\underline{v} = d\underline{r}/dt$, where $\underline{r}$ is the position vector, which locates the particle with respect to the spatial origin. Symbol: $\underline{v}$. SI unit: meter/second (m/s).

velocity four-vector (§5.7)   A **spacetime** analog of the three-dimensional **velocity vector** for a moving **particle.** Its components u^μ are the **derivatives** with respect to **proper time** of the spacetime coordinates of the particle: $u^\mu = dx^\mu/d\tau$, where τ is the proper time. These four components are related to the ordinary **velocity** of the particle as follows: $u^0 = \gamma$, $u^i = \gamma v_i$, $i = 1, 2, 3$, where γ is the **relativistic** factor **gamma** (which also depends on $\underline{v}$).

virtual particle   A **particle** that plays a transient role in some of the **virtual processes** that contribute to the **transition probability** for a physical process. The virtual particle can't be said to exist at any actual time, but the fact of its possible existence allows real processes to take place that could not do so otherwise. An example is the intermediate W particle that allows the **beta decay** of the **neutron** to occur [fig. 19.6.4(a)].

virtual process   Processes involved in the possible histories that contribute to the **transition probability** for a physical process. A virtual process may violate the **conservation** of **energy** and still play an active role in allowing a real transition to take place. The reality of virtual processes and virtual particles is arguable, but the math works very well.

wave (§1.3, §10.1, §11.1)   An extended spatial pattern that varies in time. Most waves, no matter how irregular, can be understood (by **Fourier analysis**) as **superpositions** of simple **sine waves.** See **traveling waves** and **standing waves.**

wave equation (§14.1)   The **partial differential equation** that describes the **dynamics** of any **wave** motion.

wave function (§14.1)   The **complex** function, often called the Schrödinger wave function, that is responsible for the wavelike behavior of **particles** in **quantum theory** and provides a complete specification of the **quantum state** of a particle or **system** of particles. Also known, in early days, as a **matter wave.** The **wave equation** that describes the wave function is the **Schrödinger equation.** Symbol: ψ

wavelength (§11.1)   The spatial distance over which a **sine wave** repeats itself. The wave $y = A \sin kx$ has wavelength $\lambda = 2\pi/k$, since when x increases by this amount the argument kx increases by just 2π, giving one full cycle of the sine wave. Symbol: λ. SI unit: meter (m).

wave mechanics (Chapter 14)   An early name for **quantum theory** in the form involving the **Schrödinger wave equation,** to be distinguished from *matrix mechanics,* a formulation of quantum theory that involves the manipulation of certain arrays of numbers called *matrices.* The two were devised independently and were later found to be equivalent.

wave number **1.** (§12.1)   The reciprocal of the **wavelength** of a **spectral line** (equal to number of waves per unit length). Written $1/\lambda$ in this book. SI unit: $1/$meter (m^{-1}). **2.** (§11.1)   (Loosely) the **angular wave number,** equal to $2\pi/\lambda$ (number of radians of phase per unit length).

wave–particle duality (Part II Introduction)   The distinctive characteristic of **quantum theory** that **waves** behave like **particles** and particles behave like waves. This duality springs from the fact that the equation governing the time evolution of a **system** of particles is a **wave equation** (the **Schrödinger equation**), while the measurements whose **probabilities** are predicted using the wave function are normally particle-type measurements: position, **momentum, energy,** and so on.

wave pulse (§10.1, §14.3)   A localized **wave** pattern (often called a *wave packet*) that vanishes except in some particular region in space. The region where the pulse is approximately located can itself move through space with a fairly well-defined **velocity** (the **group velocity**), although the pulse also typically spreads as it travels.

wave speed (§11.2)   The **speed** with which the individual **waves** move in a traveling wave; also called **phase velocity.** See also **group velocity.**

weak force **Weak interaction;** used in parallel with the other categories of **forces: strong force, electromagnetic** force, and **gravitational** force.

weak interactions The fundamental interactions responsible for radioactive processes such as **beta decay.** According to the **Standard Model,** weak interactions are mediated by a **gauge field** whose **quanta** are the **intermediate vector bosons** W^+, W^-, and Z^0. This gauge field is intertwined with the **electromagnetic field** in a partially **unified gauge theory** known as the **electroweak** theory.

weak isospin A modified form of **isospin,** partially conserved, that together with **electric charge** provides the sources for the **electroweak gauge theory** in the **Standard Model.** The weak isospin **multiplets** are made up of **quarks** and **leptons** in various combinations. The violated **conservation** of weak isospin is associated in the Standard Model with a **spontaneously broken symmetry.**

weak neutral current A term referring to interactions involving the neutral **intermediate vector boson** Z^0. Such interactions are predicted by the **electroweak gauge theory** but were extremely difficult to verify because they are masked by similar processes involving **strong interactions.**

width **1.** The spread in the **wave number** (definition 1) of a **spectral line** that is due to a variety of physical causes, such as interatomic collisions, **Doppler shifts,** and finite **width** (definition 2) of the **energy levels** responsible. Referred to specifically as *linewidth.* **2.** The spread in the energy of any energy level that is due to its finite **lifetime**—an example of the **uncertainty principle.** **3.** The spread in the **rest mass** of an unstable particle that is due to its finite lifetime (see definition 2).

work (§5.1) The amount of **energy** expended by a person, machine, or physical **system** in producing some change in another system. Mechanical work is normally defined as (**force** exerted) × (distance moved) and is often used as the starting point for a definition of energy. Note that work would not be a meaningful idea if energy were not **conserved.** Symbol: W. SI unit: joule (J).

work function (§10.3) A term for the minimum amount of **energy** that must be supplied in order to remove an **electron** from the surface of a conducting material. Often given as a **potential,** in volts, equal to the energy needed divided by e, the **electron** charge.

world line (§2.2) A path in **spacetime** representing the motion of a **particle** over an extended period of time. Each point on the world line represents a particular time together with the location of the particle at that time. A straight world line corresponds to motion at constant **velocity.**

zero-point energy The **ground state energy** of a **quantum state,** with particular reference to the fact that quantities like **kinetic energy** and **potential energy** have nonzero average values in this state.

Index

Law of cosines, 389
Law of universal gravitation, 122, 174, 176
Laws of motion, 2, 109–110, 118–121, 414
Laws of thermodynamics, 240–241. *See also*
    Second law of thermodynamics
Lenard, Philipp, 265, 266
Length–mass equivalence in general relativ-
    ity, 191
Lepton, 11
Light
  as electromagnetic radiation, 16, 261
  particles of, 261–269. *See also* Photons
  speed of. *See* Speed of light
Lightlike vector, 104, 105–106
Light path, curvature of, 193–194
Light waves. *See also* Electromagnetic waves;
    Wave(s)
  interference of, 25–27
Light-year, 51, 54
Limits of integration, 95
Line, straight
  graph of, 395
  slope of, 395–396
Linear algebra, 335
Local Group of galaxies, 201
Local symmetry, 9, 174
Logarithmic scale, 301
Logarithms, natural, 249
  base of (e), 96
Loop, in standing wave, 297
Lorentz contraction, 43–45, 72–74
  Lorentz transformation and, 81, 85
Lorentz invariance, 18
Lorentz transformation, 81–87, 168, 172

Magnetic field, 301, 303–305
Magnetic moment, of electron, 321
Marsden, Ernest, 308
Mass, 111, 413. *See also* Massless particles
  of electron (m_e), 420
  energy and, equivalence of, 157–158,
    160–164
  equivalence of length and, 191
  gravitational, 160–161
  gravity and, 124, 161
  inertia and, 121, 131, 161
  of neutron (m_n), 420
  nonrelativistic, 131
  of proton (m_p), 315, 420
  relativistic, 132
    equivalence to total particle energy, 157
    of massless particle, 133

  rest, 132
    equivalence to rest energy, 157
    equivalence to invariant length of
      energy–momentum four-vector, 168,
      172
  total conversion of, to energy? 163–164
  units of, 413, 417, 418
  weight and, 120, 161
Massless particles, 132–133
  energy of, 155
Math facts, 387–408
Matrix, 352n, 393
Matter wave, 260, 338. *See also* Wave function
Maxwell, James Clerk, 1, 2, 15, 225n
Maxwell's baby, 225–230, 249
Maxwell's demon, 225n
Maxwell's equations, 16–17, 382
Medium, 22
  ether as, 2, 17n
Megaparsec, 200
Memory, 231–234
Metric, 185–186
  indefinite, 210
  spacetime structure and, 216
Metric tensor, 186
Metric units, prefixes for, 419
Michelson interferometer, 21–22, 27–30
Michelson–Morley experiment, 16–18, 20–
    21, 31–38
  principle of relativity and, 40–42, 44
Microscopic randomness, 244–246, 248–
    249. *See also* Disorder
Millikan, Robert, 268
Minkowski (spacetime) diagrams, 52–56
MKS units, 416
Modulus, 341, 402
Momentum, 110–111, 412–413
  angular, 314, 319–320
    quantization of, 314–315, 319–322,
      345–346
  change in, 119–120
  conservation of
    Newton's third law and, 134–138
    space displacement symmetry,
      Noether's theorem, and, 110
    as spatial components of energy–mo-
      mentum four-vector, 145, 170–171
  force and, 111, 118–121, 412–414
  relativistic, 130–133
  units of, 111, 413
  wavelength and, 270. *See also* De Broglie
    wavelength
Momentum eigenstate, 344n, 356–357